Fred Gifford

Path Analysis: A Primer

by

Ching Chun Li

Books by the same author

1948 *Introduction to Population Genetics.* National Peking University Press.

1949 *Heredity and its Variability* (by T. D. Lysenko). Chinese translation, New China Book Co.

1953 *Soviet Genetics and World Science* (by Julian Huxley). Chinese translation, Taipei, Taiwan.

1955 *Population Genetics.* University of Chicago Press.

1959 *Numbers from Experiments.* Boxwood Press.

1961 *Human Genetics, principles and methods.* McGraw-Hill Book Co.

1964 *Introduction to Experimental Statistics.* McGraw-Hill Book Co.

1976 *First Course in Population Genetics.* Boxwood Press.

Path Analysis
—a primer

Ching Chun Li, Ph.D.

University Professor of Biometry
Graduate School of Public Health
University of Pittsburgh

Pacific Grove, California

© 1975
by
The Boxwood Press

Second printing, with corrections, 1977

No part of this book may be reproduced in any form without written permission of the Copyright owner.

Distributed by:

The Boxwood Press
183 Ocean View Blvd.
Pacific Grove, CA 93950

408—375-9110

Library of Congress Cataloging in Publication Data:

Li, Ching-chun
   Path Analysis

   Includes bibliographical references and index.
   1. Regression analysis. 2. Biology—Statistical methods.
1. Title.
QA278.2.L5   519.5'36'024573   75—19232
ISBN: 0-910286-41-8 *Cloth edition*
ISBN: 0-910286-40-X *Paper edition*

Printed in U.S.A.

Dedicated to

Professor Sewall Wright

on his 85th birthday

PREFACE

Geneticists and breeders have long known and have been using the method of path coefficients developed in 1920-21 by the pioneer population geneticist Sewall Wright. The method has only recently gained recognition in other disciplines and has been used by research workers since the mid-1960's. Most students in the social sciences, however, are caught unprepared, because the method of path coefficients is not to be found in conventional statistics textbooks. It would be unrealistic to expect these students to go to textbooks on population genetics to search for a basic and systematic introduction to the subject. The present volume, truly a primer, is intended to fill this need. It is an exposition of the method of path coefficients and it touches very little on any particular subject matter, except for the three chapters on the application of path coefficients to human genetics. The main purpose of these three chapters is to explain the meaning of heritability, a term that has been increasingly used by psychologists, educationalists, and social scientists. As science progresses, it is difficult to see how social research can be completely divorced from biological considerations.

The primer requires no particular proficiency in statistics or mathematics on the part of the reader; everything is explained in the most elementary manner; but it does assume that the reader is already familiar with the usual statistical terms. The primer may be suitable either as a textbook or as a supplement for a second course in statistics. Chapters 3 and 4 provide the necessary background in linear regression; Chapters 5 and 6 deal with the method of path coefficients. The rest are applications. The reader must be familiar with the elements of population genetics (Chapter 7) before he learns heritability (Chapters 8 and 9). At the end of each chapter, except the first and the last, a few exercises are given. These exercises are not braintwisters; they are very simple, frequently with answers provided. They are intended to give the student an opportunity to review the contents of the chapter and to verify some of its statements by numerical calculations. The student will discover that if he does all the exercises, he will not only get more out of that chapter, but he will be better prepared to proceed to the next.

The writer is a geneticist, with little training in the social sciences. In preparing the primer, he received much encouragement from his friends in other disciplines. Particularly he wishes to thank Professors Otis Dudley Duncan and Arthur S. Goldberger, who introduced him to the new literature dealing with path analysis in the social sciences. Their generous help is deeply appreciated.

C. C. Li
University of Pittsburgh

December 21, 1974

NOTE TO SECOND PRINTING

A few references, cited in the text but missing from the list of references, have been added. Typographical errors—the known ones—have been corrected. "Boxes" have been added to Fig. 298 for greater clarity. A few short sentences have been added where space permits. The only significant additions are three notes added at the ends of three chapters (Ex. 4, p. 74; Ex. 5, p. 99; Ex. 5, p. 134). Otherwise, there is no revision. There is no change in pagination.

C. C. Li

July 1, 1977

CONTENTS

Chapter		Page
1	Introduction	1
2	Identities and Equations	17
3	Linear Regression and Correlation	37
4	Multiple Regression and Correlation	75
5	Standardized Variables; Path Coefficients	100
6	Path Analysis of Causal Systems	135
7	Elements of Population Genetics	187
8	Quantitative Traits; Heritability	214
9	Inbreeding and Assortative Mating	266
10	Applications in Bio-social Sciences	299
	References	338
	Index	343

Chapter 1

INTRODUCTION

THE METHOD of *path coefficients* is a form of structured linear regression analysis with respect to standardized variables in a closed system (formally complete). In such an analysis, a diagram will be most helpful, if not indispensable, to specify the exact nature of a proposed structure, according to which the subsequent analysis is to be made. Hence, a *path analysis* and its corresponding *path diagram* always go hand in hand. The meaning of these general remarks, if not clear at the present moment, will be made explicit in Chapters 5 and 6, in which the method of path analysis is developed.

Whenever we propose or assume a structure about the interrelationships among a number of variables, it is unavoidable to involve some degree of subjective judgment. Two investigators, given the same data on the same variables, may very well come up with two different understandings of the variables and propose two different path diagrams for analysis. Hence, path analysis is not a fixed and routine method of handling data that can be preprogrammed (the program comes free with the computing equipment), as is the case with many traditional statistical procedures.

Since it is possible to analyze a set of data in several different ways and for different purposes, it is a moot question whether to say one is correct or wrong, at least, pending further evidence. What is required of path analysis, however, is that its results must be consistent throughout the structure and compatible with the observed data on all variables involved in the structure.

Since path analysis depends on structure, and structure in turn depends on the cause-and-effect relationship among the variables, we shall first say a few words about the way these terms will be used in this book. There are a number of formal definitions as to what constitutes a cause and what an effect. For instance, one may think that a cause must be doing something to lead to something else (effect). While this is clearly one type of cause-and-effect relationship, we shall not limit ourselves to that type only. Nor shall we enter into philosophical discussions about the nature of cause-and-effect. We shall simply use the words "cause" and "effect" as statistical terms similar to independent and dependent variables, or exogenous and endogenous variables.

The study of the effect of rainfall (X) on wheat yield (Y) is one of the earliest applications of linear regression in the history of statistics. The equation to be fitted is of the form $Y = bX$. In such a case, we say that rainfall (X) is a cause and the wheat yield (Y) is an effect. This is such a clear case of cause and effect that there seems no argument over the use of these words. The fact that rainfall affects plant growth is indisputable. Rainfall also affects the growth of trees. Now, suppose that we have cut down a 300-year-old tree and examine its history of growth as measured by the width (W) of its annual rings. From the measurements of the annual rings, we have a fairly reliable indication of the amount of rainfall for the last 300 years. Using the method of linear regression, an equation of the form $X = cW$ may be found, from which the rainfall (X) of 1776 may be estimated from the ring measurement (W) of 1776. In this formulation, the growth of a tree (W) becomes a cause and the rainfall (X) becomes an effect, in spite of the fact that it was actually the rainfall that influenced the growth of trees. This example brings up clearly one important facet of path analysis: it depends on the purpose or viewpoint of the investigator; his viewpoint leads to the formulation of a "causal scheme" and the analysis will then be made accordingly. This point will be discussed again in Chapter 6.

Beginning with Chapter 5, the reader will find a number of path diagrams throughout the remainder of the book. These diagrams are the qualitative representations of a set of structural equations relating the variables under study. These path diagrams are to be read according to certain rules which are based on analytical results. Such diagrams are not to be confused with popular, semi-scientific or purely qualitative diagrams seen on other occasions. The structural equations implied by a path diagram are all linear equations. In practice, if the relationship among the variables is non-linear, some type of transformation may be employed to render it linear or at least approximately so. Path analysis as outlined in this book is not applicable to non-linearly related variables. Also, the equations are all recursive without feedback loops. Wright (1960b) did treat the problem of feedback by path coefficients, but it is too complicated to fit into a primer.

INTRODUCTION

Now we come to the question: "why path coefficients?". With respect to linearly related variables, a correlation coefficient describes and measures the degree of association between two variables. No attempt has been made to "explain" the correlation from any point of view. A multiple regression equation of the type $\hat{Y} = b_1 X_1 + \ldots + b_k X_k$ has a viewpoint; that is, expresses Y as determined by $X_1, \ldots, X_k$, but also without interpretation. The multiple regression equation is fitted by minimizing the discrepancy between the actually observed Y and the value given by the equation $\hat{Y} = b_1 X_1 + \ldots$ obtained from the method of least squares. This is an ideal job for computing machines. The usefulness of such an equation is that it gives the best or closest prediction of Y based on the information in $X_1, \ldots, X_k$. The traditional statistics derived from the method of least squares are all aimed at the best prediction, whatever the meaning of the X's. When prediction is our sole purpose, we will continue to use multiple regression equations. The path method, on the other hand, is not so much concerned with prediction as to the proposal of a *plausible interpretation* of the relationships between the variables. In other words, path analysis is concerned with erecting a causal structure compatible with the observed data.

To illustrate the distinction between prediction and plausible interpretation, I find the following example instructive as well as interesting. In an early study of the degree of acceptance or rejection of contraceptive devices by young housewives in Taiwan, the research team (consisting of social scientists, biostatisticians, physicians, and nurses) included many variables in their study. The list includes background variables, adaptation variables, and health variables, of both wife and husband. Important ones are age, education, occupation, income, dwelling, health, number of children living, etc. When a multiple regression of acceptance rate on all these variables was fitted, they found that the most important single factor was the number of electric gadgets (electric fans, electric toasters, electric irons, etc.) possessed by the family. If prediction is the purpose, then the number of electric gadgets owned by a family is probably the best single indicator of acceptance of contraceptive devices. But we can hardly assert that electric gadgets constitute an important cause for acceptance of contraceptives. For, if it were true, the success of the family planning program would be assured by simply distributing free toasters and irons (which are much cheaper than raising a child). If the same set of data were presented to a path analyst, his first concern would be to propose a plausible explanation of the phenomenon by constructing a causal scheme, that is, by arranging the variables in some order to indicate their relationships. In this particular case, probably we should regard the acceptance of contraceptives (Y) and the number of electric gadgets (X) in a family as two consequences of the same causal variables such as education, income, etc. It is the people with more education and higher income who tend to accept contraceptives. It is the same people who can afford the new electric appliances. Let W be the aggregate of the variables such as education, occupation, income, etc. The difference between the multiple regression type of analysis and the path diagram that would be proposed for such a case may be roughly indicated by the following two diagrams:

multiple regression   a plausible explanation

Exact methods of reading such diagrams will be given in detail in Chapters 5 and 6. For the time being we shall be contented with intuitive understanding of the visual differences between the two diagrams. In the left diagram, the acceptance of contraceptives (Y) is expressed as a linear function of all other variables, regardless of their meaning, so that a close prediction of Y may be made. In the right diagram, the number of electric products (X) and the acceptance of contraceptives (Y) are regarded as two consequences of the same set of variables W.

Usually a path diagram is a network with directions specified. Therefore, a path analysis enables us to evaluate the direct effect of one cause on an effect and its indirect effect via other causes. It amounts to a subdivision of an observed correlation. Since a correlation coefficient is a standardized covariance, and a covariance may be decomposed into various components, it follows that a correlation coefficient may also be decomposed into various components. A considerable portion of path analysis is nothing more than the decomposition of correlations according to specified structures. These general remarks may not be clear to the reader at the present moment, but he will have ample opportunity to learn and practice the technique of subdividing correlations in later chapters.

In studies in epidemiology and the social sciences, not to mention economics, there are usually a large number of variables to be considered simultaneously. A multiple regression equation involving 30 to 50 variables is not infrequent in certain types of problems. However, our ability to understand and interpret the interrelationships among a large number of variables is quite limited. The physical forces in Nature are also large in number and complicated in interactions; but the physical scientists have the great advantage of experimentation by controlling or eliminating the influences of variables other than those they want to study. They study only a few variables at a time. Biologists are in a similar position except that the "other" variables are not so easily controlled. Hence, a special branch of applied statistics, known as the "design of experiments" involving combinations of several factors, has been developed to cope with multifactorial studies. The social scientists and economists cannot do "experiments" (in the narrow sense) and have to rely on observed events, and hence the interpretation is never as clear-cut as those in the experimental sciences. So much for the justification of studying a large number of variables simultaneously.

But it is not true that nothing can be done about it. Not too long ago, psychology was a purely descriptive subject. Psychologists used to merely observe and describe certain types of people and their behavior. The interpretation was arbitrary and ad hoc, and could often be contradicted by merely observing two more persons. But we now have experimental psychology. The principles of their design of experiments are no different from those in the biological sciences. A more extreme example is geology, the science about rocks, including diamonds. Geologists study the formation and the relationships between the various types of rocks. But now we have experimental geology to test if the traditional theories on the formation of rocks are correct or need revision. It is unnecessary to belabor the point that there is a need for some type of experimental or semi-experimental study in the social sciences. When there is a will, there is a way. When social scientists make up their minds to do experimental studies, they will in due time be able to produce "diamonds;" the artificial diamonds, though not quite good enough for "jewelry," are nevertheless "diamonds" and are good enough for all industrial purposes. After all, the physicists cannot produce a real rainbow, but they can produce a miniature one in the laboratory very easily. And that is convincing enough.

There is another aspect concerning the number of variables to be considered simultaneously. A study of 20 variables is not necessarily more enlightening or fruitful than a study of 10 variables. As a matter of fact, to understand or to explain a phenomenon (any phenomenon), we are looking for the *minimum* number of factors that can account for it. If there are several possible explanations, we choose the one with the least number of factors or assumptions. This is known as the *principle of parsimony* in formulating scientific hypotheses. When two factors are sufficient, we will never introduce a third. In other words, we are always seeking the *simplest* explanation of a phenomenon. A battery of 100 factors that can explain anything and everything in the world is of no use to science. The 14th century philosopher William of Occam is reputed to have said: "entities are not to be multiplied without necessity" *(entia non sunt multiplicanda praeter necessitatem)*—a maxim since known as "Occam's razor." In a study we are not considering all events in a system; rather we are merely studying an isolated subsystem with a small number of variables. The principle of parsimony requires that the inferential route from one event to another be as short as possible. With wise use of Occam's razor we can obtain better understanding with fewer variables rather than with more variables.

When dealing with a small number of variables, each of them must be relevant and play a role in the system. Occam's razor is not intended to cut off the critical entities. In studying the factors influencing the success or achievement (as measured by some criteria) of individuals, social scientists usually include such factors as father's education, father's occupation, father's income, etc. Most of the time it is found that these factors account for very little of the individual's achievement. The characteristics of the individuals themselves (such as their own education) would account for more, but very few of their own characteristics are included in the studies. Hence, the *residual component* (due to unknown factors, to be explained in later chapters) is large, say .80 to .90. This means that some critical factors are not included in the study. It is possible that the missing factors are biological in nature, as most of the social variables have been included. If so, the inclusion of biological factors in social research would make the path analysis more complete and the explanation more plausible. Should the inclusion of biological variables not improve the situation, let it be shown to be so by the results of study.

A scientific hypothesis (in fact, any statement at all), to be useful, must be susceptible of being disproved by obtainable evidence. A statement that cannot be disproved under all circumstances amounts to saying nothing at all. The scientific method, similar to statistical tests, is a procedure seeking evidence to *disprove* a hypothesis. It is essentially a process of elimination of false statements, and therefore it is a self-correcting process. This is our only reliable road to follow toward scientific truth. Thus, we see that preconceived ideas and doctrines, however noble or evil, have no place in science. In the process of seeking evidence to disprove a hypothesis, a true scientist would not hesitate even if the hypothesis were his own. It is irrelevant whose hypothesis it is. He is only interested in how true it is. A scientist has no vested interest to defend, as a businessman does. Daily experience in the real world, however, shows that not everyone follows the rule of scientific argument. (I will skip the details.)

I have ardently supported the "Infant Learning Program," because I thought it could do no harm but might do some good for preschool children. The basic philosophy behind the program is that all children, at birth, have the same learning abilities; they need only the proper stimuli to develop those abilities. The purpose of the infant learning program is to provide the needed stimuli to the children. An investigator in charge of the infant learning program related to me some very encouraging stories. For instance, a three-year old child drew all the time and could draw almost everything he saw; a four-year child played the piano beautifully. When I asked him to what extent these events are consistent with the philosophy that "learning abilities are born equal," he suddenly recalled that I am a geneticist and reassured me: "whatever the reason, it is *not* genetic." I was not thinking about genetic factors at that time. I was merely interested in what would constitute a disproof of the hypothesis that children are born equal in learning abilities.

A blind refusal to accept disproof of a hypothesis would only serve to prolong the wrong course and delay the self-correcting mechanism toward scientific progress. It would be a disservice to the infant program, because it fails to recognize and adjust to the many different natural abilities of the children. Many behavior traits of people, not to mention physical and biochemical traits, have been demonstrated to have a genetic component. To say that learning ability is the one exception is like saying that all celestial bodies are spherical in shape except the earth which is flat. The late Professor Jack Schultz (1973), in discussing the relationships between human values and human genetics, had this to say: "The relation between biological and cultural heredity may be viewed as a very special case of gene action, now perhaps susceptible to analysis. Being so related to gene action, behavior patterns would appear ultimately as part of the system of biological heredity. On this view, our newborn babe—he is still with us—already has its capacities for learning set in very much the same way as its capacity for antibody formation.... The uniqueness of an individual's immunological experience is probably second only to that of his mental experience." This view says that every child is unique in its learning abilities (because of its unique combination of genes)—exactly the opposite of the contention that children are born equal in their learning abilities.

In the early stages of planning for writing this primer, I wrote to a few friends in the social sciences for suggestions and guidance, and received much encouragement. One of them generously advised: "There seems to be a need for such a volume, judging by the kinds of inquiries I receive.... One thing that would be most helpful to sociologists would be to have some explanation of models that involve the concept of heritability, with special reference to I.Q. Not many are acquainted with Sewall Wright's use of Ms. Burks' data in this connection, and for most of us, the very notion of heritability remains murky. This is a case where your background in another field will be an advantage rather than a handicap in writing for social science students." It is this letter that prompted the production of this primer. I followed his advice closely and gained a perfect crutch for what I have written in the next nine chapters: very little about the social sciences. Students in those fields will learn from their own professors of sociology, not from a biologist. But there are three chapters on genetics:

INTRODUCTION

one on the elements of the genetics of random mating populations; followed by one on quantitative inheritance, and then one on inbreeding and assortative mating populations. Maybe I am overdoing it, but I do not know how else I can make the concept of heritability understandable. There are quite a number of expository articles on quantitative inheritance and heritability. They are either too advanced for me (and thus also for the social science students), or so much involved in subject matter (such as certain traits of fruit flies, mice, corn, etc.) that a human geneticist, let alone a sociologist, would have great difficulty and little motivation to follow it through. In presenting the concept of heritability in this volume, I have written as simply as I can, at the sacrifice of brevity and generality. My main concern is its understandability by students not in the field of population genetics. Whether this purpose has been accomplished or not, remains to be seen. I hope I will continue to get advice and guidance from my friends in the social sciences.

The question of heritability arises directly from the simple fact that part of the observed variation of a trait is due to environmental conditions and is not heritable. The old concept of inheritance of acquired characters has long been disproved beyond a reasonable doubt. Lysenko's theory that farm crops can be educated, trained, and induced to acquire certain abilities from the environment and that these abilities (e.g., cold resistance), are then transmitted from one generation to the next, has also been disproved. Soviet scientists had never accepted Lysenko's theory in the first place, despite the fact that his theory was said to be based on "dialectical materialism" and had the full support of the Central Committee of the Communist Party.

Probably most of the superficial traits of an individual are not heritable; they are either due to environmental effects or acquired through training (which is also regarded as environmental effects in the sense that they are not genetically determined). Let σ_G^2 be the variance due to genetic factors, σ_E^2 be the variance due to environmental factors, and $\sigma_T^2 = \sigma_G^2 + \sigma_E^2$ be the total (directly observed) variance of a trait. The fraction $h^2 = \sigma_G^2 / \sigma_T^2$ is called the heritability of the trait under consideration. This is a very rough introduction to the subject. At the present time, it is merely an arbitrary definition. Refinement and explanation will be given in Chapter 8, wherein the relationship between this fraction and the correlation between parent and child will be shown. Only then can one appreciate the usefulness and meaning of the fraction h^2. The point I wish to make here is that the very notion of heritability implies the recognition of environmental effects on the trait. Otherwise, there will be no heritability to speak of ($h^2 = 1$ all of the time). The impression of some psychologists that those who speak of heritability are "hereditarians" and that they ignore the environmental effects is clearly due to misunderstanding. A study of heritability is a study of the effects of both environment and heredity.

There are many excellent textbooks, old and new, on general genetics; but few, if any, of them are written for social science students. Instead of blaming the geneticists, I think that perhaps some biologically-minded social scientists should write one addressed to their own students. Take the problem of variations due to environmental and genetical factors. This is not an abstract conception; it may be demonstrated in a clear-cut manner with suitable experimental materials in exactly the same way as we demonstrate certain physical and chemical properties in a laboratory. It was due to his effort to distinguish the environmental from the hereditary variation that the Danish botanist Johannsen (1857-1927) established the "pure-line theory" in genetics. He was not satisfied with the biometrical school at that time which treated all variations alike. He wanted to isolate and show the distinction between inherited and environmentally-produced variations. Their properties are different. Johannsen worked with beans and the metrical trait to be measured was the size of the seed. The biometrical school merely deals with the magnitude of the differences rather than with their causes. Johannsen found that selection for large or small beans was without effect within pure lines, and that two different lines might be only slightly different in size (with much overlapping), but would maintain this slight difference generation after generation. The variations observed within pure lines are due to environmental conditions and are not heritable. The differences between lines are genetic and heritable. When we study his experimental results and interpretations, we see clearly what is environmental variation and what is genetic variation. Such key genetic terms as *gene*, *genotype*, and *phenotype*, were thus proposed by Johannsen (1909, *Elemente der exakten Erblichkeitslehre)* and have helped to clarify our

INTRODUCTION

thinking. These terms will be explained in Chapter 7. This is not the place to give detailed descriptions of the historical experiments. Suffice it to say here that the environmental and hereditary variation can be isolated and evaluated separately with suitable experimental materials. The fact that this cannot be done with human beings does not nullify its validity, just as the fact that we cannot backcross human beings does not nullify Mendel's law of heredity. It is regretable that "modern" genetics textbooks devote many pages to describing the chemical structure of deoxyribonucleic acid and related molecules, but do not devote a few pages to reviewing experiments and evidence concerning hereditary and environmental variations. Where should a social science student go for a solid understanding of the subject if the textbooks on general genetics have ignored one of the essential problems of heredity?

The writer is a geneticist, not a hereditarian. Being a geneticist has prevented him from being a hereditarian. The classification of people as "environmentalists" or "hereditarians" is unsatisfactory. Although they would differ on one subject, they are very much alike in all other aspects. I would put them in one group rather than in two different groups, because they are more alike than different. They both have a fixed doctrine to believe, a fixed message to sell, a fixed viewpoint toward humanity, and a fixed interpretation of human variation. Consequently, they both ignore evidence contradictory to their beliefs and exaggerate the evidence that seems favorable. They argue blindly and illogically, not following any established rules in science. They often change the subject and shift to irrelevant issues. Frequently they miss the point made by the other side, because they are not listening. These characteristics are incompatible with scientific inquiry. A scientist has no doctrine to preach or vested interest to defend. Like a wondering child, he merely wants to find out something and increase his knowledge. Our knowledge and understanding of human variation is limited indeed and can stand considerable increase.

It is not clear what the "hereditarians" and "environmentalists" are trying to prove. Concerning two varieties of barley, a seed company claims that variety A is better than variety B. A fertilizer salesman claims that his product can increase the yield of barley. There is no connection, let alone contradiction, between the two claims. Are we to call the breeders of the seed companies the hereditarians and the manufacturers of fertilizers the environmentalists? The best yield comes from a particular variety in a compatible environment.

We shall close this introductory chapter with some comments on the history of path analysis. My first encounter with the name Sewall Wright was in 1937 when I was a graduate student, taking a course in "advanced" genetics and using Th. Dobzhansky's *Genetics and the Origin of Species* (1937) as a textbook. This book taught me population genetics and led me to Wright's "Evolution in mendelian populations" (1931) which was repeatedly cited. The long paper proved to be absolutely abstruse to a student who knew little more than Mendel's laws. I thought reading his previous papers would help. While my classmates were looking for new literature, I was constantly searching for the old. Finally, I reached the article which may be used as a starting point: Wright's *Correlation and causation* (1921). Even today, this 1921 paper may still be used as a starting point for learning about path coefficients, because it requires knowledge only of linear regression and partial correlation.

While statisticians may accept "correlation and causation" (1921) as the first systematic presentation of the path method, geneticists may well go back one year to Wright's "The relative importance of heredity and environment in determining the piebald pattern of guinea-pigs" (1920). In the second half of the 1920 paper we see a comprehensive path diagram, the explicit use of path coefficients, and the first few basic theorems of path analysis. The coat pattern of the guinea-pigs was not denoted by a letter such as P (for phenotype, as one would use today), but was actually represented by realistic drawings like the following:

INTRODUCTION

The legend reads: "Diagram illustrating the *casual* relations between litter mates.... The small letters stand for the various path coefficients" [italics mine]. The misprint *casual* gave me considerable uneasiness at first, since the dictionary says it means "coming by chance, without design." After some calm thinking I finally decided that it must have meant *causal*. Then Wright (1920) proceeds to explain:

> The path coefficient, measuring the importance of a given path of influence from cause to effect, is defined as the ratio of the variability of the effect to be found when all causes are constant except the one in question, the variability of which is kept unchanged, to the total variability. Variability is measured by the standard deviation.
>
> It can be shown that the squares of the path coefficients measure the degree of determination by each cause. If the causes are independent of each other, the sum of the squared path coefficients is unity. If the causes are correlated, terms representing joint determination must be recognized.

The two basic formulas given in 1920 are (with respect to his Fig. 6)—

determination: $\quad a^2 + b^2 + c^2 + 2bc\, r_{BC} = 1$

correlation: $\quad r_{XY} = bb' + cc' + br_{BC}c' + c\, r_{BC}b'$

The reader will learn the meaning of these expressions in Chapters 5 and 6.

Then Wright gave genetic applications of these formulas which the reader will learn in Chapter 8. The 1920 paper gave no derivations; its author says: "In a forthcoming paper, a method of estimating the degree to which a given effect is determined by each of a number of causes will be discussed at some length." Apparently he refers to the 1921 paper on correlation and causation which we mentioned before.

If the reader is historically minded, he may even want to go back a little further. The embryo of path analysis is to be found in Wright's "On the nature of size factors" (1918). In this paper Wright was largely concerned with the degree of determination of a dependent variance. It was soon found that the square root of determination is more convenient to use than the coefficient of determination itself. Hence, we may say 1918-20 is the gestation period and 1921 is the actual birthyear of path analysis.

There are five other papers by Wright in 1921, published under the general title "Systems of mating." This series of five papers is perhaps of little interest to general statisticians and social scientists, but it is a "must" for population geneticists. All the basic results on the path relations between relatives and the results of continued inbreeding are given there. The third paper of the series deals with assortative mating by the method of path coefficients. Our Chapter 9 depends heavily upon this pioneer work on assortative mating, although we deal with equilibrium populations only, with minimum use of path coefficients. However, Moran (1961), says: "It is possible to extend the [path] theory to deal with mutation but selection (and similarly assortative mating) seems to lie outside the scope of the method." More up-to-date applications of path coefficients to assortative mating may be found in Wright's volume II, 1969.

The most spectacular success of the path method is in deducing the consequences of continued inbreeding systems. For instance, for continued regular brother-sister mating, the key equation is simply

$$m = r'_{00}$$

which says that the correlation between two mates (m) in one generation is equal to the correlation between brother and sister (r_{00}) of the previous generation (indicated by a prime). Substituting path coefficients, we get in merely two or three lines a recurrence equation relating the proportion of heterozygotes (H) in the population, viz., $H_{t+2} = \frac{1}{2} H_{t+1} + \frac{1}{4} H_t$, where t is the number of generations of inbreeding. Other methods would involve fairly difficult algebra and finding the eigenvalues of matrices. With the path method, Wright (1921) worked out the consequences of many regular and irregular systems of inbreeding. Most animal breeders are still using his results as guidance for inbreeding programs. This topic however, is out of the scope of the present volume.

Enough has been said about the early history of path analysis. A detailed account of Wright's contributions would constitute a book of moderate length and we obviously cannot mention everything, even if we limit ourselves to path analysis. However, we shall mention one more of his early publications which is of considerable interest to economists, especially econometricians. That is his analysis of corn and hog prices (1925) by the path method. This is probably the very first use of structural

INTRODUCTION 15

equations in economic analysis. The story of Schultz and Wright, as reviewed by Goldberger (1974), will be found at the end of Chapter 10, where some practical considerations and examples of application are given to wind up the book.

The next major item that is of special interest to the social sciences and educational disciplines is Wright's path analysis and estimation of heritability of the intelligence scores of true and foster children, the paper being entitled "Statistical methods in biology" (1931). This is another first of its kind. Not only is the subject matter of general interest, but there is a great deal to learn from the methodological point of view. Some details of this historical example will be given in Chapter 10.

A major exposition of path analysis as a statistical method is Wright's "The method of path coefficients" (1934) in the *Annals of Mathematical Statistics*, vol. 5. This paper is self-contained and may also serve as a starting point to learn the method for those who have had the background in statistics. The reader is assumed to be familiar with multiple regression and multiple correlation. Our Chapters 3 and 4 provide such a necessary background. Wright showed that certain equations concerning path coefficients are nothing more than the standardized normal equations. Hence, the reader must be familiar with conventional normal equations before he can recognize their standardized form.

Now, just a few sentences about how the path-coefficient method has suddenly commanded so much attention by social scientists. My first book, *An introduction to population genetics* (1948, National Peking University Press) has a chapter on the method of path coefficients, followed by two chapters of genetic applications, all based on Wright, of course. That book was merely my ten-year (1937-47) progress report to my colleagues who wanted to know why I had not produced a new variety of cotton or wheat for China. (I used to hybridize American and Chinese cottons and do selection of cereal crops.) As late as 1954, Tukey still finds: "The only extensive account outside of Wright's papers known to the writer is in Chapters 11 ff. of the book by Li (1948)." Shortly after that, however, Kempthorne's book (1957) appeared; it also has a section on path coefficients and is also limited to genetic applications. Apparently, neither Li (1955) nor Kempthorne (1957) had any direct means of communication with the social scientists or vice versa. Then, who built the bridge? My interpretation: it is not due to any effort on the part of the geneticists; it is due to the demand from the social scientists themselves. When the atmosphere for interpretative analysis is ripe, they will discover the path method, one way or another. There will be no way to hide it. Geneticists will recall that in 1900 when the scientific atmosphere for plant hybridization was ripe, not one but three biologists rediscovered Mendel's law simultaneously. In the present case, the popularization of the path method in the social sciences is largely due to Duncan's *Path analysis: sociological examples* (1966). However, Boudon's "dependence analysis" (1965) actually appeared one year earlier. The so-called dependence analysis is path analysis. Boudon says: "I use the word 'dependence' where Wright uses 'path,' ... it [path] obscures the logic of the analysis." We shall not quibble about terminology. I would like to point out that Blalock has been concerned with causal and interpretative analysis for years. In the appendix of his *Causal inferences in nonexperimental research* (1961) he mentioned the method of path coefficients and gave three references to Wright. It is possible that the discovery of Wright was made after the book was written; be that as it may, that was in 1961, several years before Boudon and Duncan. In the last 7 or 8 years (since 1966), the phenomenon in the social sciences was very much like that in the few years after 1900 in the biological world—everybody was hybridizing something and everybody confirmed the "3 : 1 ratio." But social variables are more complicated than unit-factor inheritance. This is why, at the end of Chapter 10, this book is closed with a warning rather than with a cheer.

Chapter 2

IDENTITIES AND EQUATIONS

BEFORE WE proceed to the study of linear regression and path coefficients it would be well for us to clear up some purely algebraic matters which are by themselves not a part of statistics, although we shall make use of them frequently. Algebraic identities and equations are two different kinds of mathematical expressions and they must be distinguished even by a nonmathematical reader. First we shall consider some examples of identities.

Algebraic Identities

1. If a, b, c, d are four numbers, any four numbers at all, we know from elementary algebra that

$$(a+b+c+d)^2 = a^2 + b^2 + c^2 + d^2 + 2ab + 2ac + 2ad + 2bc + 2bd + 2cd \quad (1)$$

which may be easily verified by substituting arbitrary values for the letters and carrying out the arithmetic. The right-hand side of the expression above consists of two kinds of terms. One is $a^2 + b^2 + c^2 + d^2$ which is the sum of squares of the individual numbers. The other is $2(ab + ... + cd)$, twice the sum of all the possible products of two numbers if we write the numbers of a product in the order of the alphabet ($2ab$ instead of $ab + ba$; etc.). Expression (1) is always true, no matter what the four numbers are. In other words, the left and right sides of expression (1) are merely two different ways of writing the same quantity. Expressions of this nature are called *algebraic identities*.

2. An algebraic identity may be simple or complex. Frequently an identity may be generalized. The expression (1) is an identity for four numbers. It is also true for three numbers by omitting d from the expression. It is also true for five, six, or more numbers. Hence, if $y_1, y_2, ..., y_n$ are n numbers, the square of their sum will be

$$(y_1 + ... + y_n)^2 = y_1^2 + ... + y_n^2 + 2y_1 y_2 + ... + 2y_{n-1} y_n$$

In shorthand, it may be written

$$(\Sigma y_i)^2 = \Sigma y_i^2 + 2\Sigma y_i y_j \qquad \text{where} \quad (i<j) \qquad (2)$$

The summation is understood to be taken with respect to $i = 1, 2, ..., n$; and the condition $i < j$ means that we write $2y_3 y_5$ instead of $y_3 y_5 + y_5 y_3$, etc. Identities will enable us to simplify algebraic expressions.

3. As additional examples of identities we may consider the following two:

$$(a - b)(c - d) = ac + bd - bc - ad \qquad (3)$$

$$(a - b)(c - d) = (a - b)c - (a - b)d \qquad (4)$$

These show that an algebraic identity may be written in more than one way. In expression (3) we write the two positive terms first and then the two negative terms. In expression (4) we preserve the quantity $(a - b)$ and multiply it by c and $-d$. Despite their different appearances, the quantities in (3) and (4) are of course always equal.

From one identity we may obtain others by assigning a particular value to the general symbol. For instance, if $c = a$ and $d = b$, then identity (3) becomes

$$(a - b)(a - b) = (a - b)^2 = a^2 + b^2 - 2ab \qquad (5)$$

The square of a quantity is merely a special type of multiplication. In the following paragraphs there will be an application of each of the three identities shown above.

4. Since the left and right sides of an identity are always equal, it follows that the sum of two identities will also be an identity. More generally, if there are a series of identities, their sum will be another identity. As an illustration, consider four (or more if you wish) pairs of numbers (x_1, y_1), (x_2, y_2), (x_3, y_3), (x_4, y_4) and let $\bar{x} = \Sigma x/n$ and $\bar{y} = \Sigma y/n$ be the mean values of the x's and the y's respectively. By identity (3) we write for each product

IDENTITIES AND EQUATIONS

$$(x_1 - \bar{x})(y_1 - \bar{y}) = x_1 y_1 + \bar{x}\bar{y} - \bar{x} y_1 - \bar{y} x_1$$
$$(x_2 - \bar{x})(y_2 - \bar{y}) = x_2 y_2 + \bar{x}\bar{y} - \bar{x} y_2 - \bar{y} x_2$$
$$(x_3 - \bar{x})(y_3 - \bar{y}) = x_3 y_3 + \bar{x}\bar{y} - \bar{x} y_3 - \bar{y} x_3 \quad (6)$$
$$(x_4 - \bar{x})(y_4 - \bar{y}) = x_4 y_4 + \bar{x}\bar{y} - \bar{x} y_4 - \bar{y} x_4$$

We may generalize the situation by adding $(x_5 - \bar{x})(y_5 - \bar{y}), ..., (x_n - \bar{x})(y_n - \bar{y})$. In summing, we note that $\bar{x} \Sigma y = n\bar{x}\bar{y}$ and similarly $\bar{y} \Sigma x = n\bar{x}\bar{y}$. Hence the sum of these n identities is

$$\Sigma(x_i - \bar{x})(y_i - \bar{y}) = \Sigma x_i y_i + n\bar{x}\bar{y} - n\bar{x}\bar{y} - n\bar{x}\bar{y}$$
$$= \Sigma x_i y_i - n\bar{x}\bar{y} = \Sigma x_i y_i - \frac{(\Sigma x_i)(\Sigma y_i)}{n} \quad (7)$$

This is a very useful identity. Whenever the quantity $\Sigma(x_i - \bar{x})(y_i - \bar{y})$ is needed, we do not have to calculate the individual deviations $(x_i - \bar{x})$ and $(y_i - \bar{y})$ but merely calculate the products $x_i y_i$ directly and obtain their sum, from which we subtract the quantity $(\Sigma x)(\Sigma y)/n$ which is very easy to calculate.

5. In a similar way we can make use of identity (4) in obtaining the sum of products of deviations. For each product we write according to (4)

$$(x_1 - \bar{x})(y_1 - \bar{y}) = (x_1 - \bar{x}) y_1 - (x_1 - \bar{x}) \bar{y}$$
$$\cdots \qquad \cdots$$
$$(x_n - \bar{x})(y_n - \bar{y}) = (x_n - \bar{x}) y_n - (x_n - \bar{x}) \bar{y}$$

In summing, we note that the second term on the right side is $\bar{y} \Sigma(x_i - \bar{x}) = 0$, because $\Sigma(x_i - \bar{x}) = 0$. Hence the sum of these n identities is

$$\Sigma(x_i - \bar{x})(y_i - \bar{y}) = \Sigma(x_i - \bar{x}) y_i \quad (8)$$

which is of course also identical with (7), as $\Sigma(x_i - \bar{x}) y_i = \Sigma x_i y_i - n\bar{x}\bar{y}$. The new form (8) is also useful for certain problems in statistics.

6. We have already noted that the simple identity (5) concerning $(a - b)^2$ is a special case of the identity (3) concerning $(a - b)(c - d)$. Thus, by replacing $(x_i - \bar{x})$ by $(y_i - \bar{y})$ in the expressions (6), we will obtain the identity concerning the square of a deviation:

$$(y_i - \bar{y})^2 = y_i^2 + \bar{y}^2 - 2\bar{y}\, y_i$$

Summing such n identities, we obtain

$$\Sigma\, (y_i - \bar{y})^2 = \Sigma\, y_i^2 + n\bar{y}^2 - 2n\bar{y}^2 = \Sigma y_i^2 - \frac{(\Sigma y_i)^2}{n} \tag{9}$$

This is a special case of identity (7). If we replace all the x's by y's in (7) we will immediately get (9). Similarly, if we replace all the y's by x's in (7), we will obtain

$$\Sigma\, (x_i - \bar{x})^2 = \Sigma\, x_i^2 - n\bar{x}^2 = \Sigma x_i^2 - \frac{(\Sigma x_i)^2}{n} \tag{10}$$

The quantity (7) is called the *sum of products* (of deviations from mean) and the quantities (9) and (10) are called the *sum of squares* (of deviations from mean). We need these three quantities to study linear regression and correlation in the following chapters. For a geometrical representation of the identities (7) and (9), the reader is referred to Li (1955; 1964b).

7. Perhaps another form for the sum of products and the sum of squares would help us to gain further insight into these quantities. The new identity may be illustrated by four pairs of numbers. Instead of using the deviations from mean, we may use the differences between all possible values of x and of y. For four numbers there are six possible differences; thus

$$\begin{aligned}
(x_1 - x_2)(y_1 - y_2) &= x_1 y_1 + x_2 y_2 - x_2 y_1 - x_1 y_2 \\
(x_1 - x_3)(y_1 - y_3) &= x_1 y_1 + x_3 y_3 - x_3 y_1 - x_1 y_3 \\
(x_1 - x_4)(y_1 - y_4) &= x_1 y_1 + x_4 y_4 - x_4 y_1 - x_1 y_4 \\
(x_2 - x_3)(y_2 - y_3) &= x_2 y_2 + x_3 y_3 - x_3 y_2 - x_2 y_3 \\
(x_2 - x_4)(y_2 - y_4) &= x_2 y_2 + x_4 y_4 - x_4 y_2 - x_2 y_4 \\
(x_3 - x_4)(y_3 - y_4) &= x_3 y_3 + x_4 y_4 - x_4 y_3 - x_3 y_4
\end{aligned} \tag{11}$$

IDENTITIES AND EQUATIONS

It is seen that among the positive terms, each term of the type $x_i y_i$ appears three times. In general, if there are n pairs, each $x_i y_i$ term will appear $n-1$ times. Among the negative terms, if we collect them according to y, we shall have, for instance, $(x_2 + x_3 + x_4) y_1$. If we collect them according to x, we shall have, for instance, $(y_2 + y_3 + y_4) x_1$. The terms are symmetrical with respect to x and y. In taking the sum of the six identities in (11), let us add the following quantity which is zero:

$$x_1 y_1 + x_2 y_2 + x_3 y_3 + x_4 y_4 - x_1 y_1 - x_2 y_2 - x_3 y_3 - x_4 y_4 = 0$$

so that the negative terms will be $(x_1 + x_2 + x_3 + x_4) y_1 + \ldots + (x_1 + x_2 + x_3 + x_4) y_4$. Then the sum of the identities of (11) becomes in the general case under the condition $i<j$,

$$\Sigma (x_i - x_j)(y_i - y_j) = n \Sigma x_i y_i - (\Sigma x_i)(\Sigma y_i) = n \left\{ \Sigma x_i y_i - \frac{(\Sigma x_i)(\Sigma y_i)}{n} \right\} \quad (12)$$

which is exactly n times the sum of products of deviations from the mean.

8. In exactly the same way a similar relationship may be established for the sum of squares. Or one may simply replace the x's in (11) and (12) by the corresponding y's and then find that

$$\Sigma (y_i - y_j)^2 = n \Sigma y_i^2 - \Sigma y_i^2 - 2 \Sigma y_i y_j = n \left\{ \Sigma y_i^2 - \frac{(\Sigma y_i)^2}{n} \right\} \quad (13)$$

recalling identity (2) that $\Sigma y_i^2 + 2 \Sigma y_i y_j = (\Sigma y_i)^2$ where $i<j$. In conclusion, *the sum of squares of all pair-wise differences is exactly n times the ordinary sum of squares of deviations from the mean.* The last two quantities, (12) and (13), are not frequently encountered in statistics; but we shall use them to elucidate the geometrical meaning of a regression coefficient in the next chapter.

Linear Equations

9. *An equation is a statement of a condition that a certain number should satisfy.* For instance, we want to know what value should the

unknown number b take such that

$$14.5\,b + 29 = 0. \tag{14}$$

The condition (14) is called an *equation.* The unknown b can no longer take any value whatever as it does in identities, but must take a particular value to satisfy the requirement (14). It is called a *linear equation* because it involves only b, and does not involve higher powers of b or products with other unknowns. The particular value of b that satisfies (14) is called the *solution* of the equation. In this simple case, $b = -29/14.5 = -2$ is the solution, and there are no other solutions. Thus we say the solution is *unique*.

10. Most of the time we face a set of *simultaneous equations* to be solved. For instance, the following is a set of two linear equations with two unknowns: a and b.

$$\begin{array}{rl} \text{(i)} & 10\,a + 25\,b = 62 \\ \text{(ii)} & 25\,a + 77\,b = 126 \end{array} \tag{15}$$

In high school we learned that we may multiply the first equation by 25 and multiply the second equation by 10 so that the coefficients of a in both equations are 250. Then by subtraction the unknown a may be eliminated, resulting in a simple equation for b only. In most equations that arise from practical problems, however, the coefficients are large numbers, so that if a calculator is available, it is more convenient to divide the first and second equations by 10 and 25, respectively. Thus, the two equations become

$$\begin{array}{rl} \text{(i)}' & a + 2.50\,b = 6.20 \\ \text{(ii)}' & a + 3.08\,b = 5.04 \end{array}$$

$$\begin{array}{rl} \text{(ii)}' - \text{(i)}' & 0.58\,b = -1.16 \\ & b = -2.00 \\ \text{Substituting in (i)}' & a = 11.20 \end{array} \tag{16}$$

which is the desired solution of equations (15).

IDENTITIES AND EQUATIONS

11. *Two linear equations in two unknowns have a unique solution only when they are consistent and linearly independent.* Since a discussion of these requirements is beyond the scope of this primer, we can illustrate their meaning by looking at the following two sets of equations:

$$\left.\begin{array}{r}10a + 25b = 62 \\ 30a + 75b = 126\end{array}\right\} \qquad \left.\begin{array}{r}10a + 25b = 62 \\ 30a + 75b = 186\end{array}\right\}$$

The first set is *inconsistent*, because if $10a + 25b = 62$, then $3(10a + 25b)$ should be $3 \times 62 = 186$ and it cannot be 126. There is no solution to contradictory statements. The second set is *consistent but not independent*, because the second equation merely repeats what the first equation says; it gives us no additional condition with respect to the two unknowns. In such a case, all we can say is that $a = (62 - 25b)/10$, but there is no unique solution. Suffice it to say that the linear equations in regression problems covered in this book are consistent and linearly independent.

12. *Two linear equations in two unknowns* arise so frequently that it will be useful to derive a *general solution* at this stage. Let the two linear equations be

$$\begin{align}\text{(i)} \quad W_1 a + Z_1 b &= Y_1 \\ \text{(ii)} \quad W_2 a + Z_2 b &= Y_2\end{align} \tag{17}$$

where a and b are the unknowns to be solved and the capital letters W, Z, Y are known numbers. We proceed as we did in high school, multiplying the first equation by W_2 and the second by W_1:

$$\begin{align}\text{(i)}' \quad W_1 W_2 a + W_2 Z_1 b &= W_2 Y_1 \\ \text{(ii)}' \quad W_1 W_2 a + W_1 Z_2 b &= W_1 Y_2\end{align}$$

$$\text{(ii)}' - \text{(i)}' \quad (W_1 Z_2 - W_2 Z_1) b = W_1 Y_2 - W_2 Y_1$$

We now assume that $W_1 Z_2 \neq W_2 Z_1$. Solving,

$$b = \frac{W_1 Y_2 - W_2 Y_1}{W_1 Z_2 - W_2 Z_1} \qquad (18)$$

Most readers are familiar with solution (18). For those who encounter it for the first time, formula (18) may seem complicated and difficult to remember. For this reason we will write it in another form which requires little memory.

$$b = \frac{\begin{vmatrix} W_1 & Y_1 \\ W_2 & Y_2 \end{vmatrix}}{\begin{vmatrix} W_1 & Z_1 \\ W_2 & Z_2 \end{vmatrix}} \qquad (19)$$

The arrangement of four numbers into a 2 × 2 square bordered by two vertical bars is called a *determinant*. It is not practical to go into the subject of determinants in this primer; we use it here purely as a mnemonic device to write formulas which otherwise would be difficult to remember. The four numbers in the denominator of (19) are the four coefficients on the left side of equations (17), arranged in precisely the same way. The determinant in the numerator of (19) is obtained from the determinant in the denominator by replacing Z_1, Z_2 by Y_1, Y_2. In other words, when we are solving for b, we replace the coefficients of b by the numbers on the right side of the equations (17). *The value of a 2 × 2 determinant is the product of the two numbers in the principal diagonal minus the product of the two other numbers;* thus,

$$\begin{vmatrix} W_1 & Z_1 \\ W_2 & Z_2 \end{vmatrix} = W_1 Z_2 - W_2 Z_1 ; \qquad \begin{vmatrix} W_1 & Y_1 \\ W_2 & Y_2 \end{vmatrix} = W_1 Y_2 - W_2 Y_1 \qquad (20)$$

The solutions (18) and (19) are the same. If we solve for a, the denominator remains the same; the numerator is obtained from the denominator by replacing the coefficients of a (that is, W_1, W_2) by the numbers on the right of the equations. Thus,

IDENTITIES AND EQUATIONS

$$a = \frac{\begin{vmatrix} Y_1 & Z_1 \\ Y_2 & Z_2 \end{vmatrix}}{\begin{vmatrix} W_1 & Z_1 \\ W_2 & Z_2 \end{vmatrix}} = \frac{Y_1 Z_2 - Y_2 Z_1}{W_1 Z_2 - W_2 Z_1} \tag{21}$$

After a few minutes of practice, the reader will be able to write down the solutions of two linear equations as soon as the equations are given. For instance, the numerical solution for b of equations (15) is

$$b = \frac{\begin{vmatrix} 10 & 62 \\ 25 & 126 \end{vmatrix}}{\begin{vmatrix} 10 & 25 \\ 25 & 77 \end{vmatrix}} = \frac{(10)(126) - (25)(62)}{(10)(77) - (25)(25)} = \frac{-290}{145} = -2.00$$

in agreement with our previous solution (16). In this connection we note that there is little saving in arithmetic labor in using the formulas (18) or (19). These formulas are useful primarily for demonstrating certain relationships between various quantities we shall encounter in later chapters.

13. In a similar way a set of *three linear* (consistent and independent) *equations* may be solved. In the following, the capital letters W and Y are known numbers and the lower case a, b, c are the unknowns to be solved.

$$\begin{aligned} W_{11}\, a + W_{12}\, b + W_{13}\, c &= Y_1 \\ W_{21}\, a + W_{22}\, b + W_{23}\, c &= Y_2 \\ W_{31}\, a + W_{32}\, b + W_{33}\, c &= Y_3 \end{aligned} \tag{22}$$

The general solution may be written in terms of determinants as in the case for two equations. Let us copy down the nine W's exactly the way they appear in the equations and form the determinant

$$D = \begin{vmatrix} W_{11} & W_{12} & W_{13} \\ W_{21} & W_{22} & W_{23} \\ W_{31} & W_{32} & W_{33} \end{vmatrix} \neq 0 \qquad (23)$$

Let D_1 be the determinant formed by replacing the first column of D by the three Y's. That is, the three numbers (W_{11}, W_{21}, W_{31}) are replaced by (Y_1, Y_2, Y_3), respectively. Let D_2 be the determinant formed by replacing the second column of D by the three Y's, and D_3 be the determinant formed by replacing the third column. Then the general solution is

$$a = \frac{D_1}{D}, \quad b = \frac{D_2}{D}, \quad c = \frac{D_3}{D} \qquad (24)$$

These formulas are useful for demonstrating certain mathematical relationships but are not necessarily convenient for numerical calculation. The numerical evaluation of a determinant larger than 3 x 3 is itself a task, and we shall not go into that subject here.

14. Now we shall give the *easiest numerical procedure for solving three linear equations*. We say the "easiest," not the best nor the simplest, because the method to be illustrated does not require any new learning or memory. It is merely the high school method of elimination. There are various shortcuts and checking methods, depending on the nature of the equations, but we shall use the elementary process which always works. Consider the equations

$$\begin{aligned} 5a + 7b + 11c &= 68.3 \\ 8a + 14b + 6c &= 40.6 \\ 20a + 30b + 13c &= 108.3 \end{aligned} \qquad (25)$$

Using a calculator, divide each equation of (25) by its coefficient of a

IDENTITIES AND EQUATIONS

$$\text{(i)} \quad a + 1.40b + 2.20c = 13.660$$

$$\text{(ii)} \quad a + 1.75b + 0.75c = 5.075$$

$$\text{(iii)} \quad a + 1.50b + 0.65c = 4.415$$

$$\text{(ii)} - \text{(i)} \quad 0.35b - 1.45c = -8.585$$

$$\text{(iii)} - \text{(i)} \quad 0.10b - 1.55c = -8.245$$

The last two equations are like (15), two equations in two unknowns, which may be solved readily. To proceed systematically, divide each of the last two equations by its coefficient of b:

$$\text{(iv)} \quad b - 4.143c = -24.5286$$

$$\text{(v)} \quad b - 15.500c = -82.4500$$

$$\text{(iv)} - \text{(v)} \quad 11.357c = 57.9214;$$

solving, $\quad c = 5.10$

substituting in (v), $\quad b = -3.40$

substituting in (iii), $\quad a = 7.20$

These values of a, b, c are the desired solutions of the original equations (25). Should there be any doubt at all, substitute these values in the original equations and see that they do satisfy them. In regression problems the coefficients on the left side of the equations are symmetrical; that is, $W_{12} = W_{21}$, $W_{13} = W_{31}$, $W_{23} = W_{32}$ in equations (22). There are special and shorter methods of solving such equations but the elementary method we described above will work for such equations too. The procedure illustrated for three equations can obviously be extended to solving four linear equations with four unknowns, and so on.

Nonlinear Equations

15. While the properties of linear equations and the methods of solving them are well understood, the solution of a set of *simultaneous nonlinear equations* is far less so, except for very simple cases. There is no one standard procedure to follow because of the great variety of possible forms of nonlinearity, except that of substituting one relationship (equation) in the other. Here we shall give the simplest example of the situation by considering the following two equations:

$$x + y = 10, \qquad xy = 21 \qquad (26)$$

For such a simple case we proceed formally by substituting, say, $x = 21/y$ in the first equation which then becomes $21/y + y = 10$, or $y^2 - 10y + 21 = 0$. Solving, we find $y = 7$ or 3. Thus the two solutions of the equations (26) are

$$(x, y) = (3, 7); \qquad (x, y) = (7, 3) \qquad (27)$$

The reader will recall that for two linear (independent and consistent) equations, there is one and only one solution, but here there are two. The figure below gives a geometrical illustration of the situation. The solutions are given by the intersection points of the two equations (26) because these points satisfy both conditions. Two straight lines, if not parallel, have only one point of intersection and therefore two linear equations have only one solution.

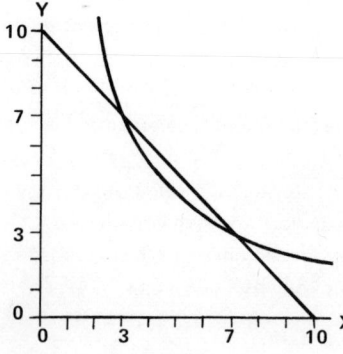

Graphs of $x + y = 10$ and $xy = 21$, showing two points of intersection: $(x, y) = (3, 7)$ and $(7, 3)$, corresponding to the two solutions of the two simultaneous equations.

IDENTITIES AND EQUATIONS

16. The geometrical representation of the two equations (26) leads immediately to further conclusions. If the straight line and the curve have only one point in common, there will be only one solution. If they have no point in common, there will be no real solution at all. Thus, the two equations $x + y = 10$ and $xy = 25$ have only one solution, viz., $(x, y) = (5, 5)$. The two equations $x + y = 10$ and $xy = 26$, or any number greater than 25, will have no solution in real numbers.

17. In the example above, one equation is linear representing a straight line and one is a curve of the second degree. When both equations are of the second degree, the diagrams in the following figure show there could be 0, 1, 2, 3, or 4 solutions corresponding to the number of intersection points of these two curves.

Graphs of two second degree curves. From left to right the successive diagrams show 0, 1, 2, 3, 4 points of intersection of the two curves. Accordingly, there are 0, 1, 2, 3, 4 solutions to the two equations.

As an exercise the student may draw the graphs of two other equations

$$x^2 + y^2 = 34, \qquad xy = 15 \qquad (28)$$

The first equation gives a circle with the center at the origin and radius $\sqrt{34} = 5.831$. The second equation gives two curves: one for positive numbers such as $(5)(3) = 15$ and one for negative numbers such as $(-5)(-3) = 15$. The reader will find that there are four solutions (that is, there are four intersection points):

$$(x, y) = (5, 3), \quad (3, 5), \quad (-5, -3), \quad (-3, -5) \qquad (29)$$

The reader's graphs will be very different from those shown on page 29. Due to the symmetry of the curves, the number of solutions is always even, viz., 0, 2, 4. Thus we see that the curves may intersect in many different ways, depending on the nature of the equations. More generally, if one equation, $f(x, y) = 0$, is of degree m and the other equation, $g(x, y) = 0$, is of degree n, then there could be as many as mn solutions. When there are three variables and three equations, the situation could be very complicated. In solving nonlinear equations, it is in general not enough to obtain merely a solution; on the other hand, it is usually a great task to obtain all possible solutions. In certain applied problems, however, the investigator usually limits himself only to those solutions that are relevant and meaningful to the problem under consideration. For instance, if we consider positive numbers only, then there are only two solutions to equations (28). If, due to the nature of the problem, only $y > x$ is meaningful, then there is only one solution. This amounts to adding further restrictions to the original equations.

18. Not all nonlinear equations are difficult to solve. Fortunately, the nonlinear equations we shall encounter in path analysis are quite simple; the solutions may be found by straightforward successive substitution. As an example, consider the following set of four equations in four unknowns (p_1, p_2, p_3, p_4):

$$
\begin{align}
\text{(i)} \quad & p_4 = -0.48 \\
\text{(ii)} \quad & p_2 + p_1 p_3 p_4 = +0.56 \\
\text{(iii)} \quad & p_2 p_4 + p_1 p_3 = -0.66 \\
\text{(iv)} \quad & p_1^2 + p_2^2 + 2 p_1 p_3 p_2 p_4 = 1.00
\end{align}
\tag{30}
$$

The solutions are not unique unless we require *a priori* that the sign of certain p's should be positive or negative. In any specific concrete problem, these decisions must be made by the investigator on the basis of his knowledge about the p's; it is not a mathematical requirement. In other words,

IDENTITIES AND EQUATIONS

the investigator is only interested in the solution that is meaningful to the problem under consideration. Let us assume that p_1 in the equations above must be a positive number on account of its biological or sociological meaning.

19. We now proceed to solve the equations (**30**). From (i) we already know that $p_4 = -0.48$. Substituting this value in (ii) and (iii) we obtain

$$\text{(ii)}' \qquad p_2 - 0.48\, p_1 p_3 = 0.56$$
$$\text{(iii)}' \qquad -0.48\, p_2 + p_1 p_3 = -0.66 \qquad (31)$$

These new equations, though nonlinear, may be regarded at this stage as two linear equations with two unknowns: p_2 and $(p_1 p_3)$, the latter as one unknown. Then, using the familiar method of solving two linear equations, we find

$$p_2 = 0.3160, \qquad p_2 p_4 = -0.1517, \qquad p_1 p_3 = -0.5083 \qquad (32)$$

The final step is substituting these numerical values in (iv) of (**30**) and transposing

$$p_1^2 = 1 - (0.3160)^2 - 2(-0.5083)(-0.1517) = 0.7459$$

yielding $p_1 = \sqrt{0.7459} = +0.8637$, discounting the possibility $p_1 = -0.8637$. This gives $p_3 = -0.5083 / 0.8637 = -0.5885$. Now, all the p's are known. Since the original equations are given in two significant figures, we may round off the answers also to two significant figures:

$$p_1 = +0.86, \qquad p_2 = +0.32, \qquad p_3 = -0.59, \qquad p_4 = -0.48 \qquad (33)$$

which may be accepted as a meaningful set of solutions to the nonlinear equations (**30**).

IDENTITIES AND EQUATIONS

Exercises

Ex. 1. Verify the identities (7), (8), (9), (10) with the following set of six pairs of numbers. The subscript i has been omitted for brevity.

x	y	$x-\bar{x}$	$y-\bar{y}$	$(x-\bar{x})^2$	$(x-\bar{x})(y-\bar{y})$	$(y-\bar{y})^2$	$(x-\bar{x})y$
2	3	-3	-4	9	12	16	-9
4	4	-1	-3	1	3	9	-4
4	6	-1	-1	1	1	1	-6
6	7	+1	0	1	0	0	+7
7	10	+2	+3	4	6	9	+20
7	12	+2	+5	4	10	25	+24
$\Sigma x = 30$ $\Sigma y = 42$		0	0	20	32	60	32
$\bar{x} = 5$ $\bar{y} = 7$		$\Sigma(x-\bar{x})$,	$\Sigma(y-\bar{y})$	$\Sigma(x-\bar{x})^2$	$\Sigma(x-\bar{x})(y-\bar{y})$	$\Sigma(y-\bar{y})^2$	$\Sigma(x-\bar{x})y$

These are the sums of products or squares of deviations from the mean calculated according to the left side of the identities. We have already verified identity (8), $\Sigma(x-\bar{x})(y-\bar{y}) = \Sigma(x-\bar{x})y = 32$ in the above. We proceed to verify identities (7), (9), (10).

x	y	x^2	xy	y^2
2	3	4	6	9
4	4	16	16	16
4	6	16	24	36
6	7	36	42	49
7	10	49	70	100
7	12	49	84	144
$\Sigma x = 30$,	$\Sigma y = 42$	$\Sigma x^2 = 170$	$\Sigma xy = 242$	$\Sigma y^2 = 354$
$(\Sigma x)^2 = 900$,	$(\Sigma y)^2 = 1764$	$\dfrac{(\Sigma x)^2}{6} = 150$	$\dfrac{(\Sigma x)(\Sigma y)}{6} = 210$	$\dfrac{(\Sigma y)^2}{6} = 294$
	subtracting,	20	32	60
		$\Sigma(x-\bar{x})^2$	$\Sigma(x-\bar{x})(y-\bar{y})$	$\Sigma(y-\bar{y})^2$

IDENTITIES AND EQUATIONS

Thus we have verified $\Sigma (x - \bar{x})(y - \bar{y}) = \Sigma xy - (\Sigma x)(\Sigma y)/n$, etc. From now on, we shall not give such detailed arithmetic in each case for our numerical examples. When n pairs of (x, y) are given, we will merely give the numerical values of quantities $\Sigma (x - \bar{x})^2$, $\Sigma (x - \bar{x})(y - \bar{y})$, $\Sigma (y - \bar{y})^2$; and it is understood that they have been obtained according to the method indicated above. However, as exercises, the reader should do his own calculation for each numerical example.

Ex. 2. If we add (or subtract) a constant quantity to (or from) each y value, then the new value is $y' = y + c$, and the new mean value will be $\bar{y}' = \bar{y} + c$, so that the new deviation will be $y' - \bar{y}' = y - \bar{y}$ = original deviation. Similar remarks may be made with respect to x. As an illustration, let us subtract *1* from each x value and subtract *2* from each y value of those in Ex. 1. The new values will be as follows (dropping the prime):

x: 1, 3, 3, 5, 6, 6; $\Sigma x = 24$, $\bar{x} = 4$

y: 1, 2, 4, 5, 8, 10; $\Sigma y = 30$, $\bar{y} = 5$

Calculate $\Sigma (x - \bar{x})^2$, $\Sigma (x - \bar{x})(y - \bar{y})$, $\Sigma (y - \bar{y})^2$, and see that they remain the same as in Ex. 1; viz., 20, 32, 60, respectively. If we multiply x by h and multiply y by k, we see that

$$\Sigma (hx - h\bar{x})^2 = h^2 \Sigma (x - \bar{x})^2$$

$$\Sigma (hx - h\bar{x})(ky - k\bar{y}) = hk \Sigma (x - \bar{x})(y - \bar{y})$$

$$\Sigma (ky - k\bar{y})^2 = k^2 \Sigma (y - \bar{y})^2$$

Ex. 3. The following is a set of simultaneous linear equations that is particularly easy to solve.

$8a$	+	$120b$	+	$80c$	=	48.0
$120a$	+	$1928b$	+	$1200c$	=	758.4
$80a$	+	$1200b$	+	$1000c$	=	544.0

IDENTITIES AND EQUATIONS

Dividing each equation by its coefficient of a,

$$
\begin{align}
\text{(i)} \quad & a + 15.00\, b + 10.0\, c = 6.00 \\
\text{(ii)} \quad & a + 16.0\dot{6}\, b + 10.0\, c = 6.32 \\
\text{(iii)} \quad & a + 15.00\, b + 12.5\, c = 6.80
\end{align}
$$

$$
\begin{align}
\text{(ii) - (i)} \quad & 1.0\dot{6}\, b = 0.32; \\
& b = 0.30 \\
\text{(iii) - (i)} \quad & 2.5\, c = 0.80; \\
& c = 0.32
\end{align}
$$

Substituting in (i) $a = 6.00 - 4.50 - 3.20 = -1.70$

In dividing the second original equation by *120*, we note that *1928/120* = *241/15* = *16(1/15)* = *16.0$\dot{6}$* which means *16.066666....* In the final solution, we have $b = 0.32/1.06 = 0.32 \times 15/16 = 0.30$ exactly.

Ex. 4. As a preliminary exercise in solving nonlinear equations, consider the following three equations in three unknowns, p_1, p_2, p_3.

(i) $p_1 p_2 = 0.50$;   (ii) $p_1 p_3 = 0.40$;   (iii) $p_2 p_3 = 0.45$

There are several different ways in which substitutions may be made. The student should try his own method first. The author simply takes the product of (ii) and (iii) and then divides it by (i), obtaining

$$p_3^2 = \frac{0.40 \times 0.45}{0.50} = 0.36$$

Limiting ourselves to positive solutions, we take $p_3 = 0.60 = 3/5$. Substituting in (ii) and (iii), we obtain $p_1 = 2/3$ and $p_2 = 3/4$, respectively. Note that if we took $p_3 = -0.60$, then p_1 and p_2 would also be negative.

IDENTITIES AND EQUATIONS

Ex. 5. Should the reader wish to gain more practice in solving nonlinear equations, the following four equations with four unknowns will provide an opportunity to do so.

(i) $\quad bs \quad + \quad e \quad\quad\quad\quad = \quad 0.49$

(ii) $\quad br \quad + \quad 0.86\,e \quad\quad = \quad 0.61$

(iii) $\quad b^2 \quad + \quad e^2 \quad + \quad 2\,bse \quad = \quad 1.00$

(iv) $\quad b/e \quad = \quad 96/29 \quad = \quad 3.31$

There is no standard procedure to follow except that of convenient substitution. In this particular case, b may be eliminated by substituting (iv), resulting in three equations with three unknowns. Thus, dividing the first two equations by e and the third equation by e^2 and using $b^2/e^2 = (3.31)^2 = 10.958$, we obtain

(i)' $\quad 3.31\,s \quad + \quad 1.00 \quad = \quad 0.49/e$

(ii)' $\quad 3.31\,r \quad + \quad 0.86 \quad = \quad 0.61/e$

(iii)' $\quad 10.958 \quad + \quad 1 \quad + \quad 6.62\,s \quad = \quad 1/e^2$

From (i)' and (iii)' we may eliminate s, resulting in a quadratic equation in e only.

2(i)' $\quad 6.62\,s \quad\quad\quad\quad = \quad 0.98/e \quad - \quad 2$

(iii)' $\quad 6.62\,s \quad\quad\quad\quad = \quad 1/e^2 \quad - \quad 11.958$

Subtracting $\quad 9.958 \quad + \quad 0.98/e \quad - \quad 1/e^2 \quad = \quad 0$

that is, $\quad\quad 9.958\,e^2 \quad + \quad 0.98\,e \quad - \quad 1 \quad = \quad 0$

solving, $\quad\quad e \quad = \quad +0.2715 \quad$ or $\quad e \quad = \quad -0.3699$

There are two solutions, of which we assume that the positive solution $e = 0.2715$ is the meaningful one. From this we immediately get $b = 3.31\,e = 0.899$. The value of s and r may be obtained from (i)' and (ii)'.

$$s = \frac{0.49/e - 1}{3.31} = 0.243, \qquad r = \frac{0.61/e - 0.86}{3.31} = 0.419$$

Finally, we round off the answers to two places of decimals:

$$b = 0.90, \qquad e = 0.27, \qquad r = 0.42, \qquad s = 0.24$$

The reader does not have to solve these equations in precisely the way indicated above; he is encouraged to solve them in some other way as an additional exercise.

Chapter 3

LINEAR REGRESSION AND CORRELATION

THE COMBINED subject of linear regression and correlation can hardly be condensed into a few pages. There are book-length treatments of linear regression and its applications. Since the reader is assumed to have had a preliminary course in statistics, all we do in this chapter is to review some of the highlights of this subject to pave the way for the study of path coefficients.

Regression of y on x

1. To avoid abstractness we begin with a set of eight $(n = 8)$ pairs of measurements: x being numbers in units of centimeters and y in units of grams. Think of x as the height of a person and y as his weight.

(i)	(1)	(2)	(3)	(4)	(5)	(6)	(7)	(8)	Total mean
x_i (cm)	2	4	5	7	10	13	14	17	$\Sigma x = 72,\ \bar{x} = 9.0\ cm$
y_i (gm)	3	2	13	10	4	11	11	10	$\Sigma y = 64,\ \bar{y} = 8.0\ gm$

(1)

The figure on page 38* gives a plot of these eight points on the x-y plane. Such a picture showing the position of the points (x_i, y_i) is called a scatter diagram. Then we wish to draw a straight line among these points to show the general trend of the positions of these points. If this is the only purpose, such a straight line may be drawn freehand by trial and error until it satisfies our visual judgement. In statistics, however, we wish to draw a straight line with some simple mathematical properties, not just a line somewhat among these points. Before we can do this we must set forth the mathematical requirements of such a straight line, following quantities calculated from data (1).

$$
\begin{array}{lll}
\Sigma x^2 = 848\ cm^2 & \Sigma xy = 656\ cm\ gm & \Sigma y^2 = 640\ gm^2 \\
(\Sigma x)^2/n = 648\ cm^2 & (\Sigma x)(\Sigma y)/n = 576\ cm\ gm & (\Sigma y)^2/n = 512\ gm^2 \\
\\
\Sigma(x-\bar{x})^2 = 200\ cm^2 & \Sigma(x-\bar{x})(y-\bar{y}) = 80\ cm\ gm & \Sigma(y-\bar{y})^2 = 128\ gm^2
\end{array}
$$

(2)

* Figures, diagrams, and tables are numbered according to the page on which they appear.

These quantities are calculated by virtue of the identities demonstrated in Chapter 2, except in one important respect. That is, the numbers here are always associated with some physical units; they are not pure numbers. For instance, $\Sigma(x-\bar{x})(y-\bar{y})$ is not a number equal to 80, but it is 80 *gm cm*.

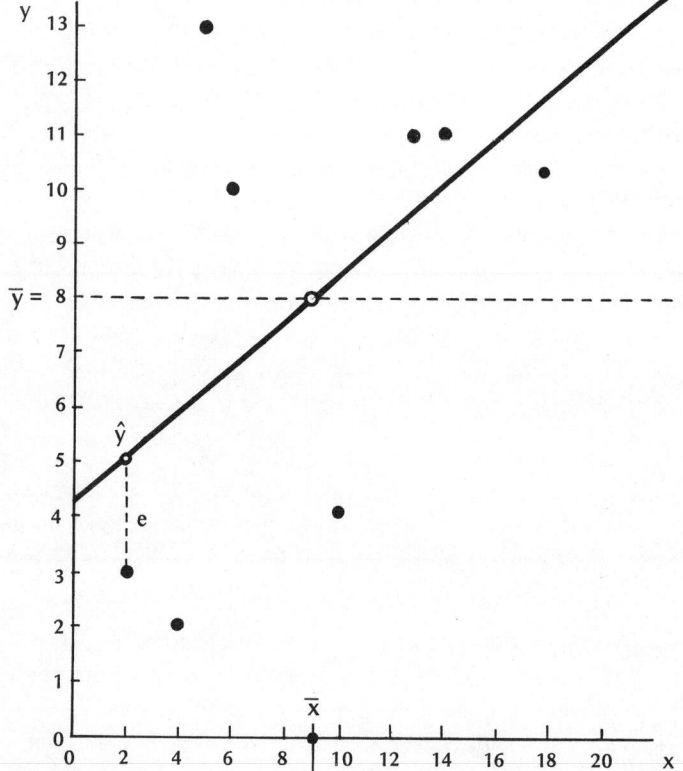

Fig. 38. The Scatter Diagram for the eight points given in (1) in the text. The dotted horizontal line is $y = \bar{y} = 8$. The equation of the solid line is $\hat{y} = a + bx = 4.4 + 0.40x$, representing the regression of y on x. The distance from point to regression line is $e = y - \hat{y}$.

2. Now let us try to draw a straight line supposedly representing the linear trend of these points. Of course, no single straight line can pass through all eight points. Some of the points will be above the line, others below the line. Let e be the vertical (parallel to the y-axis) distance from

LINEAR REGRESSION AND CORRELATION

the point to the desired straight line. For points above the straight line the distance e is positive; for those below the line, the distance e is negative, to be explained shortly. Suppose that the equation of the straight line is

$$\hat{y} = a + bx \tag{3}$$

where a and b are yet to be determined. Note the caret above y. It is frequently read as "y-hat." It is in general not equal to the actual y values (such as those shown in data (1)). It is the y-value on the straight line. In other words, when an x value is given, we calculate $\hat{y} = a + bx$, so that $(x, \hat{y})$ satisfies the linear equation (3). For this reason, $\hat{y}$ is also called the "calculated y," or the "predicted y" for given values of x. Whatever the name or terminology, the important thing to remember is that $\hat{y}$ is the y-value on the straight line; that means, $(x, \hat{y})$ are points on the straight line (collinear). Let us now consider one of the actual points, say, (x_i, y_i). For a given x_i, the calculated y-value will be $\hat{y}_i = a + bx_i$ but the actual y-value is y_i. The vertical distance from the actual point to the line is

$$e_i = y_i - \hat{y}_i \tag{4}$$

This is why we say that the distance e_i is positive for points above the line $(y_i > \hat{y}_i)$ and the distance e_i is negative for points below the line $(y_i < \hat{y}_i)$. With e and $\hat{y}$ so defined, we see that each actual y-value is equal to

$$y_i = \hat{y}_i + e_i = a + bx_i + e_i \tag{5}$$

which is the linear model for the relationship between y and x. The values of the two constants, a and b, are yet to be determined according to the mathematical properties we wish the straight line to possess.

3. The mathematical requirement for the straight line $\hat{y} = a + bx$ is that the values of a and b should be so chosen that the sum of squares of the distances e_i is as small as possible. The method of doing this is called the method of least squares. The values of a and b are chosen to minimize the following quantity:

$$Q = \Sigma e_i^2 = \Sigma (y_i - \hat{y}_i)^2 = \Sigma (y_i - a - bx_i)^2 \tag{6}$$

where the summation is over all the points $(i = 1, 2, \ldots, n)$. Setting $\partial Q/\partial a = 0$ and $\partial Q/\partial b = 0$, the two equations reduce to

$$\Sigma (y_i - a - bx_i) = \Sigma (y_i - \hat{y}) = \Sigma e_i = 0 \qquad (7)$$

$$\Sigma (y_i - a - bx_i)x_i = \Sigma (y_i - \hat{y}_i)x_i = \Sigma e_i x_i = 0 \qquad (8)$$

Readers without differential calculus may skip the operation $\partial Q/\partial a$ and $\partial Q/\partial b$, and simply accept (7) and (8) as the two equations from which the values of a and b are to be found. These two equations are called the "normal" equations, although the "normal" here has nothing to do with any normal distribution. All of the properties of the desired line to be drawn are embodied in these two normal equations. Equation (7), for instance, may be described verbally in three different ways, but they are all mathematically equivalent. First, we note that $\Sigma e_i = 0$ implies that the positive distances above the line exactly balance the negative distances below the line. Second, $\Sigma(y_i - \hat{y}_i) = \Sigma y_i - \Sigma \hat{y}_i = 0$ implies that, although any single y_i may not be equal to its corresponding $\hat{y}_i$, their totals and therefore their mean values are equal. Third, dividing the left side of (7) by n, we obtain

$$\bar{y} - a - b\bar{x} = 0, \quad \text{or} \quad \bar{y} = a + b\bar{x} \qquad (9)$$

indicating that the straight line passes through the point $(\bar{x}, \bar{y})$, because the values of $\bar{x}$ and $\bar{y}$ satisfy the equation $\hat{y} = a + bx$. These three properties are mathematically equivalent; any one of them will imply the other two. Since we already know that the desired line passes through the central point $(\bar{x}, \bar{y})$, all we need to know is one more property about the line in order to determine it uniquely; we shall determine the slope b of the line.

4. In order to solve for slope b conveniently, we rewrite the normal equations (7) and (8) as follows, noting that $\Sigma a = a + \ldots + a = a n$ and omitting subscript i for brevity.

$$a n + b \Sigma x = \Sigma y \qquad (10)$$

$$a \Sigma x + b \Sigma x^2 = \Sigma xy \qquad (11)$$

LINEAR REGRESSION AND CORRELATION

It is recognized that these are two simultaneous linear equations in two unknowns (a, b), the other quantities being known numbers such as those given in (2). Solving for slope b, we have from the methods of Chapter 2,

$$b = \frac{\begin{vmatrix} n & \Sigma y \\ \Sigma x & \Sigma xy \end{vmatrix}}{\begin{vmatrix} n & \Sigma x \\ \Sigma x & \Sigma x^2 \end{vmatrix}} = \frac{n \Sigma xy - (\Sigma x)(\Sigma y)}{n \Sigma x^2 - (\Sigma x)^2} = \frac{\Sigma (x-\bar{x})(y-\bar{y})}{\Sigma (x-\bar{x})^2} \quad (12)$$

and from (9),
$$a = \bar{y} - b\bar{x} \quad (13)$$

The last two expressions, (12) and (13), give us the numerical solutions for a and b. Substituting the numbers in (2) calculated from data (1),

$$b = \frac{80 \text{ cm gm}}{200 \text{ cm cm}} = 0.40 \frac{gm}{cm}, \qquad a = 8.0 - 0.40\,(9.0) = 4.4 \text{ gm}$$

so that the equation of the straight line is

$$\hat{y} = a + bx = 4.4 + 0.40\,x \quad (14)$$

Such a line is called the (linear) regression line of y on x. It is also known as the least square line. The slope b is called the regression coefficient of y on x, usually denoted by b_{yx}. When $x = 0$, then $y = a = 4.4$ which is the intercept of the regression line on the y-axis. The regression line in Fig. 38 is drawn this way. Note that $b = 0.40\,gm/cm$, not just 0.40. It is a rate or a slope in geometrical language. It says that when the height x changes by one centimeter, the weight y changes (or is expected to change) by $0.40\,gm$. Briefly, the change in y is $0.40\,gm$ per cm change in x. If the units of measurements for x and y are changed, the numerical appearance of b will also change. Thus, $b = 0.40\,gm/cm = 40\,gm/meter$. Also note that bx is in units of y, for $(gm/cm) \times cm = gm$.

Partition of the sum of squares of y.

5. Since we have already found the equation $\hat{y} = a + bx = 4.4 + 0.4\,x$, we can calculate the $\hat{y}$ values from it for the various x values. This is done in the upper half of Table 42. For instance, when $x = 5$, $\hat{y} = 4.4 + 2.0 = 6.4$. Remember: all the $\hat{y}$ values so calculated are on the straight line. They are collinear. Now, the original or total deviation of y, $(y - \bar{y})$, may be subdivided into two segments: $(y - \hat{y})$ and $(\hat{y} - \bar{y})$, as $(y - \hat{y}) + (\hat{y} - \bar{y}) = (y - \bar{y})$. These values are also shown in the upper portion of Table 42. For instance, $(13 - 6.4) + (6.4 - 8) = 6.6 - 1.6 = 5.0 = 13 - 8$. In terms of Fig. 38, we see that the first segment is the distance $e = y - \hat{y}$ from the point to the regression line, and the second segment, $\hat{y} - \bar{y}$, is the distance from the regression line to the horizontal line $y = \bar{y} = 8$. The lower half of Table 42 shows the corresponding squares of the numbers in the upper half.

Table 42. Subdivision of the deviations of y and subdivision of the sum of squares of deviations of y by linear regression of y on x.

$$\hat{y} = a + bx = 4.4 + 0.4\,x$$

x	y	$\hat{y}$	$\bar{y}$	$(y - \hat{y})$	$(\hat{y} - \bar{y})$	$(y - \bar{y})$
2	3	5.2	8	− 2.2	− 2.8	− 5
4	2	6.0	8	− 4.0	− 2.0	− 6
5	13	6.4	8	+ 6.6	− 1.6	+ 5
7	10	7.2	8	+ 2.8	− 0.8	+ 2
10	4	8.4	8	− 4.4	+ 0.4	− 4
13	11	9.6	8	+ 1.4	+ 1.6	+ 3
14	11	10.0	8	+ 1.0	+ 2.0	+ 3
17	10	11.2	8	− 1.2	+ 3.2	+ 2
72	64	64.0	64	0	0	0
Σx	Σy	$\Sigma \hat{y}$	$\Sigma \bar{y}$	$\Sigma (y - \hat{y})$	$\Sigma (\hat{y} - \bar{y})$	$\Sigma (y - \bar{y})$
	y^2	$\hat{y}^2$	$\bar{y}^2$	$(y - \hat{y})^2$	$(\hat{y} - \bar{y})^2$	$(y - \bar{y})^2$
	9	27.04	64	4.84	7.84	25
	4	36.00	64	16.00	4.00	36
	169	40.96	64	43.56	2.56	25
	100	51.84	64	7.84	0.64	4
	16	70.56	64	19.36	0.16	16
	121	92.16	64	1.96	2.56	9
	121	100.00	64	1.00	4.00	9
	100	125.44	64	1.44	10.24	4
	640	544.00	512	96.00	32.00	128
	Σy^2	$\Sigma \hat{y}^2$	$\Sigma \bar{y}^2$	$\Sigma (y - \hat{y})^2$	$\Sigma (\hat{y} - \bar{y})^2$	$\Sigma (y - \bar{y})^2$

LINEAR REGRESSION AND CORRELATION

From the sums of such squares shown at the bottom of Table 42, it does not take much to notice the following relationships:

$$\begin{array}{rcl} 640 - 544 & = & 96 \\ \underline{544 - 512} & = & \underline{32} \end{array} \qquad (15)$$

adding, $\quad 640 \qquad\qquad - 512 \;=\; 128$

More generally, in algebraic form,

$$\begin{array}{l} \Sigma y^2 - \Sigma \hat{y}^2 \qquad\qquad = \Sigma (y - \hat{y})^2 \\ \underline{\qquad\qquad \Sigma \hat{y}^2 - \Sigma \bar{y}^2 = \Sigma (\hat{y} - \bar{y})^2} \end{array}$$

adding, $\quad \Sigma y^2 \qquad\qquad - \Sigma \bar{y}^2 = \Sigma (y - \bar{y})^2 \qquad (16)$

A geometrical presentation of the three quantities, Σy^2, $\Sigma \hat{y}^2$, and $\Sigma \bar{y}^2$, is given in Fig. 44. Noting $\Sigma \bar{y}^2 = n\bar{y}^2 = (\Sigma y)^2/n$, we see that the last lines of (15) and (16) viz., $640 - 512 = 128$, and $\Sigma y^2 - \Sigma \bar{y}^2 = \Sigma (y - \bar{y})^2$, are not new; we have already calculated that quantity in (2) directly from data (1), as it is the total sum of squares of deviations of y, having nothing to do with the corresponding x values at all. But the other two quantities, $\Sigma y^2 - \Sigma \hat{y}^2 = \Sigma (y - \hat{y})^2 = 96$, and $\Sigma \hat{y}^2 - \Sigma \bar{y}^2 = \Sigma (\hat{y} - \bar{y})^2 = 32$, are new, on account of the introduction of the new collinear values $\hat{y} = a + bx$. In other words, the regression line has partitioned the total sum of squares of y into two components as shown numerically in (15); it will be shown that this is generally the case.

6. We shall now see that the introduction of the regression line $\hat{y} = a + bx$, determined by the two normal equations (7) and (8), splits the total sum of squares of deviations of y into two components. We begin by writing

$$(y - \bar{y}) \;=\; (y - \hat{y}) + (\hat{y} - \bar{y})$$

The expression above is always true, being an identity. Squaring and summing,

$$\Sigma (y - \bar{y})^2 = \Sigma (y - \hat{y})^2 + \Sigma (\hat{y} - \bar{y})^2 + 2 \Sigma (y - \hat{y})(\hat{y} - \bar{y}) \qquad (17)$$

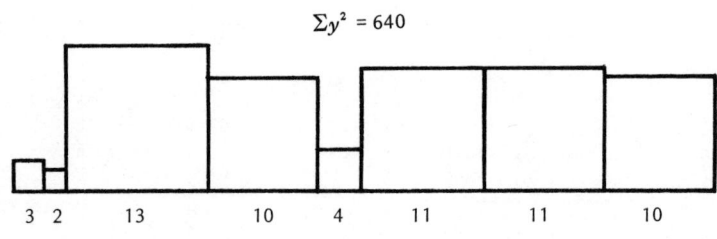
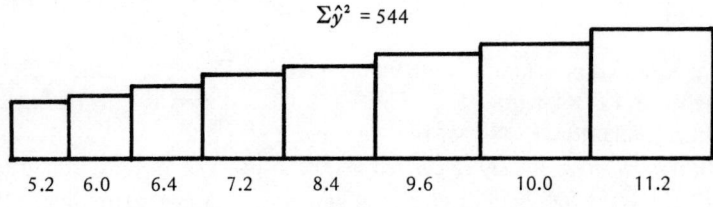
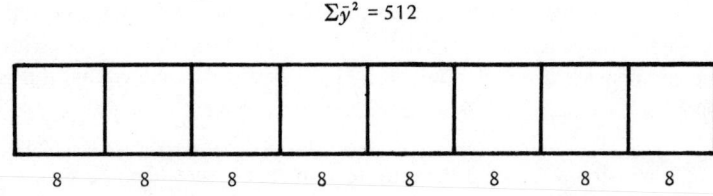

Fig. 44. A geometrical representation for the sum of squares. The first row is the area $\Sigma y^2 = 3^2 + \ldots + 10^2 = 640$; the second row is the area $\Sigma \hat{y}^2 = (5.2)^2 + \ldots + (11.2)^2 = 544$; and the last row is the area $\Sigma \bar{y}^2 = 8^2 + \ldots + 8^2 = 512$. The difference between the first two areas, $640 - 544 = 96$, is the residual ssq; the difference between the middle and the bottom areas, $544 - 512 = 32$, is the regression ssq. The difference between the first and the last area is the total $ssq = 640 - 512 = 128$.

LINEAR REGRESSION AND CORRELATION

The last term, the summation of products, is zero on account of the normal equations (i.e., the conditions under which the values of a and b are determined). Explicitly, if the reader wishes to see all the details, the sum of products is

$$\Sigma (y - \hat{y})(\hat{y} - \bar{y}) = \Sigma (y - \hat{y})\hat{y} - \Sigma (y - \hat{y})\bar{y}$$

The last term on the right is zero on account of normal equation (7), because $\bar{y}$ is a constant. The first term on the right is also zero because of normal equations (7) and (8):

$$\Sigma (y - \hat{y})\hat{y} = \Sigma (y - a - bx)(a + bx)$$
$$= a \Sigma (y - a - bx) + b \Sigma (y - a - bx) x$$
$$= 0 \quad \text{by (7)} \quad + 0 \quad \text{by (8)}$$

Hence (17) becomes

$$\Sigma (y - \bar{y})^2 = \Sigma (y - \hat{y})^2 + \Sigma (\hat{y} - \bar{y})^2 \tag{18}$$

This is a very important result of linear regression and it deserves some verbal reenforcement by introducing a system of verbal notation. Let *ssq* denote the sum of squares (of deviations). On the left of (18) is the *total ssq* of y, calculated from the y values alone, taking no account of its corresponding x values. With the introduction of the set of collinear values $\hat{y} = a + bx$ whose mean value is also $\bar{y}$, we can calculate the *ssq* for these $\hat{y}$ values and call $\Sigma (\hat{y} - \bar{y})^2$ the *regression ssq*, because such $\hat{y}$ values are calculated from x and we say it is due to regression on x. Finally, the component $\Sigma (y - \hat{y})^2 = \Sigma e^2$ is the *ssq* of the distances from the actual point to the introduced regression line. Such deviations from linearity cannot be explained by the influence of x and we call that component the *residual ssq*. With this verbal notation, the relationship (18) may be written as, term by term,

$$\text{total ssq} = \text{residual ssq} + \text{regression ssq} \tag{18'}$$

Sometimes we use both verbal and algebraic notations to facilitate reading and to provide some redundancy to guard against possible typographical errors. From (18′) it follows

$$1 = \frac{\text{residual ssq}}{\text{total ssq}} + \frac{\text{regression ssq}}{\text{total ssq}} \qquad (18'')$$

The first two relations of (16) as exhibited in Fig. 44 follow immediately from the normal equations (7) and (8). Since $\Sigma e = 0$ and $\Sigma ex = 0$, it follows that $\Sigma e(a + bx) = 0$ and that $\Sigma e\hat{y} = 0$. Hence

$$\Sigma y \hat{y} = \Sigma (\hat{y} + e)\hat{y} = \Sigma \hat{y}^2$$

and

$$\text{residual ssq} = \Sigma (y - \hat{y})^2 = \Sigma y^2 + \hat{y}^2 - 2 \Sigma y \hat{y} = \Sigma y^2 - \Sigma \hat{y}^2$$

which is the first relation shown in (16). The second relationship of (16) is even easier to see, as $\Sigma \hat{y} = \Sigma y = n\bar{y} = \Sigma \bar{y}$; thus,

$$\text{regression ssq} = \Sigma (\hat{y} - \bar{y})^2 = \Sigma \hat{y}^2 + \Sigma \bar{y}^2 - 2\bar{y} \Sigma \hat{y} = \Sigma \hat{y}^2 - \Sigma \bar{y}^2$$

This completes the justification for using the diagrams of Fig. 44 to illustrate the relationships (16).

7. To show the relationship (18) is one thing; to have a practical way of calculating the two components is another. In Table 42 the two components have been obtained by actually calculating all the $\hat{y} = a + bx$ values, each from its corresponding x value. There is nothing wrong in doing this; in fact, it is instructive to do so because it illustrates the meaning of the identity (18) clearly. The disadvantage is that it becomes very laborious when there are a large number of pairs of (x, y). There is a shortcut by which one of the components, $\Sigma (\hat{y} - \bar{y})^2$, may be obtained by one simple step without ever calculating one single $\hat{y}$. We begin with equation (3), $\hat{y} = a + bx$, where $a = \bar{y} - b\bar{x}$ as given by (9) or (13). Consequently, the equation for the regression line may also be written as

LINEAR REGRESSION AND CORRELATION

$$\hat{y} = \bar{y} + b(x - \bar{x})$$

or $\qquad (\hat{y} - \bar{y}) = b(x - \bar{x}) \qquad$ (19)

The last equation makes good sense. Since the regression line passes through the central point $(\bar{x}, \bar{y})$, equation (19) simply says that the slope of the line is $b = (\hat{y} - \bar{y})/(x - \bar{x})$, which is elementary geometry. Squaring both sides of (19) and summing, we obtain immediately

$$\text{regression ssq} = \Sigma (\hat{y} - \bar{y})^2 = b^2 \Sigma (x - \bar{x})^2 \qquad (20)$$

Since $b = \Sigma (x - \bar{x})(y - \bar{y}) / \Sigma (x - \bar{x})^2$, expression (20) may also be written

$$\text{regression ssq} = \Sigma (\hat{y} - \bar{y})^2 = b \Sigma (x - \bar{x})(y - \bar{y}) \qquad (21)$$

Referring back to the numerical values of (2) and formula (12) for b,

$$\Sigma (x - \bar{x})^2 = 200, \quad \Sigma (x - \bar{x})(y - \bar{y}) = 80, \quad \Sigma (y - \bar{y})^2 = 128, \quad b = 80/200 = 0.40$$

we arrive at

$$\text{regression ssq} = \Sigma (\hat{y} - \bar{y})^2 = b^2 \Sigma (x - \bar{x})^2 = 0.16 \times 200 = 32 \qquad (20')$$
$$= b \Sigma (x - \bar{x})(y - \bar{y}) = 0.40 \times 80 = 32 \qquad (21')$$

which agrees with the result exhibited in Table 42, without calculating the $\hat{y}$ values. Since the total sum of squares of y and the component due to influence of x are known, the other component is obtained by subtraction:

$$\text{residual ssq} = \text{total ssq} - \text{regression ssq}$$
$$\Sigma (y - \hat{y})^2 = \Sigma (y - \bar{y})^2 - b \Sigma (x - \bar{x})(y - \bar{y}) \qquad (22)$$

i.e., $\qquad \Sigma e^2 = 128 - 32 = 96$

There is no other shortcut for calculating Σe^2 to provide an independent check. The value of Σe^2 obtained by subtraction is correct only when the other two quantities are correctly calculated.

Regression of x on y.

8. With the same set of points (1) and the sum of squares and sum of products (2), we can also find a regression equation of x on y; viz.,

$$\hat{x} = c + dy \qquad (3x)$$

according to the same principle of least squares. Then d, which looks like a reversed b, is the slope giving the amount of change in height x per unit change in weight y. The label of the equation, $(3x)$, means that this expression is analogous to (3) for the regression of y on x. The whole process of fitting such a line involves merely the interchange of the letters x and y in all the previous expressions. In the following we shall give some of the expressions together with their numerical values for ease of comparison. For instance,

$$d = \frac{\Sigma (x-\bar{x})(y-\bar{y})}{\Sigma (y-\bar{y})^2} = \frac{80 \text{ cm gm}}{128 \text{ gm gm}} = 0.625 \text{ cm/gm} \qquad (12x)$$

$$c = \bar{x} - d\bar{y} = 9.0 - 0.625(8) = 4.00 \text{ cm} \qquad (13x)$$

so that the equation of the regression line of x on y is

$$\hat{x} = c + dy = 4.00 + 0.625 y \qquad (14x)$$

Note that the present d and the previous b are not only numerically different but also different in units: $d = 0.625$ cm/gm while $b = 0.40$ gm/cm. Comparing $(12x)$ with (12), we see that b and d will be numerically equal when $\Sigma (x - \bar{x})^2 = \Sigma (y - \bar{y})^2$, but this does not change the fact that they are still different in units and therefore different in meaning. Ten miles per hour and ten hours per mile are saying two entirely different things.

9. The total sum of squares $\Sigma (x - \bar{x})^2 = 200$ may be partitioned into two components by the introduction of $\hat{x} = c + dy$ in exactly the same way as we did for the total sum of squares of y. Thus, the component due to regression is

LINEAR REGRESSION AND CORRELATION

$$\Sigma\,(\hat{x}-\bar{x})^2 \;=\; d^2\,\Sigma\,(y-\bar{y})^2 \qquad =\;(0.625)^2\,(128) \;=\;50 \qquad (20x)$$

$$=\;d\,\Sigma\,(x-\bar{x})(y-\bar{y}) \;=\;(0.625)\,(80) \;=\;50 \qquad (21x)$$

The residual component is then obtained by subtraction:

$$\Sigma\,(x-\hat{x})^2 \;=\; 200 \;-\; 50 \;=\; 150 \qquad (22x)$$

This completes the calculations for the regression of x on y.

10. We have seen from the above that the regression of y on x, and the regression of x on y, have different meanings and the numerical results are quite different. However, there is one property that remains the same for both sets of calculations, and that is the relative magnitude of the two components of the total sum of squares. To illustrate, let us summarize the previous numerical results as follows:

	$\hat{y} = a + bx$		$\hat{x} = c + dy$	
	number	percentage	number	percentage
residual ssq	96	75%	150	75%
regression ssq	32	25%	50	25%
total ssq	128	100%	200	100%

It can be seen that the fraction of the total sum of squares due to regression or residual is the same in both cases. Algebraically it means that

$$\frac{\Sigma\,(\hat{y}-\bar{y})^2}{\Sigma\,(y-\bar{y})^2} \;=\; \frac{\Sigma\,(\hat{x}-\bar{x})^2}{\Sigma\,(x-\bar{x})^2} \;=\; \frac{\text{regression ssq}}{\text{total ssq}} \qquad (23)$$

The general truth of (23) will be easily seen after we introduce another measure or index for the relationship between x and y.

Correlation Coefficient

11. All of the quantities dealt with in the previous sections have physical units attached; for instance, y *gm*, y^2 *gm*^{2}, xy *cm gm*, and b *gm/cm* which measures the general linear influence of height x on weight y. Note that the sign of $b = \Sigma\ (x-\bar{x})(y-\bar{y})/\Sigma\ (x-\bar{x})^2$ is always the same as the sign of the sum of products of deviations in the numerator, because the denominator $\Sigma(x-\bar{x})^2$ is always positive. When x and y have the same trend of increasing and decreasing (that is, the larger the x, the larger the corresponding y, and the smaller the x, the smaller the corresponding y), then the pair of deviations, $(x_i - \bar{x})$ and $(y_i - \bar{y})$, tends to be both positive or both negative, so that the product $(x_i - \bar{x})(y_i - \bar{y})$ is positive. Conversely, when large x's are associated with small y's and small x's are associated with large y's, most of the pairs of deviations, $(x_i - \bar{x})$ and $(y_i - \bar{y})$, will have opposite signs (one positive and one negative), so that the product $(x_i - \bar{x})(y_i - \bar{y})$ will be negative. Hence, the sign of the sum $\Sigma\ (x - \bar{x})(y - \bar{y})$ tells us where most of the pairs of deviations are. When it is positive, we say x and y are positively correlated; when it is negative, we say x and y are negatively correlated. The sign of b tells us the positive or negative nature of the correlation but it measures the relationship in a directional sense (from x to y) and in concrete physical units. We would need a measurement or an index to indicate the degree of correlation between x and y in a symmetrical manner and without physical units.

12. The coefficient of correlation between x and y is defined as

$$r\ =\ r(x, y)\ =\ r_{xy}\ =\ \frac{\Sigma\ (x - \bar{x})(y - \bar{y})}{\sqrt{\Sigma(x-\bar{x})^2 \cdot \Sigma(y-\bar{y})^2}} \qquad (24)$$

The usual notation is r_{xy}; since we are considering x and y only, we may simply write r without ambiguity. It is seen that the expression is perfectly symmetrical with respect to x and y. Consequently, $r(x, y) = r(y,x)$. For our numerical example **(1)**, the correlation coefficient is

$$r\ =\ \frac{80}{\sqrt{200 \times 128}} = \frac{80}{160} = \frac{1}{2} \qquad (24')$$

LINEAR REGRESSION AND CORRELATION

It is a pure number, as the units $cm\ gm/\sqrt{cm^2\ gm^2} = cm\ gm/cm\ gm$ cancel each other out. The correlation coefficient gives a symmetrical measurement of the degree of linear association of x and y. When the positive and negative products of deviations are of equal magnitude so that the sum $\Sigma\ (x-\bar{x})(y-\bar{y}) = 0$, then $r = 0$, and we say that x and y are (linearly) uncorrelated.

13. The correlation coefficient has many pleasing and useful properties, a few of which shall be described. First of all, we note that its numerical (absolute) value cannot exceed unity; that is,

$$-1 \leqslant r \leqslant +1, \qquad r^2 \leqslant 1 \qquad (25)$$

The demonstration of $r^2 \leqslant 1$ is given here and that $-1 \leqslant r \leqslant 1$ in Ex. 4. In view of definition (24) and its square, we are now required to show that the denominator $\geqslant$ the numerator:

$$\left\{\Sigma\ (x_i - \bar{x})^2\right\} \left\{\Sigma\ (y_i - \bar{y})^2\right\} \geqslant \left\{\Sigma\ (x_i - \bar{x})(y - \bar{y})\right\}^2 \qquad (25')$$

To shorten the writing we denote the deviations by a capital letter; $X_i = x_i - \bar{x}$ and $Y_i = y_i - \bar{y}$. Then the inequality above may be written

$$(X_1^2 + X_2^2 + \ldots)(Y_1^2 + Y_2^2 + \ldots) \geqslant (X_1 Y_1 + X_2 Y_2 + \ldots)^2$$

Expanding,

$$X_1^2 Y_1^2 + X_2^2 Y_2^2 + X_1^2 Y_2^2 + X_2^2 Y_1^2 + \ldots \geqslant X_1^2 Y_1^2 + X_2^2 Y_2^2 + 2 X_1 Y_2 X_2 Y_1 + \ldots$$

Cancelling the terms that appear on both sides,

$$(X_1 Y_2)^2 + (X_2 Y_1)^2 + \ldots \geqslant 2(X_1 Y_2)(X_2 Y_1) + \ldots$$

analogous to $\qquad A^2 + \quad B^2 + \ldots \geqslant 2AB + \ldots$

which is always true because $(A - B)^2 = A^2 + B^2 - 2AB \geqslant 0$. Thus we have proved the inequality (25). As a quick and easy exercise you should write out all the terms for $n = 3$, viz., $(X_1^2 + X_2^2 + X_3^2)(Y_1^2 + Y_2^2 + Y_3^2) \geqslant (X_1 Y_1 + X_2 Y_2 + X_3 Y_3)^2$ and see that the inequality holds.

Origin and Units

14. An inspection of the definition (24) shows that the value of r will remain unchanged when x and y are transformed into

$$x' = k_1 + b_1 x; \qquad y' = k_2 + b_2 y \qquad (26)$$

where the b's and k's are constants. While the k's may be positive or negative, we shall assume that the b's are positive (see paragraph 19). The transformation (26) amounts to a shifting of origin and change of concrete units (e.g., centimeters into inches and grams into ounces). The addition of a constant to x and/or y does not change the deviations $(x - \bar{x})$ and/or $(y - \bar{y})$. The constant multiplier of x and/or y will cancel each other out from the numerator and denominator of (24). In order to avoid the small size of the subscripts, we may write $r(x,y)$ for r_{xy}. Thus, we see that the following four correlations are equal:

$$r(x', y') = r(x', y) = r(x, y') = r(x, y) \qquad (27)$$

Any transformation of variables of the type (26) has no effect on correlation. In other words, the correlation coefficient is independent of origin and units. If we measure the heights of school children each standing on a stool two feet high, and measure their weights each with a five-pound pack on the back, the correlation coefficient between such weights and such heights will remain the same as that between weights and heights measured without the stool, or without the pack, or without either. Nor does it matter whether these children are measured in terms of inches and pounds, or in terms of kilograms and meters.

15. When will the correlation coefficient be exactly unity? Under usual circumstances the points (x, y) are not collinear and $|r| < 1$. When the series of points under consideration are exactly collinear (all of them lying on a straight line), then the correlation coefficient will be unity (or -1). For instance, consider the collinear points $(x, \hat{y})$, where $\hat{y} = a + bx$. Then

LINEAR REGRESSION AND CORRELATION

$$\Sigma(x-\bar{x})(\hat{y}-\bar{y}) = \Sigma(x-\bar{x})(a+bx-a-b\bar{x}) = b\Sigma(x-\bar{x})^2$$

and $\quad \Sigma(\hat{y}-\bar{y})^2 \quad = b^2 \Sigma(x-\bar{x})^2$

Hence,

$$r(x,\hat{y}) = \frac{\Sigma(x-\bar{x})(\hat{y}-\bar{y})}{\sqrt{\Sigma(x-\bar{x})^2 \Sigma(\hat{y}-\bar{y})^2}} = \frac{b\Sigma(x-\bar{x})^2}{\sqrt{\Sigma(x-\bar{x})^2 \cdot b^2 \Sigma(x-\bar{x})^2}} = 1 \quad (28)$$

We shall not do this kind of longhand algebra every time. A much simpler way of seeing this result is to recall the general relations (26) and (27). Since $\hat{y} = a + bx$, it may be considered as a transformed x' of the type (26). Then the correlation between x and $\hat{y}$ will be the same as the correlation between x and x'. But correlation is independent of origin and units; the correlation between x and x' is the same as that between x and itself. Thus,

$$r(x,\hat{y}) = r(x, a+bx) = r(x, x') = r(x, x) = 1 \quad (29)$$

Similarly, $\quad r(\hat{x}, y) = r(c+dy, y) = r(y', y) = r(y, y) = 1$

The correlation between a variable and itself is always unity by definition (24). We shall make use of this fact frequently in path analysis.

16. Consider the square of the correlation coefficient once more.

$$r^2 = \frac{\{\Sigma(x-\bar{x})(y-\bar{y})\}^2}{\Sigma(x-\bar{x})^2 \cdot \Sigma(y-\bar{y})^2} \quad (30)$$

Recalling (12) and (12x), $b = \Sigma(x-\bar{x})(y-\bar{y})/\Sigma(x-\bar{x})^2$ and $d = \Sigma(x-\bar{x})(y-\bar{y})/\Sigma(y-\bar{y})^2$, we see that

$$r^2 = bd, \qquad r = \sqrt{bd} \quad (31)$$

Since most of the time we will use either regression of y on x or regression of x on y, a more useful expression for r^2 may be obtained by substituting either b only or d only. Thus, recalling (21), substituting b, $\quad r^2 = \dfrac{b \sum (x-\bar{x})(y-\bar{y})}{\sum (y-\bar{y})^2} = \dfrac{\sum (\hat{y}-\bar{y})^2}{\sum (y-\bar{y})^2} = \dfrac{\text{regression ssq}}{\text{total ssq}}$ (32)

substituting d, $\quad r^2 = \dfrac{d \sum (x-\bar{x})(y-\bar{y})}{\sum (x-\bar{x})^2} = \dfrac{\sum (\hat{x}-\bar{x})^2}{\sum (x-\bar{x})^2} = \dfrac{\text{regression ssq}}{\text{total ssq}}$ (33)

Thus we have proved the relationship (23). In our numerical example, this fraction has been found to be 25% or ¼ for both regressions. Now, we not only know that the fraction of the total sum of squares that is due to regression is the same for both regressions, but that they are both equal to r^2. In our numerical example, it has been found that $r = ½$ and hence $r^2 = ¼$. This is a very important property of the correlation coefficient and worthwhile repeating, although in a somewhat different form:

$$\text{regression ssq} = r^2 \sum (y - \bar{y})^2 = \sum (\hat{y} - \bar{y})^2 \qquad (34)$$

that is, r^2 of the total sum of squares is due to regression. It follows that $(1 - r^2)$ of the total sum of squares is the residual:

$$\text{residual ssq} = (1 - r^2) \sum (y - \bar{y})^2 = \sum (y - \hat{y})^2 = \sum e^2 \qquad (35)$$

17. We have defined the correlation coefficient in (24) symmetrically with respect to x and y, without considering linear regression explicitly. When we introduce the regression $\hat{y} = a + bx$, however, the correlation $r_{xy} = r(x, y)$ may be considered from a different viewpoint. Instead of saying that it is the correlation between x and y, we can equally well say that it is the correlation between $\hat{y}$ and y, for in view of (26) and (27),

$$r(x, y) = r(a + bx, y) = r(\hat{y}, y) \qquad (36)$$

The reader may confirm this either by longhand algebra or verify it numerically with the values of $\hat{y}$ and y listed in Table 42. The concept that the correlation $r(x, y)$ is the correlation $r(\hat{y}, y)$ is important as it leads to the definition of multiple correlation in the next chapter when we study multiple linear regression. Having taken the viewpoint of $r(\hat{y}, y)$, then by definition it is equal to

LINEAR REGRESSION AND CORRELATION

$$r(x, y) = r(\hat{y}, y) = \frac{\Sigma (\hat{y}-\bar{y})(y-\bar{y})}{\sqrt{\Sigma (\hat{y}-\bar{y})^2 \Sigma (y-\bar{y})^2}} \qquad (37)$$

And also remember that the mean of the $\hat{y}$'s is $\bar{y}$. The numerator of (37) is equal to

$$\Sigma (\hat{y}-\bar{y}) y = \Sigma(\hat{y}-\bar{y})(\hat{y} + e) = \Sigma (\hat{y}-\bar{y})\hat{y} = \Sigma (\hat{y}-\bar{y})^2 \qquad (38)$$

because $\hat{y} \Sigma e = 0$ and $\Sigma \hat{y} e = \Sigma (a + bx)e = 0$ by virtue of the normal equations (7) and (8). In other words, the regression line splits the original y value into two uncorrelated parts, $\hat{y}$ and e. On substitution of the numerator (38), the correlation (37) becomes

$$r(x, y) = r(\hat{y}, y) = \frac{\Sigma (\hat{y}-\bar{y})^2}{\sqrt{\Sigma (\hat{y}-\bar{y})^2 \Sigma (y-\bar{y})^2}} = \frac{\sqrt{\Sigma (\hat{y}-\bar{y})^2}}{\sqrt{\Sigma (y-\bar{y})^2}} \qquad (39)$$

which is, of course, the square root of (32). The present derivation of (39) gives a new interpretation of r, viz., $r(x, y) = r(\hat{y}, y)$. In analysis of variance the form (32) or (34) is most useful. In path analysis, as we shall see later, we use an equivalent form of (39) frequently.

18. Let us recall our linear model (5) once more, viz.,

$$y = a + bx + e = \hat{y} + e$$

where $e = y - \hat{y}$. The numerical values of these variables have been listed in Table 42. We have shown in previous paragraphs that $r(y, \hat{y}) = r(y, x)$ and $r(\hat{y}, e) = 0$ because $\Sigma \hat{y} e = 0$. The reader who has never done so before should take this opportunity to verify these numerically with the numbers in Table 42. There remains one correlation we have to calculate; that is the correlation between the actual y and the residual distance e. Since $\Sigma e = 0$ and $\bar{e} = 0$, the correlation by definition is

$$r(y, e) = \frac{\Sigma (y - \bar{y}) e}{\sqrt{\Sigma (y - \bar{y})^2 \Sigma e^2}} \qquad (40)$$

But the numerator is $\Sigma (y - \bar{y}) e = \Sigma y e = \Sigma (\hat{y} + e) e = \Sigma e^2$. Hence the correlation (40) becomes

$$r(y, e) = \frac{\Sigma e^2}{\sqrt{\Sigma (y-\bar{y})^2 \Sigma e^2}} = \frac{\sqrt{\Sigma e^2}}{\sqrt{\Sigma (y-\bar{y})^2}} = \sqrt{\frac{\text{residual ssq}}{\text{total ssq}}} \quad (41)$$

which is similar in form to (39). In our numerical example, $r(y, e) = \sqrt{96/128} = \sqrt{0.75} = 0.866$. In view of (41), (39), (32) and (18), we have

$$\begin{aligned} r^2_{y\hat{y}} + r^2_{ye} &= \frac{\text{regression ssq}}{\text{total ssq}} + \frac{\text{residual ssq}}{\text{total ssq}} = 1 \\ &= (0.50)^2 + (0.886)^2 = 1 \end{aligned} \quad (42)$$

where $\hat{y}$ and e are uncorrelated. In path analysis we have frequent occasions to make use of this simple fact and extend it to more complicated cases. Also, (42) implies

$$r_{y\hat{y}} = \sqrt{1 - r^2_{ye}} \quad \text{and} \quad r_{ye} = \sqrt{1 - r^2_{y\hat{y}}} \quad (43)$$

19. In the last few paragraphs since the transformation (26) was introduced, we have tacitly assumed that the original set of x and y values exhibits positive correlation and hence the regression coefficient b is also positive. Under these circumstances it is obviously true that $r(x, y) = r(a + bx, y) = r(\hat{y}, y)$ as stated in (36). Now, suppose that the original values of x and y have a negative correlation and thus a negative regression coefficient. The numerical example in Table 57 shows such a situation. From the preliminary calculations shown at the bottom of the table, we obtain

$$b = b_{yx} = \frac{\Sigma (x - \bar{x})(y - \bar{y})}{\Sigma (x - \bar{x})^2} = \frac{-80}{200} = -0.40$$

$$r = r_{yx} = r(x, y) = \frac{-80}{\sqrt{200 \times 128}} = -0.50$$

and the regression equation

$$\hat{y} = a + bx = (\bar{y} - b\bar{x}) + bx = 11.60 - 0.40 x$$

LINEAR REGRESSION AND CORRELATION

Table 57. Negative correlation between x and y, but the correlation between $\hat{y} = a + bx$ and y is always positive; $|r(x, y)| = r(\hat{y}, y)$. $\hat{y} = 11.60 - 0.40x$.

x	y	xy	$\hat{y}$	$y\hat{y}$	$\hat{y}^2$
2	13	26	10.8	140.4	116.64
4	14	56	10.0	140.0	100.00
5	3	15	9.6	28.8	92.16
7	6	42	8.8	52.8	77.44
10	12	120	7.6	91.2	57.76
13	5	65	6.4	32.0	40.96
14	5	70	6.0	30.0	36.00
17	6	102	4.8	28.8	23.04
72	64	496	64.0	544.0	544.00

$$\Sigma(x - \bar{x})^2 = \Sigma x^2 - (\Sigma x)^2/n = 848 - 648 = 200$$
$$\Sigma(y - \bar{y})^2 = \Sigma y^2 - (\Sigma y)^2/n = 640 - 512 = 128$$
$$\Sigma(x - \bar{x})(y - \bar{y}) = \Sigma xy - (\Sigma x)(\Sigma y)/n = 496 - 576 = -80$$
$$\Sigma(y - \bar{y})(\hat{y} - \bar{y}) = \Sigma(\hat{y} - \bar{y})^2 = 544 - 512 = 32$$

These $\hat{y}$'s are the best linear estimates of the observed y's; hence $\hat{y}$ and y are always positively correlated, indicating the degree of success of estimating y based on the regression on x. The calculated $\hat{y}$'s are also shown in Table 57. The sum of squares due to regression is

$$\text{regression ssq} = b^2 \Sigma(x-\bar{x})^2 = (-0.40)^2\, 200 = 32$$
$$= b \Sigma(x-\bar{x})(y-\bar{y}) = (-0.40)(-80) = 32$$

which may be independently verified by using the individual $\hat{y}$'s in Table 57. This gives the reader one more opportunity to review what we have covered in the last few paragraphs, especially (37), (38), (39). Briefly, $\Sigma y\hat{y} = \Sigma \hat{y}^2 = 544$ and

$$\Sigma(y-\bar{y})(\hat{y}-\bar{y}) = \Sigma y\hat{y} - (\Sigma y)^2/n = 544 - 512 = 32$$
$$\Sigma(\hat{y} - \bar{y})^2 = \Sigma \hat{y}^2 - (\Sigma y)^2/n = 544 - 512 = 32$$

The correlation between $\hat{y}$ and y is thus

$$r(y, \hat{y}) = \frac{\Sigma(y-\bar{y})(\hat{y}-\bar{y})}{\sqrt{\Sigma(y-\bar{y})^2\, \Sigma(\hat{y}-\bar{y})^2}} = \frac{32}{\sqrt{128 \times 32}} = \sqrt{\frac{32}{128}} = \tfrac{1}{2}$$

In conclusion, no matter whether x and y are positively or negatively correlated, the correlation between $\hat{y}$ and y is always positive and equal numerically to that between x and y. That is, in the case of negative correlation, our previous formula (36) should be modified to $|\, r(x,y)\, |$ = $r(\hat{y}, y)$. This does not affect any relationships that involve only the square of the correlation coefficient.

Population parameters

20. Throughout the chapter so far, there have been described merely the various relationships among certain quantities calculated from the n given points (x, y). If the eight points in (1) are actually a random sample from a certain population, then we have to find the sampling errors of the various statistics calculated from the sample. For a statistical treatment of that aspect of the subject, the reader may refer to a textbook of statistics (e.g., Brownlee, 1965, and many others). To prepare us for path analysis, however, we shall, at least initially, make believe that we are dealing with a population, a large population. To do this, we regard the eight distinct points of (1) as constituting the whole population; the frequency of the first pair (x, y) = (2, 3) is $f_1 = 1/8$; the frequency of the second pair is $f_2 = 1/8$; etc., where $f_1 + \ldots + f_n = 1$. If the population size is 8 million, then there are one million pairs of (2, 3), etc. These f's need not be equal for all the pairs of (x, y), but in this particular numerical example, $f_i = f = 1/8$. We shall subsequently give another example in which the individual f's are not equal; $\Sigma f_i = 1$, $f_i \geq 0$ is the only requirement.

21. First we need definitions for a few basic population parameters, from which other parameters may be defined later. Sometimes a numerical value is attached along with a definition in order to specify its meaning clearly. These numerical values are those derived from the numerical examples (1) and (2). Since we have to write $\hat{y}$ as a subscript frequently, it is more convenient to use a letter without the caret. We shall use $g = \hat{y}$, just as we have already used e to denote $y - \hat{y}$. The definition of a *mean* needs no comment: $\bar{x} = \Sigma f_i x_i$ and $\bar{y} = \Sigma f_i y_i$, where $\Sigma f_i = 1$. The definition of a *variance* is, for that of x and y respectively,

$$\begin{aligned} \text{Var } x = V(x) = V_x = \sigma_x^2 = \Sigma f_i (x_i - \bar{x})^2 = E(x - \bar{x})^2 \\ \text{Var } y = V(y) = V_y = \sigma_y^2 = \Sigma f_i (y_i - \bar{y})^2 = E(y - \bar{y})^2 \end{aligned} \quad (44)$$

LINEAR REGRESSION AND CORRELATION

In our numerical example, from (2), $V(x) = 200/8 = 25$ and $V(y) = 128/8 = 16$. It is the average value of the squared deviations in a population. Note that the denominator is $n = 8$, not $n - 1 = 7$, because we are treating the data as a population in which $f_i = 1/8$. The positive square root of a variance is defined as the *standard deviation* to be denoted by σ. Thus

$$\sigma_x = \sqrt{25} = 5, \qquad \sigma_y = \sqrt{16} = 4 \qquad (45)$$

The various forms of expression in (44) need a few words of explanation. The first four forms, $Var\ y = V(y) = V_y = \sigma_y^2$, are merely different notations for the variance of y. We shall not use $Var\ y$ or V_y very often, but we shall use $V(y)$ and σ_y^2 and use them interchangeably. Convenience is almost as important as consistency, and that is why both notations, $V(y)$ and σ_y^2, are in common use, depending upon the occasion. The next form, $\Sigma f_i(y_i - \bar{y})^2 = \Sigma f(y - \bar{y})^2$, is the real definition of a variance; it tells us what a variance is and how it should be calculated. This form says that the variance of y is the average value of the quantities $(y - \bar{y})^2$. For numerical calculation it is more convenient to use the following identities:

$$\begin{aligned}\sigma_x^2 &= \Sigma f_i(x_i - \bar{x})^2 = \Sigma f_i x_i^2 - \bar{x}^2 \\ \sigma_y^2 &= \Sigma f_i(y_i - \bar{y})^2 = \Sigma f_i y_i^2 - \bar{y}^2\end{aligned} \qquad (46)$$

The last form, $E(y - \bar{y})^2$, is merely an abbreviation of $\Sigma f_i(y_i - \bar{y})^2$ and says the same thing, of course, without writing out the specific f's. It denotes the *expected* value or the population average of the quantities $(y - \bar{y})^2$. The symbol E is also known as an "operator," it tells us to do something; and $E(\ldots)$ tells us to calculate the average value of $(\ldots)$. We shall use the E-operator quite freely in later chapters, as it is the simplest way of deriving certain relationships.

22. The *covariance* between the variables x and y is defined as

$$\begin{aligned}Cov(x, y) = \sigma_{xy} &= E(x - \bar{x})(y - \bar{y}) = \Sigma f_i(x_i - \bar{x})(y_i - \bar{y}) \qquad (47) \\ &= \Sigma f_i x_i y_i - \bar{x}\bar{y}\end{aligned}$$

where $\Sigma f_i = 1$, as usual. Due to the symmetry of the definition (47),

$$\text{Cov}(x, y) = \text{Cov}(y, x) = \sigma_{xy} = \sigma_{yx} \qquad (48)$$

In our numerical example, $\sigma_{xy} = 80/8 = 10$. With these basic parameters defined, the other quantities will just be written out in terms of them without further explanation, except for indicating their source. Writing $g = \hat{y} = a + bx$ for convenience, we list the relationships as follows:

                                                                                                      Source

(i) regression model: $\quad y = (a + bx) + e = g + e \qquad (5)$

(ii) regression coefficient: $\quad b = b_{yx} = \sigma_{xy} / \sigma_x^2 \qquad (12)$

(iii) partition of variance: $\quad \sigma_y^2 = \sigma_g^2 + \sigma_e^2 \qquad (18)$

(iv) due to regression: $\quad \sigma_g^2 = b^2 \sigma_x^2 = b \sigma_{xy} \qquad (20, 21)$

(v) correlation coefficient: $\quad r = r_{xy} = r_{yx} = \dfrac{\sigma_{xy}}{\sigma_x \sigma_y} \qquad (24)$

(vi) covariance: $\quad \sigma_{xy} = \sigma_{yx} = r_{xy} \sigma_x \sigma_y \qquad (24)$

(vii) conversion formula: $\quad r_{yx} = b_{yx} \dfrac{\sigma_x}{\sigma_y} \qquad (12)$

(viii) squared correlation: $\quad r^2 = \dfrac{b \sigma_{xy}}{\sigma_y^2} = \dfrac{\sigma_g^2}{\sigma_y^2} \qquad (32)$

(ix) another form of (iv): $\quad \sigma_g^2 = r^2 \sigma_y^2 \qquad (34)$

(x) reduction in variance: $\quad \sigma_e^2 = \sigma_y^2 (1 - r^2) \qquad (35)$

(xi) alternative form of r: $\quad r(x, y) = r(g, y) = \sigma_g / \sigma_y \qquad (39)$

(xii) other correlations: $\quad r(g, e) = 0; \quad r(y, e) = \sigma_e / \sigma_y \qquad (41)$

(xiii) complete determination: $\quad r_{yg}^2 + r_{ye}^2 = \dfrac{\sigma_g^2}{\sigma_y^2} + \dfrac{\sigma_e^2}{\sigma_y^2} = 1 \qquad (42)$

(xiv) inter-relationship: $\quad r_{yg} = \sqrt{1 - r_{ye}^2} \text{ and } r_{ye} = \sqrt{1 - r_{yg}^2} \qquad (43)$

LINEAR REGRESSION AND CORRELATION

The reader should acquaint himself with all the expressions listed above. They are not independent relationships; they are written out explicitly so that they will look familiar. Some of these results will be treated again when the occasion arises in a different context. Introducing still another useful notation, following Wright (1968, I), y is identified as the variable 0, and x as the variable 1; then expression (viii) for the squared correlation may be written as

$$r_{01}^2 = \frac{\sigma_g^2}{\sigma_y^2} = \frac{\sigma_{0(1)}^2}{\sigma_0^2}$$

where $\sigma_{0(1)}^2$ denotes the variance of the y-values determined by x. The expression (x) for reduction of variance may be written as

$$\sigma_e^2 = \sigma_{0.1}^2 = \sigma_0^2 (1 - r_{01}^2)$$

where $\sigma_{0.1}^2$ denotes the variance of the y-values that are uncorrelated with x. The advantage of the new notation is that it may be extended to cases involving more variables.

23. An example of a bivariate population. Having written the relationships in terms of population parameters, we are now ready for a numerical analysis of a population to illustrate the application of these formulas. The first three columns of Table 61 show the composition of a bivariate population: 36% of the (x, y) pairs are $(2, 40)$, etc. In such a case we really cannot talk of the number of points, but we can use the frequency f_i of points, where $\Sigma f_i = 1$. The means are calculated in the usual way; e.g., $\bar{x} = \Sigma f_i x_i = 0.36(2) + 0.48(1) + 0.16(0) = 1.20$. The variances are calculated according to (44) or (46). The covariance is obtained according to (47), viz.,

$$\sigma_{xy} = \Sigma f x y - \bar{x}\bar{y} = 45.6 - 32(1.2) = 7.2 \qquad (49)$$

Table 61. Linear regression and correlation in a bivariate population.

f	x	y	fxy	$g = a + bx$	$e = y - g$
0.36	2	40	28.8	44	-4
0.48	1	35	16.8	29	$+6$
0.16	0	5	0	14	-9
1.00	$\bar{x} = 1.20$	$\bar{y} = 32$	45.60	$\bar{g} = 32$	$\bar{e} = 0$
	$\sigma_x^2 = 0.48$ $\sigma_y^2 = 144$		$\sigma_{xy} = 7.20$ $\sigma_g^2 = 108$		$\sigma_e^2 = 36$

All the other quantities are then easily obtained:

$$b = b_{yx} = \frac{\sigma_{xy}}{\sigma_x^2} = \frac{7.20}{0.48} = 15 \qquad (50)$$

$$a = \bar{y} - b\bar{x} = 32 - 15(1.2) = 14 \qquad (51)$$

so that the equation of the regression line of y on x is

$$\hat{y} = g = a + bx = 14 + 15x \qquad (52)$$

The values of g in Table 61 are calculated from this equation. Then the e values are obtained by subtraction: $e = y - g$. Note that $\bar{y} = \bar{g}$ and $\bar{e} = 0$. The variance of g may be calculated from the g values directly; that is,

$$\sigma_g^2 = \Sigma f_i g_i^2 - \bar{g}^2 = 1132 - 1024 = 108 \qquad (53)$$

or by the shortcuts

$$\sigma_g^2 = b^2 \sigma_x^2 = 225 \times 0.48 = 108 \qquad (54)$$

$$= b\sigma_{xy} = 15 \times 7.20 = 108$$

Similarly, the variance of e may be directly calculated from the e values in Table 61 or simply by subtraction: $\sigma_e^2 = \sigma_y^2 - \sigma_g^2 = 144 - 108 = 36$. Then the correlation between x and y may be obtained in various ways. By definition it is

$$r(x, y) = \frac{\sigma_{xy}}{\sigma_x \sigma_y} \quad \frac{7.20}{\sqrt{0.48 \times 144}} = 0.866 \qquad (55)$$

Or we can make use of the fact that

$$r(x, y) = r(g, y) = \frac{\sigma_g}{\sigma_y} = \sqrt{\frac{108}{144}} = \sqrt{0.75} = 0.866 \qquad (56)$$

As to the partition of the total variance of y, we see that

LINEAR REGRESSION AND CORRELATION

$$r_{gy}^2 = \frac{\sigma_g^2}{\sigma_y^2} = \frac{108}{144} = 0.75 \quad \text{due to regression} \tag{57}$$

$$1 - r_{gy}^2 = \frac{\sigma_e^2}{\sigma_y^2} = \frac{36}{144} = 0.25 \quad \text{due to residual}$$

The correlation between y and e is

$$r(e, y) = \sqrt{1 - r_{gy}^2} = \frac{\sigma_e}{\sigma_y} = \frac{6}{12} = 0.50 \tag{58}$$

The reader may verify that $r(g, e) = 0$.

Further Notes on Correlation

24. Independent and uncorrelated variables. The reader may have noticed that when the sum of products of deviations $\Sigma (x - \bar{x})(y - \bar{y}) = 0$ and thus the correlation coefficient $r(x, y) = 0$, we simply say x and y are (linearly) uncorrelated and do not say that x and y are independent (or independently distributed). There is an important technical difference between these two situations, as illustrated in Table 64. If $f_i(x)$ is the frequency of x_i, and $f_j(y)$ is the frequency of y_j, of the two marginal variables, then independent distribution means that the frequency of the pair (x_i, y_j) in the body of the table is equal to the product of the respective marginal frequencies, viz.,

$$f_{ij}(x_i, y_j) = f_i(x_i) \cdot f_j(y_j)$$

The top table of Table 64 fulfills this condition; hence x and y are independently distributed. In such a situation the correlation $r(x, y)$ is always zero. Therefore independent distribution implies no correlation. The remaining four tables of Table 64 show no linear trend of the covariation of x and y and hence $\Sigma (x - \bar{x})(y - \bar{y}) = 0$ and $r(x, y) = 0$; but obviously they are not independently distributed. In such a case we simply say x and y are uncorrelated, meaning linearly uncorrelated. In our analysis of linearly related variables, most of the time we merely assume a certain covariance is zero and do not require independent distribution.

Table 64. The difference between independent and uncorrelated variables. The common denominator 30 for all entries is omitted in the following tables; thus the entry 9 means 9/30. The table on the top shows independent distribution of the two marginal variables. The other four tables show that the two variables are linearly uncorrelated but they are not independently distributed.

r = 0

1	3	2	6
3	9	6	18
1	3	2	6
5	15	10	30

uncorrel but not indep

	6		6
5	3	10	18
	6		6
5	15	10	30

2		4	6
1	15	2	18
2		4	6
5	15	10	30

	6		6
2	9	7	18
3		3	6
5	15	10	30

3		3	6
2	9	7	18
	6		6
5	15	10	30

25. Correlations are not transitive. Let us consider three variables denoted by x_1, x_2, x_3, as shown in Table 65. Also, let $r_{12} = r(x_1, x_2)$, etc. The mean, sum of squares, and two sums of products are calculated and listed in Table 65. For the example in the upper half of the table, we have the correlations

$$r_{12} = \frac{39}{\sqrt{110 \times 22}} = 0.793; \qquad r_{23} = \frac{20}{\sqrt{22 \times 64}} = 0.533$$

LINEAR REGRESSION AND CORRELATION

Table 65. The three pair-wise correlations among three variables. For brevity, deviations are denoted by capital letters: $X_1 = (x_1 - \bar{x}_1)$, etc.

	First example				mean	sum of squares	sum of products
x_1:	3	7	10	14	16	10	$\Sigma X_1^2 = 110$	
								$\Sigma X_1 X_2 = 39$
x_2:	8	7	12	11	12	10	$\Sigma X_2^2 = 22$	
								$\Sigma X_2 X_3 = 20$
x_3:	12	4	14	12	8	10	$\Sigma X_3^2 = 64$	

	Second example				mean	sum of squares	sum of products
x_1:	3	7	10	14	16	10	$\Sigma X_1^2 = 110$	
								$\Sigma X_1 X_2 = 39$
x_2:	8	7	12	11	12	10	$\Sigma X_2^2 = 22$	
								$\Sigma X_2 X_3 = 32$
x_3:	4	8	12	12	14	10	$\Sigma X_3^2 = 64$	

But $r_{13} = 0$, as $\Sigma(x_1 - \bar{x}_1)(x_3 - \bar{x}_3) = \Sigma x_1 x_3 - n\bar{x}_1\bar{x}_3 = 500 - 500 = 0$. This shows that from the knowledge of r_{12} and r_{23}, it is impossible to know the value of r_{13} without further assumption. In particular, $r_{13} \neq r_{12} r_{23}$ in general. In other words, the correlations are not transitive. Should x_1 and x_3 be also correlated, then their correlation r_{13} should be calculated separately and directly from the data on x_1 and x_3, not deduced from r_{12} and r_{23}. A second example is provided in the lower half of Table 65. Here we have

$$r_{12} = \frac{39}{\sqrt{110 \times 22}} = 0.793; \quad r_{23} = \frac{32}{\sqrt{22 \times 64}} = 0.853$$

The sum of products for x_1 and x_3 is calculated separately.

$$\Sigma(x_1 - \bar{x}_1)(x_3 - \bar{x}_3) = \Sigma x_1 x_3 - n\bar{x}_1\bar{x}_3 = 580 - 500 = 80,$$

so that

$$r_{13} = \frac{80}{\sqrt{110 \times 64}} = 0.953$$

The Scattering of Points

26. Let us consider the data in Table 61 once more. The points in that population assume only three distinct positions: $(x, y) = (2, 40), (1, 35), (0, 5)$, with frequencies $f = 0.36, 0.48, 0.16$, respectively. Now suppose that for the given condition $x = 2$, not all the y's assume the value 40 but they assume a series of values, e.g., $50, 46, 34, 35$, with frequencies $0.04, 0.12, 0.12, 0.08$, respectively, so that the mean of such values is 40. Then the original point $(2, 40)$ has been split into a series of points with mean position at $(2, 40)$. The new series of y values can be given a new symbol z where $z = y + \in$ and the average value of the $\in$'s is zero for each given value of x. Table 66 and Fig. 67 give the full picture of the scattering of points for the population of Table 61. Note that the mean of the z's in the first group $(x = 2)$ is $y = 40$; the mean of the z's in the second group $(x = 1)$ is $y = 35$; and that of the last group $(x = 0)$ is $y = 5$, so that the general mean of the z's in the entire population is $\bar{z} = \bar{y} = 32$, the same as that in Table 61. Also note that the extent of the scattering for the three x-groups need not be the same or symmetrical. The spread in the second group is much larger than that in the third group. The important feature is that the scattering of y is about the mean of y of each group.

Table 66. The scattering of a point (x, y) into a series of points $(x, y + \in)$ with mean position at the original point. The sum of the $\in$'s is zero for each x-group. This is an extension of the data in Table 61. See Fig. 67.

	f	x	y	z	fyz	fxz	$\in$	$f\in^2$
0.36	0.04	2	40	50	80.0	4.00	10	4.00
	0.12	2	40	46	220.8	11.04	6	4.32
	0.12	2	40	34	163.2	8.16	−6	4.32
	0.08	2	40	35	112.0	5.60	−5	2.00
0.48	0.04	1	35	43	60.2	1.72	8	2.56
	0.12	1	35	39	163.8	4.68	4	1.92
	0.20	1	35	34	238.0	6.80	−1	0.20
	0.08	1	35	31	86.8	2.48	−4	1.28
	0.04	1	35	28	39.2	1.12	−7	1.96
0.16	0.08	0	5	8	3.2	0	3	0.72
	0.08	0	5	2	0.8	0	−3	0.72
	1.00	$\bar{x}=1.2$	$\bar{y}=32$	$\bar{z}=32$	1168.0	45.60	$\in = 0$	24.00
	$\sigma_x^2 = 0.48$	$\sigma_y^2 = 144$	$\sigma_z^2 = 168$		$\sigma_{yz}=144$	$\sigma_{xz}=7.20$		$\sigma_\in^2 = 24.00$
			$\sigma_z^2 = \sigma_y^2 + \sigma_\in^2$		$\sigma_{yz}=\sigma_y^2$	$\sigma_{xz}=\sigma_{xy}$		

LINEAR REGRESSION AND CORRELATION

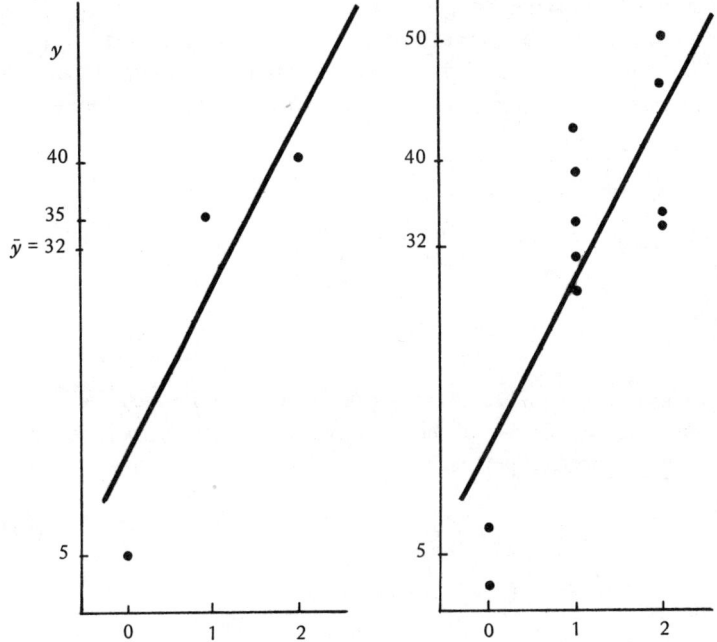

Fig. 67. Two scatter diagrams with the same regression line. *Left:* the points are (x, y) of Table 61. Although only three points are shown in the diagram, we must remember that the frequencies of these points are *0.36, 0.48, 0.16*. *Right:* the scattering of each y value into a series of values which are denoted by z in the text; the mean of each array of z's is the original y value. The equation of the regression line is (52), $\hat{y} = 14 + 15x$ in both diagrams.

27. Variance and covariance involving z. The introduction of the new variable z obviously does not affect the mean, variance, and covariance of x and y, which we have studied in paragraph 22. Now we need only to study the variance and covariances involving z. The variance of z may be calculated directly from the numerical values of Table 66.

$$\sigma_z^2 = \Sigma f z^2 - \bar{z}^2 = 1192 - (32)^2 = 168 \qquad (59)$$

Or we may use the fact that $z = y + \in$, where y and $\in$ are uncorrelated. Hence

$$\sigma_z^2 = \sigma_y^2 + \sigma_\in^2 = 144 + 24 = 168 \qquad (60)$$

The $\in$ values and $\sigma_\in^2 = \Sigma f \in^2$ are given in the last two columns of Table 66. There are two covariances involving z, viz., $Cov(y, z)$ and $Cov(x, z)$. Let us calculate $Cov(y, z)$ first, as shown in the fyz column of Table 66. Note that for each fixed x-group (y also constant), the sum of product is

$$(x \text{ and } y \text{ fixed}), \quad \Sigma fyz = y\Sigma fz = y\Sigma f(y + \in) = f_x y^2$$

where $f_x = \Sigma f$ is the total frequency of that x-group. For example, for the $x = 2$ group the sum of product is (see Table 66):

$$80.0 + 220.8 + 163.2 + 112.0 = 576 = 0.36(40)^2$$

This is so because for each fixed x-group (that is also for each fixed y-group), the sum $\Sigma f \in = 0$. Since this is true for every x-group, the covariance in the entire population is

$$\sigma_{yz} = \Sigma fyz - \bar{y}\bar{z} = \Sigma fy^2 - \bar{y}^2 = \sigma_y^2 \quad (61)$$

$$= 1168 - 32(32) = 144$$

The covariance (x, z) may be calculated in a similar way, using the same fact that for each fixed x-group, $\Sigma f \in = 0$. Thus,

$$(x \text{ and } y \text{ fixed}), \quad \Sigma fxz = x\Sigma fz = x\Sigma f(y + \in) = f_x xy$$

e.g., $\quad 4.00 + 11.04 + 8.16 + 5.60 = 28.80 = 0.36 \times 2 \times 40$

Therefore, for the entire population, the covariance is

$$\sigma_{xz} = \Sigma fxz - \bar{x}\bar{z} = \Sigma fxy - \bar{x}\bar{y} = \sigma_{xy} \quad (62)$$

$$= 45.60 - 1.2(32) = 7.20$$

The last result is important. Since $\sigma_{xz} = \sigma_{xy}$, it follows that the regression slope on x is the same whether we consider the original points (x, y) or the scattered points (x, z). That is $b = b_{yx} = b_{zx} = 15$. The intercept is

LINEAR REGRESSION AND CORRELATION

also the same: $a = \bar{y} - b\bar{x} = \bar{z} - b\bar{x} = 14$. In other words, the regression line remains the same for the points (x, y) and the points (x, z), as emphasized in Fig. 67. Because it is the same regression line, the calculated value of z for a given x is the same as the calculated y for the same given x. That is,

$$\hat{z} = \hat{y} = g = a + bx \tag{63}$$

Then the variance due to regression will also remain the same for

$$\sigma_{\hat{z}}^2 = \sigma_{\hat{y}}^2 = \sigma_g^2 = b^2 \sigma_x^2 = b\sigma_{xy} = b\sigma_{xz} \tag{64}$$

This leads to the conclusions in the next paragraph.

28. Correlations among the three variables x, y, z. Using the last expression (64), or some of the previous expressions, we see that

$$\frac{\sigma_{xy}}{\sigma_x} = \frac{b\sigma_{xy}}{b\sigma_x} = \frac{\sigma_g^2}{\sigma_g} = \sigma_g \tag{65}$$

Now we shall summarize all of the previous results by arranging the variables in the following order and calculating their correlations:

$$
\begin{array}{cccc}
0 & 1 & 2 & 3 \\
x & g = a + bx, & y = g + e, & z = y + \epsilon
\end{array} \tag{66}
$$

where g and e are uncorrelated and so are y and ϵ. The correlation $r(x, y) = r(g, y)$ has already been calculated (55, 56) but will be given here again for comparison under the new symbol r_{12}.

$$r_{12} = r(g,y) = r(x,y) = \frac{\sigma_{xy}}{\sigma_x \sigma_y} = \frac{\sigma_g^2}{\sigma_y} = \sqrt{\frac{108}{144}} = 0.866$$

$$r_{23} = r(y,z) = \frac{\sigma_{yz}}{\sigma_y \sigma_z} = \frac{\sigma_y^2}{\sigma_y \sigma_z} = \frac{\sigma_y}{\sigma_z} = \sqrt{\frac{144}{168}} = 0.926 \tag{67}$$

$$r_{13} = r(x,z) = \frac{\sigma_{xz}}{\sigma_x \sigma_z} = \frac{\sigma_{xy}}{\sigma_x \sigma_z} = \frac{\sigma_g}{\sigma_z} = \sqrt{\frac{108}{168}} = 0.802$$

LINEAR REGRESSION AND CORRELATION

It is seen that

$$r_{13} = r_{12}\, r_{23} = \frac{\sigma_g}{\sigma_y}\frac{\sigma_y}{\sigma_z} = \frac{\sigma_g}{\sigma_z} \quad (68)$$

$$= 0.866 \times 0.926 = 0.802$$

Thus, under certain circumstances the correlations do have a kind of chain property. A knowledge of linear regression and correlation is essential to the development of path analysis. One should be familiar with the contents of this chapter before going on to multiple regression and multiple correlation in the next chapter.

Exercises

Ex. 1. Given the following six pairs of (x, y), find the equation $\hat{y} = a + bx$ of the linear regression line of y on x. Then verify expression (18) for the partition of the sum of squares of y. Follow the format of Table 42, or use a slightly simplified table by omitting certain columns as shown in the following:

	x	y	$\hat{y}$	$(y-\hat{y})$	$(\hat{y}-\bar{y})$	$(y-\hat{y})^2$	$(\hat{y}-\bar{y})^2$	$\hat{y}^2$
	2	3	2.2	+.8	−4.8	.64	23.04	4.84
	4	4						
	4	6						
	6	7						
	7	10						
	7	12						
Total	30	42	42.0	0	0	8.80	51.20	345.20
mean	5	7	7					

Verify $\quad r = \dfrac{\Sigma(x-\bar{x})(y-\bar{y})}{\sqrt{\Sigma(x-\bar{x})^2\ \Sigma(y-\bar{y})^2}} = \sqrt{\dfrac{\Sigma(\hat{y}-\bar{y})^2}{\Sigma(y-\bar{y})^2}} = 0.92376$

LINEAR REGRESSION AND CORRELATION

Also verify
$$\Sigma(\hat{y}-\bar{y})^2 = b^2 \Sigma (x-\bar{x})^2$$
$$= r^2 \Sigma (y-\bar{y})^2$$
$$= b \, \Sigma(x-\bar{x})(y-\bar{y}) = 51.20$$

Hint: Find the three basic values first: $\Sigma (x-\bar{x})^2$, $\Sigma (y-\bar{y})^2$, $\Sigma (x-\bar{x})(y-\bar{y})$.

Partial Ans. $\hat{y} = 1.60\, x - 1$.

Ex. 2. The slope of the regression line of y on x has been shown to be

$$b = b_{yx} = \frac{\Sigma (x-\bar{x})(y-\bar{y})}{\Sigma (x-\bar{x})^2} \quad \text{i.e. (12)}$$

purely as a consequence of the normal equations (10) and (11) which are in turn derived from the least square conditions (7) and (8). How do we justify that the expression (12) actually has the geometrical meaning of a slope? Li (1959, 1964, 1964b) gives the following justification. Let $P_i = (x_i, y_i)$ and $P_j = (x_j, y_j)$ be any two points. Then the slope of the straight line connecting these two points is

$$b_{ij} = \frac{y_i - y_j}{x_i - x_j}$$

Not all such slope should be of equal importance, as the slope determined over a short base is less accurate than that determined over a long base. Hence a kind of weight should be attached to each of such b_{ij}'s. Li proposed that the weight should be (or proportional to) the square of the base over which the slope is determined (Fig. 72, left). That is, the weight for b_{ij} is $w_{ij} = (x_i - x_j)^2$. Then the expression (12) is simply the weighted mean of the b_{ij}'s. And,

$$b = \frac{\Sigma w_{ij}\, b_{ij}}{\Sigma w_{ij}} = \frac{\Sigma (x_i - x_j)(y_i - y_j)}{\Sigma (x_i - x_j)^2} \quad (12')$$

where the summation is overall possible $n(n-1)/2$ pairs of points. The expression (12') above is the same as (12) by virtue of the identities established in Chapter 2. Alternatively, we may find the slope from any point (x_i, y_i) to the fixed central point $(\bar{x}, \bar{y})$:

$$b_i = \frac{y_i - \bar{y}}{x_i - \bar{x}}, \quad \text{with weight} \quad w_i = (x_i - \bar{x})^2$$

as shown in Fig. 72, right. Then the regression coefficient is the weighted mean of such slopes:

$$b = \frac{\Sigma w_i b_i}{\Sigma w_i} = \frac{\Sigma (x_i - \bar{x})(y_i - \bar{y})}{\Sigma (x_i - \bar{x})^2}$$

Thus the expression for b does have a geometrical meaning.

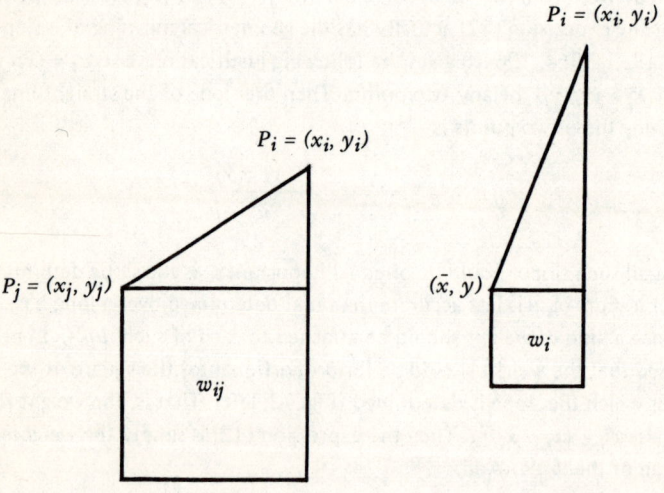

Fig. 72. The weight w associated with a slope is the square of the base over which the slope is calculated. *Left:* the slope between any two points P_i and P_j. *Right:* the slope between a point (x_i, y_i) and the fixed point $(\bar{x}, \bar{y})$.

LINEAR REGRESSION AND CORRELATION

Ex. 3. In the following table the population of x and y is the same as that in Tables 61 and 66, but the scattering pattern of z is different. Hence all the statistics involving x and y will remain the same as before, but the variance of z and $\in$ will be different.

f_x	f	x	y	z	fyz	fxz	$\in$	$f\in^2$
	0.09	2	40	48	172.8	8.64	8	5.76
0.36	0.18	2	40	38	273.6	13.68	−2	0.72
	0.09	2	40	36	129.6	6.48	−4	1.44
	0.12	1	35	43	180.6	5.16	8	7.68
0.48	0.24	1	35	33	277.2	7.92	−2	0.96
	0.12	1	35	31	130.2	3.72	−4	1.92
	0.04	0	5	13	2.6	0	8	2.56
0.16	0.08	0	5	3	1.2	0	−2	0.32
	0.04	0	5	1	0.2	0	−4	0.64
	1.00	$\bar{x}=1.2$	$\bar{y}=32$	$\bar{z}=32$	1168.0	45.60	$\bar{\in}=0$	$\sigma^2_\in=22.00$

(i) Note that the distribution of $\in$ is the same in three x-groups. Hence the variance of $\in$ may be calculated from any one of the x-groups instead of from the entire table as we did above. Calculate the variance of $\in$ from the following distribution and see $\sigma^2_\in = 22.0$.

$\in:$	8	−2	−4	Total
$f:$	1	2	1	4

(ii) Calculate the variance of z directly from the listed values in the table. Then verify $\sigma^2_z = \sigma^2_{\hat{y}} + \sigma^2_\in = 144 + 22 = 166$.

(iii) Verify: the covariance of (y, z) is the same as variance of y. That is,

$$\Sigma fyz = \Sigma fy^2 \quad \text{implies} \quad \sigma_{yz} = \sigma^2_y$$

Can you see why?

(iv) Also verify: $\sigma_{xz} = \sigma_{xy}$. Does this mean that the slope of the regression line of y on x is the same as that of z on x?

(v) Finally, calculate the three correlations $r(x, y)$, $r(y, z)$, $r(x, z)$, and see if $r(x, y)\, r(y, z) = r(x, z)$.

Ex. 4. Write $\sigma_x^2 = \Sigma\,(x - \bar{x})^2/n$ and $\sigma_y^2 = \Sigma\,(y - \bar{y})^2/n$, where n is the number of pairs of (x, y). If we let

$$U = \frac{x - \bar{x}}{\sigma_x}, \qquad V = \frac{y - \bar{y}}{\sigma_y}$$

then, by definition (24), the correlation coefficient may be written

$$r = r(x,y) = E(U\,V)$$

where the operator E means "the average value of." Also, $E(U^2) = 1$ and $E(V^2) = 1$, on account of the definitions of U and V, which are known as the "standardized" variables (Chapter 5). Now, we have $E(U \pm V)^2 \geqslant 0$, which is always true. Expanding,

$$E(U^2) + E(V^2) \pm 2\,E(UV) \geqslant 0$$

that is, $\qquad\qquad 2 \pm 2\,r \quad\ \geqslant 0$

When $1 + r \geqslant 0$, we have $r \geqslant -1$. When $1 - r \geqslant 0$, we have $1 \geqslant r$. The final conclusion is $-1 \leqslant r \leqslant 1$.

Chapter 4

MULTIPLE REGRESSION AND CORRELATION

WE HAVE thought of y as weight and x as height of people and have obtained a regression equation $\hat{y} = a + bx$, so that for a given height (x) we can calculate the "expected" weight $(\hat{y})$, expected from the linear equation. The equation is such that the amount by which our prediction is off, $e = y - \hat{y}$, yields the smallest sum of squares; and in this sense we say that $\hat{y}$ is the best predictor. The weight, however, may also be predicted from other related factors, such as diet or daily food intake. Let x_1 be measurement of height and x_2 be measurement of average daily food intake. Then we wish to find a linear equation of the type $\hat{y} = a + b_1 x_1 + b_2 x_2$, which presumably would give us an even more accurate prediction for weight y. A regression equation involving two or more variables on the right ("independent" variables) is called a multiple (linear) regression equation. Its general form is $\hat{y} = a + b_1 x_1 + \ldots + b_k x_k$, where k is the number of independent variables on the right side of the equation. The case for two independent variables is of particular importance because it illustrates the general principle of the matter without much mathematics and because it often arises in biological and social research for no other reason than the fact that each child has two parents. So we shall treat the simple case in some detail. Once understood, the case with three or more independent variables will follow the same pattern.

Regression on two variables.

1. The data. To avoid abstractness we follow the procedure of the last chapter by presenting a set of data first before dealing with the method of finding the regression equation.

(weight)	y:	8	4	21	49	26	33	84	55	$\Sigma y = 280$,	$\bar{y} = 35$	
(height)	x_1:	10	9	20	17	11	18	18	17	$\Sigma x_1 = 120$,	$\bar{x}_1 = 15$	(1)
(diet)	x_2:	3	5	6	8	11	14	15	18	$\Sigma x_2 = 80$,	$\bar{x}_2 = 10$	

In addition to the sums and means indicated above, we will also need the various sums of squares and sums of products, all directly calculated from the data above.

$$\Sigma y^2 = 14768, \quad \Sigma x_1^2 = 1928, \quad \Sigma x_2^2 = 1000 \quad (2)$$
$$\Sigma x_1 x_2 = 1280, \quad \Sigma x_1 y = 4696, \quad \Sigma x_2 y = 3560$$

Note that these are uncorrected (or direct) sums of squares and sums of products to fulfill our immediate need in paragraph 4. The corrected sums of squares and products (i.e., of deviations from mean) will be calculated later.

2. The regression equation. The model (assumption) is that each observed value of weight y is a linear function of height x_1 and diet x_2, plus an uncorrelated residual term:

$$y = a + b_1 x_1 + b_2 x_2 + e \quad (3)$$

The corresponding regression equation to be found is

$$\hat{y} = a + b_1 x_1 + b_2 x_2 \quad (4)$$

Geometrically, the equation (4) represents a plane in a three dimensional space, just as $\hat{y} = a + bx$ represents a straight line in a two dimensional space. Each set of observations (y, x_1, x_2) represents a point in a three dimensional space. Data (1) give us the positions of eight points in space; the central point is $(\bar{y}, \bar{x}_1, \bar{x}_2) = (35, 15, 10)$ The plane represented by equation (4) lies among these points; some of the points will be above the plane, while others will be below the plane. The distance (along the y-axis) between an actual point to the plane is $e = y - \hat{y}$. The plane (4) to be found should be such that the sum of squares of these distances is a minimum. That is, Σe^2 is to be minimized. This is again the principle of least squares which we have used for the simple case $\hat{y} = a + bx$. If we let x_2 be a constant, then the plane becomes a straight line on the y-x_1 plane with slope b_1. Similarly, if x_1 is held constant, the plane becomes a

MULTIPLE REGRESSION AND CORRELATION

straight line on the y-x_2 plane with slope b_2. Hence, b_1 and b_2 are called partial regression coefficients. At this stage, a few words should be said about the notation employed above. For convenience we have used simply b_1 and b_2, while the more complete notation should be

$$b_1 = b_{01.2} \quad \text{and} \quad b_2 = b_{02.1}$$

where the subscript o represents the dependent variable y. The symbol $b_{01.2}$, for instance, then means the partial regression coefficient (slope) of y on x_1, holding x_2 constant. In the next few paragraphs, we shall use b_1 and b_2 for brevity; this should not cause confusion as we are using them in the context of equations (3) and (4). In later developments, when the simple regression coefficients are also involved, we shall use the full notation $b_{01.2}$.

3. The normal equations.
The quantity to be minimized is

$$Q = \Sigma e^2 = \Sigma(y - \hat{y})^2 = \Sigma(y - a - b_1 x_1 - b_2 x_2)^2 \quad (5)$$

Taking the partial derivatives $\partial Q/\partial a$, $\partial Q/\partial b_1$, $\partial Q/\partial b_2$, and setting them equal to zero, we obtain the following normal equations:

$$\Sigma(y - a - b_1 x_1 - b_2 x_2) = \Sigma(y - \hat{y}) = \Sigma e = 0$$

$$\Sigma(y - a - b_1 x_1 - b_2 x_2)x_1 = \Sigma(y - \hat{y})x_1 = \Sigma e x_1 = 0 \quad (6)$$

$$\Sigma(y - a - b_1 x_1 - b_2 x_2)x_2 = \Sigma(y - \hat{y})x_2 = \Sigma e x_2 = 0$$

These are the basic equations. The first equation of (6) shows that the distances from actual y to plane $\hat{y}$ add up to zero; this means that the positive distances above the plane cancel the negative distances below the plane. Hence $\Sigma y = \Sigma \hat{y}$; in other words, the mean of the $\hat{y}$ values lying on the plane is equal to $\bar{y}$. The remaining two equations of (6) show that the distance e is uncorrelated with x_1 or x_2. For the convenience of numerical solution the normal equations (6) are usually written in the form shown below:

$$na + b_1 \Sigma x_1 + b_2 \Sigma x_2 = \Sigma y$$

$$a \Sigma x_1 + b_1 \Sigma x_1^2 + b_2 \Sigma x_1 x_2 = \Sigma x_1 y \quad (7)$$

$$a \Sigma x_2 + b_1 \Sigma x_1 x_2 + b_2 \Sigma x_2^2 = \Sigma x_2 y$$

from which the values of a, b_1, b_2 may be found readily.

4. Numerical example. Substituting the numerical values obtained in (1) and (2), the normal equations (7) become

$$8a + 120 b_1 + 80 b_2 = 280$$

$$120 a + 1928 b_1 + 1280 b_2 = 4696 \quad (8)$$

$$80 a + 1280 b_1 + 1000 b_2 = 3560$$

Following the usual procedure of solving linear equations (Chapter 2), the reader should be able to obtain the solutions for a, b_1, b_2, in a few minutes; he will find that

$$a = -25, \quad b_1 = 2, \quad b_2 = 3 \quad (9)$$

so that the regression equation is

$$\hat{y} = -25 + 2 x_1 + 3 x_2 \quad (10)$$

From the first normal equation of (7) we see that the value of a is

$$a = \bar{y} - b_1 \bar{x}_1 - b_2 \bar{x}_2 \quad (11)$$

Hence the point $(y, x_1, x_2) = (\bar{y}, \bar{x}_1, \bar{x}_2)$ satisfies the regression equation (4). In other words, the regression plane passes through the central point $(\bar{y}, \bar{x}_1, \bar{x}_2)$. We shall make use of (11) later to simplify our normal equations. Before proceeding further on the subject, it would be well to calculate the individual values of $\hat{y}$ and e as is done in Table 79 to verify some

MULTIPLE REGRESSION AND CORRELATION

of the properties of regression. Since e is uncorrelated with x_1 or x_2, it follows that it is also uncorrelated with $\hat{y}$. The other quantities shown in Table 79 are for later use; A = uncorrected sum of squares and C = correction term; so that $A - C = ssq$.

Table 79. Multiple regression of y on x_1 and x_2. The regression equation is $\hat{y} = -25 + 2x_1 + 3x_2$.

x_1	x_2	y	$\hat{y}$	e	$y\hat{y}$	ye	$\hat{y}e$
10	3	8	4	+ 4	32	32	16
9	5	4	8	− 4	32	− 16	− 32
20	6	21	33	− 12	693	− 252	− 396
17	8	49	33	+ 16	1617	784	528
11	11	26	30	− 4	780	− 104	− 120
18	14	33	53	− 20	1749	− 660	− 1060
18	15	84	56	+ 28	4704	2352	1568
17	18	55	63	− 8	3465	− 440	− 504
Total 120	80	280	280	0	13072	1696	0
mean 15	10	35	35	0			
A: 1928	1000	14768	13072	1696	13072	1696	0
−C: −1800	− 800	− 9800	− 9800	0	− 9800	0	0
ssq: 128	200	4968	3272	1696	3272	1696	0
$\sigma_1^2 = 16$	$\sigma_2^2 = 25$	$\sigma_y^2 = 621$	$\sigma_{\hat{g}}^2 = 409$	$\sigma_e^2 = 212$	$\sigma_{yg} = 409$	$\sigma_{ye} = 212$	$\sigma_{ge} = 0$
$\sigma_1 = 4$	$\sigma_2 = 5$	$\sigma_y = 24.92$	$\sigma_g = 20.22$	$\sigma_e = 14.56$			

5. Partition of the total sum of squares.

The total sum of squares $\Sigma(y - \bar{y})^2$ may be subdivided into two components by the use of the regression equation $\hat{y} = a + b_1 x_1 + b_2 x_2$ in exactly the same manner as in the simple regression case, viz.:

$$\begin{array}{rcccc}
\text{total ssq} & = & \text{regression ssq} & + & \text{residual ssq} \\
\Sigma(y - \bar{y})^2 & = & \Sigma(\hat{y} - \bar{y})^2 & + & \Sigma(y - \hat{y})^2 \\
\text{Table 79} \quad 4968 & = & 3272 & + & 1696
\end{array} \quad (12)$$

The product term, $\Sigma(y-\hat{y})(\hat{y}-\bar{y})$, vanishes, because it is equal to

$$\Sigma(y-\hat{y})(a+b_1x_1+b_2x_2) = a\Sigma e + b_1\Sigma ex_1 + b_2\Sigma ex_2 = 0$$

by virtue of the normal equations (6). If we consider data (1) as a population, we may describe the same fact in terms of variances $(g=\hat{y})$:

$$\sigma_y^2 = \sigma_g^2 + \sigma_e^2 \tag{13}$$

Table 79   621   =   409   +   212

The reader, if he wishes, may construct a table exactly like Table 42 with the values of $y, \hat{y}, \bar{y}$. He will find that the three "areas" (similar to Fig. 44) are

$$\begin{aligned}\Sigma y^2 &= 14768 \\ &\searrow \Sigma y^2 - \Sigma \hat{y}^2 = 1696 = \text{residual ssq} \\ \Sigma \hat{y}^2 &= 13072 \\ &\searrow \Sigma \hat{y}^2 - \Sigma \bar{y}^2 = 3272 = \text{regression ssq} \\ \Sigma \bar{y}^2 &= 9800\end{aligned} \tag{14}$$

To summarize, all the relationships regarding the sum of squares of $y, \hat{y}, \bar{y}$ remain the same as those in the simple regression case.

Sum of squares due to regression

6. In order to calculate the multiple correlation we need to know the sum of squares due to regression. Its value in our numerical example, *regression ssq = 3272* as given in (12) and (14), was obtained from the individual values of $\hat{y}$ calculated in Table 79. As in the case of simple regression, this quantity may be calculated by a shorter method without explicitly calculating the individual values of $\hat{y}$. Recalling (11), $a = \bar{y} - b_1\bar{x}_1 - b_2\bar{x}_2$. Substituting in the regression equation (4), and transposing the term $\bar{y}$ to the left side, we obtain a form in terms of deviations from the means:

$$\hat{y} - \bar{y} = b_1(x_1 - \bar{x}_1) + b_2(x_2 - \bar{x}_2)$$

MULTIPLE REGRESSION AND CORRELATION

or, briefly,
$$\hat{Y} = b_1 X_1 + b_2 X_2 \quad (15)$$

where the capital letters denote the deviations from mean. Squaring both sides and summing, we obtain the regression ssq:

$$\Sigma (\hat{y} - \bar{y})^2 = \Sigma \hat{Y}^2 = b_1^2 \Sigma X_1^2 + 2 b_1 b_2 \Sigma X_1 X_2 + b_2^2 \Sigma X_2^2 \quad (16)$$

which is analogous to $b^2 \Sigma (x - \bar{x})^2$ for the case of simple regression. From (1) and (2), we have the corrected sum of squares and products:

$$\Sigma X_1^2 = \Sigma x_1^2 - n \bar{x}_1^2 = 1928 - 1800 = 128$$

$$\Sigma X_1 X_2 = \Sigma x_1 x_2 - n \bar{x}_1 \bar{x}_2 = 1280 - 1200 = 80$$

$$\Sigma X_2^2 = \Sigma x_2^2 - n \bar{x}_2^2 = 1000 - 800 = 200 \quad (17)$$

and from (9), $b_1 = 2$, $b_2 = 3$. Substituting, we obtain the regression ssq:

$$\Sigma (\hat{y} - \bar{y})^2 = 4(128) + 12(80) + 9(200) = 3272 \quad (16n)$$

correctly. Thus the regression ssq may be obtained directly from the sum of squares of the x-variables together with the partial regression coefficients.

7. Normal equations in a reduced form. The normal equations (6) and (7) will take a simpler form if expressed in terms of the corrected sum of squares and products. If we regard (15) as the regression equation to be found, then the quantity to be minimized would be

$$Q = \Sigma e^2 = \Sigma (Y - \hat{Y})^2 = \Sigma (Y - b_1 X_1 - b_2 X_2)^2 \quad (18)$$

Setting $\partial Q / \partial b_1$ and $\partial Q / \partial b_2$ equal to zero, we obtain the normal equations

$$\Sigma (Y - b_1 X_1 - b_2 X_2) X_1 = 0$$
$$\Sigma (Y - b_1 X_1 - b_2 X_2) X_2 = 0 \quad (19)$$

Transposing,

$$b_1 \Sigma X_1^2 + b_2 \Sigma X_1 X_2 = \Sigma X_1 Y$$
$$b_1 \Sigma X_1 X_2 + b_2 \Sigma X_2^2 = \Sigma X_2 Y \qquad (20)$$

Normal equations (19) and (20) are the simplified versions of (6) and (7) by eliminating the parameter a and working directly with the corrected sum of squares and products.

8. Another form for regression ssq. Using the normal equations (20), we may rewrite the regression ssq (16) in an even simpler form:

$$\Sigma(\hat{y}-\bar{y})^2 = b_1 [b_1 \Sigma X_1^2 + b_2 \Sigma X_1 X_2] + b_2 [b_1 \Sigma X_1 X_2 + b_2 \Sigma X_2^2]$$

by (20), $= b_1 \Sigma X_1 Y + b_2 \Sigma X_2 Y \qquad (21)$

which, the reader will recognize, is analogous to $b\Sigma(x-\bar{x})(y-\bar{y}) = b\Sigma XY$ for the case of simple regression. From the numerical values of (1) and (2), we calculate the sums of products of deviations from mean:

$$\Sigma X_1 Y = \Sigma x_1 y - n\bar{x}_1\bar{y} = 4696 - 4200 = 496$$
$$\Sigma X_2 Y = \Sigma x_2 y - n\bar{x}_2\bar{y} = 3560 - 2800 = 760 \qquad (22)$$

Substituting in (21), we obtain

$$\text{regression ssq} = 2(496) + 3(760) = 3272 \qquad (21n)$$

in agreement with our previous result. Expressions (16) and (21) represent two forms of subdivision of the regression ssq. When there are more than two x-variables, the subdivision (21) is more convenient to use than the subdivision (16).

Multiple Correlation Coefficient

9. The correlation coefficient between paired values such as (y, x) has been defined in the last chapter. Now we have triplets of numbers (y, x_1, x_2). One may ask what should be the correlation between y on the one hand and x_1, x_2 on the other. Such a correlation (to be defined) is called the multiple correlation coefficient between y and x_1, x_2, and is usually denoted by the capital letter $R = R(y; x_1, x_2) = R_{y(x_1 x_2)} = R_{0(12)}$, where the subscript 0 refers to the dependent variable y. It is defined as the ordinary correlation coefficient between y and $\hat{y}$, where $\hat{y} = a + b_1 x_1 + b_2 x_2$ as determined by the least square method.

$$R(y; x_1, x_2) = r(y, \hat{y}) = \frac{\Sigma (y - \bar{y})(\hat{y} - \bar{y})}{\sqrt{\Sigma (y - \bar{y})^2 \cdot \Sigma (\hat{y} - \bar{y})^2}} \quad (23)$$

Thus defined, the multiple correlation is always positive or zero $(R \geq 0)$. We shall calculate its numerical value first and then show its algebraic relationships with some other quantities. The individual values of $\hat{y}$ have already been given in Table 79. Then the simple correlation coefficient between the two columns of numbers $(y$ and $\hat{y})$ may be readily obtained. From Table 79, we see that the sum of products and the sum of squares of $\hat{y}$ have the same value:

$$\Sigma (y - \bar{y})(\hat{y} - \bar{y}) = \Sigma y \hat{y} - n \bar{y}^2 = 13072 - 9800 = 3272$$
$$\Sigma (\hat{y} - \bar{y})^2 = \Sigma \hat{y}^2 - n \bar{y}^2 = 13072 - 9800 = 3272 \quad (24)$$

Substituting in (23) we have, writing $R_{0(12)} = R(y; x_1, x_2)$,

$$R_{0(12)} = r(y, \hat{y}) = \frac{3272}{\sqrt{4968 \times 3272}} = \sqrt{\frac{3272}{4968}} = \sqrt{0.6586} = 0.81 \quad (25)$$

10. To show the two quantities of (24) to be equal in general, we need only to show that $\Sigma y \hat{y} = \Sigma \hat{y}^2$, which the reader may recall, is true for simple regression. The proof here is the same:

$$\Sigma y\hat{y} = \Sigma(\hat{y}+e)\hat{y} = \Sigma\hat{y}^2 + \Sigma e\hat{y} = \Sigma\hat{y}^2 \quad (26)$$

because $\Sigma e\,\hat{y} = \Sigma e(a + b_1 x_1 + b_2 x_2) = 0$ by virtue of the normal equations (6). Using the relationship $\Sigma(y-\bar{y})(\hat{y}-\bar{y}) = \Sigma(\hat{y}-\bar{y})^2$, we may rewrite the definition (23) as

$$R^2_{0(12)} = r^2_{y\hat{y}} = \frac{\Sigma(\hat{y}-\bar{y})^2}{\Sigma(y-\bar{y})^2} = \frac{\text{regression ssq}}{\text{total ssq}} = \frac{\sigma^2_g}{\sigma^2_y} \quad (27)$$

and $\quad R_{0(12)} = r_{y\hat{y}} = \dfrac{\sigma_g}{\sigma_y} = \dfrac{\sigma_{0(12)}}{\sigma_0} \quad (28)$

Also, on account of (12) and (13), we may rewrite (27) as

$$R^2_{0(12)} = r^2_{y\hat{y}} = 1 - \frac{\text{residual ssq}}{\text{total ssq}} = 1 - \frac{\sigma^2_e}{\sigma^2_y} \quad (29)$$

$$= 1 - \frac{\sigma^2_{0.12}}{\sigma^2_0}$$

In the expression above, $\sigma^2_{0(12)}$ is the variance of the part of y that is determined by x_1 and x_2, while $\sigma^2_{0.12}$ is the variance of the part of y that is uncorrelated with x_1 and x_2. Since $\sigma^2_y = \sigma^2_g + \sigma^2_e$, or $\sigma^2_0 = \sigma^2_{0(12)} + \sigma^2_{0.12}$, we have

$$\frac{\text{regression ssq}}{\text{total ssq}} + \frac{\text{residual ssq}}{\text{total ssq}} = \frac{\sigma^2_{0(12)}}{\sigma^2_0} + \frac{\sigma^2_{0.12}}{\sigma^2_0} = 1 \quad (30)$$

Readers who have had previous knowledge of multiple correlation should carefully note that in following Wright (1954), we have used the symbol $R_{0(12)}$ for $R(y; x_1, x_2)$, not the conventional $R_{0.12}$, in order to be consistent with our notation for $\sigma_{0(12)}$ and $\sigma_{0.12}$. Consider $y = \hat{y} + e$, where $\hat{y}$ and e are uncorrelated. Then $R_{0(12)} = r(y, \hat{y})$ is the correlation between y and $\hat{y}$, while $R_{0.12} = r(y, e)$ is the correlation between y and e. Hence the expression (30) amounts to

$$R^2_{0(12)} + R^2_{0.12} = 1 \quad (31)$$

Note that all of these formulas involving multiple correlation remain the same as those for simple correlation when expressed in terms of $\hat{y}$ and $e = y - \hat{y}$.

11. Components of R^2. In the previous two sections, (23) – (31), we have used the regression ssq as one number *(3272)*. But the regression ssq itself consists of subcomponents as indicated by (16) and (21). Hence the value of $R^2 = R^2_{0(12)}$ may be similarly subdivided into components. Dividing the regression ssq (16) by the total ssq of y, we have

$$R^2_{0(12)} = b_1^2 \left(\frac{\sigma_1}{\sigma_0}\right)^2 + 2b_1 b_2 r_{12} \left(\frac{\sigma_1}{\sigma_0}\right)\left(\frac{\sigma_2}{\sigma_0}\right) + b_2^2 \left(\frac{\sigma_2}{\sigma_0}\right)^2 \quad (32)$$

$$= 0.1031 + 0.1932 + 0.3623 = 0.6586$$

Similarly, dividing the regression ssq (21) by the total ssq, we obtain, remembering that the covariance $\sigma_{10} = \sigma_{01} = r_{10} \sigma_1 \sigma_0$,

$$R^2_{0(12)} = \frac{b_1 \sigma_{10}}{\sigma_0^2} + \frac{b_2 \sigma_{20}}{\sigma_0^2}$$

$$= b_1 r_{10} \left(\frac{\sigma_1}{\sigma_0}\right) + b_2 r_{20} \left(\frac{\sigma_2}{\sigma_0}\right) \quad (33)$$

$$= 0.1997 + 0.4589 = 0.6586$$

These expressions will look more symmetrical if we replace the abbreviated notations b_1 and b_2 by b_{01} and b_{02}. Diagrammatic interpretations of the expressions (32) and (33) will be given when we come to path analysis.

Determinants; explicit forms.

12. The numerical values of the partial regression coefficients are in practice always obtained from the normal equations (7) or their equivalents (20). In this section we shall write out explicit expressions for the partial regression coefficients, not for the sake of convenience in computation but to facilitate future comparisons with some other related quantities.

The reduced normal equations (20) permit us to write out explicit expressions for b_1 and b_2 in various ways. An exercise at the end of the chapter shows that they can be expressed in terms of simple regression coefficients. The following gives the expressions in terms of simple correlation coefficients and standard deviations. To simplify the matter, we divide the normal equations (20) by n, so that $\Sigma X_1^2/n = \sigma_1^2$, and $\Sigma X_1 X_2/n = Cov(x_1, x_2) = \sigma_{12} = r_{12}\sigma_1\sigma_2$, etc. Then equations (20) become

$$b_1 \sigma_1^2 \qquad + \qquad b_2 r_{12} \sigma_1 \sigma_2 \;=\; r_{01} \sigma_0 \sigma_1$$
$$b_1 r_{12} \sigma_1 \sigma_2 + \qquad b_2 \sigma_2^2 \qquad\;=\; r_{02} \sigma_0 \sigma_2 \tag{34}$$

Divide the first equation by σ_1 and the second by σ_2; then solve for the b's by the method described in Chapter 2. The reader will find that

$$b_1 = b_{01.2} = \frac{\begin{vmatrix} r_{01}\sigma_0 & r_{12}\sigma_2 \\ r_{02}\sigma_0 & \sigma_2 \end{vmatrix}}{\begin{vmatrix} \sigma_1 & r_{12}\sigma_2 \\ r_{12}\sigma_1 & \sigma_2 \end{vmatrix}} = \frac{\sigma_0 \begin{vmatrix} r_{01} & r_{12} \\ r_{02} & 1 \end{vmatrix}}{\sigma_1 \begin{vmatrix} 1 & r_{12} \\ r_{12} & 1 \end{vmatrix}} \tag{35}$$

$$= \frac{\sigma_0}{\sigma_1} \cdot \frac{r_{01} - r_{02} r_{12}}{1 - r_{12}^2} \tag{35'}$$

Similarly,

$$b_2 = b_{02.1} = \frac{\sigma_0 \begin{vmatrix} 1 & r_{01} \\ r_{12} & r_{02} \end{vmatrix}}{\sigma_2 \begin{vmatrix} 1 & r_{12} \\ r_{12} & 1 \end{vmatrix}} \tag{36}$$

MULTIPLE REGRESSION AND CORRELATION

When the two x-variables are uncorrelated $(r_{12} = 0)$, the partial regression coefficients reduce to simple regression coefficients:

$$b_1 = b_{01.2} = r_{01} \frac{\sigma_0}{\sigma_1}; \qquad b_2 = b_{02.1} = r_{02} \frac{\sigma_0}{\sigma_2} \qquad (37)$$

These are the simple regression coefficients of y on x_1 without considering x_2, and of y on x_2 without considering x_1, respectively.

13. Now let us consider the expression for $R^2_{0(12)}$ in (33) once more. The first term contains σ_1/σ_0 and the second term consists of σ_2/σ_0. On the other hand, expressions (35) and (36) show that b_1 has the factor σ_0/σ_1 and b_2 has the factor σ_0/σ_2. Substitution of (35) and (36) in (33) therefore would yield an expression for R^2 involving simple correlations only, all the factors involving standard deviations being cancelled out. For this simple case, it is an easy matter to write out the full expression. Thus, substituting (35) and (36) in (33) and cancelling, we have

$$R^2_{0(12)} = \frac{r_{10}\begin{vmatrix} r_{01} & r_{12} \\ r_{02} & 1 \end{vmatrix}}{\begin{vmatrix} 1 & r_{12} \\ r_{12} & 1 \end{vmatrix}} + \frac{r_{20}\begin{vmatrix} 1 & r_{01} \\ r_{12} & r_{02} \end{vmatrix}}{\begin{vmatrix} 1 & r_{12} \\ r_{12} & 1 \end{vmatrix}} \qquad (37)$$

This expression may seem intractable. Actually, it may be put into a very succinct form that can be readily generalized. To do this, let us begin with the following determinant in which $r_{00} = r_{11} = r_{22} = 1$.

$$\Delta = \begin{vmatrix} 1 & r_{01} & r_{02} \\ r_{01} & 1 & r_{12} \\ r_{02} & r_{12} & 1 \end{vmatrix} \qquad (38)$$

Expanding the symmetrical determinant Δ in terms of the elements of the first row (or first column) we have

$$\Delta = 1 \begin{vmatrix} 1 & r_{12} \\ r_{12} & 1 \end{vmatrix} - r_{01} \begin{vmatrix} r_{01} & r_{12} \\ r_{02} & 1 \end{vmatrix} + r_{02} \begin{vmatrix} r_{01} & 1 \\ r_{02} & r_{12} \end{vmatrix} \quad (39)$$

The interchange of two columns or two rows of a determinant changes the sign of the determinant. Thus, changing the sign of the last determinant on the right, we have

$$\Delta = 1 \begin{vmatrix} 1 & r_{12} \\ r_{12} & 1 \end{vmatrix} - r_{01} \begin{vmatrix} r_{01} & r_{12} \\ r_{02} & 1 \end{vmatrix} - r_{02} \begin{vmatrix} 1 & r_{01} \\ r_{12} & r_{02} \end{vmatrix} \quad (40)$$

Let the first determinant on the right be

$$\Delta_{00} = \begin{vmatrix} 1 & r_{12} \\ r_{12} & 1 \end{vmatrix} \quad (41)$$

which is obtained by deleting the first row and the first column of the full determinant Δ. Now, if we divide (40) by Δ_{00} and compare the result with the expression (37) for R^2, we see that

$$\frac{\Delta}{\Delta_{00}} = 1 - R_{0(12)}^2$$

or
$$R_{0(12)}^2 = 1 - \frac{\Delta}{\Delta_{00}} \quad (42)$$

which is the most succinct form for R^2, and the expression remains true when there are more than two x-variables.

14. A numerical verification of the formula (42) may help the reader to get familiar with it. From the sum of squares and products given in (17),

$$r_{12} = \frac{80}{\sqrt{128 \times 200}} = \frac{10}{4 \times 5} = \frac{1}{2}$$

From the sum of squares of y in Table 79 and the sums of products in (22), we have

$$r_{01} = \frac{496}{\sqrt{128 \times 4968}} = \frac{62}{\sqrt{16 \times 621}} = \frac{15.5}{\sqrt{621}}$$

$$r_{02} = \frac{760}{\sqrt{200 \times 4968}} = \frac{95}{\sqrt{25 \times 621}} = \frac{19.0}{\sqrt{621}}$$

Hence, $\Delta_{00} = 1 - r_{12}^2 = 1 - (\frac{1}{2})^2 = 3/4$

and

$$\Delta = \begin{vmatrix} 1 & \frac{15.5}{\sqrt{621}} & \frac{19}{\sqrt{621}} \\ \frac{15.5}{\sqrt{621}} & 1 & \frac{1}{2} \\ \frac{19}{\sqrt{621}} & \frac{1}{2} & 1 \end{vmatrix} = \frac{159}{621}$$

so that (referring to Table 79)

$$\frac{\Delta}{\Delta_{00}} = \frac{159}{621} \times \frac{4}{3} = \frac{212}{621} = \frac{\sigma_e^2}{\sigma_y^2} = \frac{\sigma_{0.12}^2}{\sigma_0^2}$$

$$= \frac{\text{residual ssq}}{\text{total ssq}}$$

and $R_{0(12)}^2 = 1 - \frac{212}{621} = \frac{409}{621} = 0.6586$

It is seen that (42) is consistent with (27) and (29).

15. The expressions (35) and (36) for the partial regression coefficients show that they can also be expressed in terms of smaller determinants derived from the full determinant Δ of (38). In addition to Δ_{00} already defined in (41), let

$$\Delta_{01} = \begin{vmatrix} r_{01} & r_{12} \\ r_{02} & 1 \end{vmatrix}, \quad \Delta_{02} = \begin{vmatrix} r_{01} & 1 \\ r_{02} & r_{12} \end{vmatrix} \tag{43}$$

where Δ_{01} is the determinant obtained from Δ by deleting the rows and columns involving r_{01} (i.e., first row and second column); and Δ_{02} is derived from Δ by deleting the rows and columns involving r_{02} (i.e., first row and third column). Then, we see that (35) and (36) may be written in the succinct form below:

$$b_{01.2} = \frac{\sigma_0}{\sigma_1} \frac{\Delta_{01}}{\Delta_{00}}, \quad b_{02.1} = \frac{-\sigma_0}{\sigma_2} \frac{\Delta_{02}}{\Delta_{00}} \tag{44}$$

If we substitute these in (33), we will of course obtain the formula for R^2 in (42) again, using (39): $\Delta = \Delta_{00} - r_{01} \Delta_{01} + r_{02} \Delta_{02}$

Regression on three variables.

16. The reader must be familiar with the situation of regression on two x-variables described above before he studies regression on three x-variables. In fact, when he works out the case for two x-variables in detail, he will find that regression on three variables is such a straightforward extension that there is practically nothing new. All we have to do is to write out a few key expressions. The model is

$$y = a + b_1 x_1 + b_2 x_2 + b_3 x_3 + e \tag{45}$$

and the linear regression equation to be found is

$$\hat{y} = a + b_1 x_1 + b_2 x_2 + b_3 x_3 \tag{46}$$

MULTIPLE REGRESSION AND CORRELATION

subject to the least squares requirement. That is, the values of a and the b's should be such that the following sum of squares is a minimum:

$$Q = \Sigma(y - a - b_1 x_1 - b_2 x_2 - b_3 x_3)^2 = \Sigma(y - \hat{y})^2 = \Sigma e^2 \quad (47)$$

Differentiating Q with respect to a and the b's and equating these partial derivatives to zero, we obtain the basic normal equations. It will be found that $\partial Q/\partial a = 0$ yields the solution

$$a = \bar{y} - b_1 \bar{x}_1 - b_2 \bar{x}_2 - b_3 \bar{x}_3 \quad (48)$$

so that the regression equation (46) may be expressed in terms of deviations from mean. Let $\hat{Y} = \hat{y} - \bar{y}$ and $X_i = x_i - \bar{x}_i$, $i = 1, 2, 3$. Then the regression equation (46) becomes

$$\hat{Y} = b_1 X_1 + b_2 X_2 + b_3 X_3 \quad (49)$$

The abbreviated notation b_1 stands for $b_{01.23}$, the partial regression coefficient (slope) of y on x_1, holding x_2 and x_3 constant, and similarly for b_2 and b_3. The reduced normal equations, analogous to (20), are

$$\begin{aligned} b_1 \Sigma X_1^2 + b_2 \Sigma X_1 X_2 + b_3 \Sigma X_1 X_3 &= \Sigma X_1 Y \\ b_1 \Sigma X_1 X_2 + b_2 \Sigma X_2^2 + b_3 \Sigma X_2 X_3 &= \Sigma X_2 Y \quad (50) \\ b_1 \Sigma X_1 X_3 + b_2 \Sigma X_2 X_3 + b_3 \Sigma X_3^2 &= \Sigma X_3 Y \end{aligned}$$

from which the numerical values of the b's may be found readily.

17. Using the calculated values $\hat{y}$ and the normal equations, the total sum of squares of y may be subdivided into two parts:

$$\begin{aligned} \text{total ssq} &= \text{regression ssq} + \text{residual ssq} \\ \Sigma(y - \bar{y})^2 &= \Sigma(\hat{y} - \bar{y})^2 + \Sigma(y - \hat{y})^2 \quad (51) \end{aligned}$$

since $\hat{y}$ and $e = y - \hat{y}$ are uncorrelated. The sum of squares due to regression (on x_1, x_2, x_3) is

$$\text{regression ssq} = \Sigma(\hat{y} - \bar{y})^2 = \Sigma(b_1 X_1 + b_2 X_2 + b_3 X_3)^2 = b_1^2 \Sigma X_1^2 + \ldots \quad (52)$$

Expanding and using the normal equations (50), it becomes

$$\text{regression ssq} = \Sigma(\hat{y} - \bar{y})^2 = b_1 \Sigma X_1 Y + b_2 \Sigma X_2 Y + b_3 \Sigma X_3 Y \quad (53)$$

which, obviously, is a simple extension of the case for two x-variables.

18. The equations (50) – (53), when divided by n, may be rewritten in terms of variances and covariances or correlations. Thus, the normal equations (50) become

$$\begin{aligned}
b_1 \sigma_1^2 \; + \; & b_2 \sigma_{12} \; + \; & b_3 \sigma_{13} \; & = \; \sigma_{10} \\
b_1 \sigma_{12} \; + \; & b_2 \sigma_2^2 \; + \; & b_3 \sigma_{23} \; & = \; \sigma_{20} \\
b_1 \sigma_{13} \; + \; & b_2 \sigma_{23} \; + \; & b_3 \sigma_3^2 \; & = \; \sigma_{30}
\end{aligned} \quad (50^*)$$

and (51) becomes

$$\sigma_y^2 \; = \; \sigma_g^2 \; + \; \sigma_e^2 \quad (51^*)$$

or

$$\sigma_0^2 \; = \; \sigma_{0(123)}^2 \; + \; \sigma_{0.123}^2$$

where the first term on the right is the variance of the part of y that is determined by x_1, x_2, x_3, while the second term is the variance of the part of y that is uncorrelated with the x's. Then (52) and (53) give further subdivision:

$$\begin{aligned}
\sigma_{0(123)}^2 \; &= \; b_1^2 \sigma_1^2 \; + \; 2 b_1 b_2 r_{12} \sigma_1 \sigma_2 \; + \; \ldots & (52^*) \\
&= \; b_1 \sigma_{10} \; + \; b_2 \sigma_{20} \; + \; b_3 \sigma_{30} & (53^*)
\end{aligned}$$

MULTIPLE REGRESSION AND CORRELATION

19. The multiple correlation coefficient between y and x's is, as before defined as the simple correlation between y and $\hat{y} = a + b_1 x_1 + b_2 x_2 + b_3 x_3$. Analogous to **(23)**,

$$R_{0(123)} = r(y, \hat{y}) = \frac{\Sigma (y - \bar{y})(\hat{y} - \bar{y})}{\sqrt{\Sigma (y - \bar{y})^2 \cdot \Sigma (\hat{y} - \bar{y})^2}} \quad (54)$$

But

$$\Sigma (y - \bar{y})(\hat{y} - \bar{y}) = \Sigma (\hat{y} - \bar{y})^2 \quad (55)$$

Therefore,

$$R_{0(123)} = \frac{\sqrt{\Sigma (\hat{y} - \bar{y})^2}}{\sqrt{\Sigma (y - \bar{y})^2}} \quad (56)$$

$$R^2_{0(123)} = \frac{\text{regression ssq}}{\text{total ssq}} = 1 - \frac{\text{residual ssq}}{\text{total ssq}} \quad (57)$$

$$\frac{\sigma^2_{0(123)}}{\sigma^2_0} = 1 - \frac{\sigma^2_{0.123}}{\sigma^2_0} \quad (58)$$

If we consider $y = g + e = y_{(123)} + y_{0.123}$, where the two components are uncorrelated, then $r^2_{yg} + r^2_{ye} = 1$, or

$$R^2_{0(123)} + R^2_{0.123} = 1 \quad (59)$$

All of these expressions remain the same as those for two x-variables except for replacing the subscripts $0(12)$ and 0.12 by $0(123)$ and 0.123, respectively.

20. Analogous to **(38)** we define the following symmetrical determinant:

$$\Delta = \begin{vmatrix} 1 & r_{01} & r_{02} & r_{03} \\ r_{10} & 1 & r_{12} & r_{13} \\ r_{20} & r_{21} & 1 & r_{23} \\ r_{30} & r_{31} & r_{32} & 1 \end{vmatrix} \quad (60)$$

in which $r_{ij} = r_{ji}$; and $r_{ii} = 1$, $i = 0, 1, 2, 3$. Again, let Δ_{00} be the minor determinant obtained from Δ above by deleting the first row and the first column; and let Δ_{0i} be the minor determinant obtained from Δ by deleting the first row and the column headed by r_{0i}. Then the partial regression coefficients are

$$b_1 = \frac{\sigma_0}{\sigma_1} \frac{\Delta_{01}}{\Delta_{00}}, \qquad b_2 = \frac{-\sigma_0}{\sigma_2} \frac{\Delta_{02}}{\Delta_{00}}, \qquad b_3 = \frac{\sigma_0}{\sigma_3} \frac{\Delta_{03}}{\Delta_{00}} \qquad (61)$$

and the (squared) multiple correlation coefficient is, as before,

$$R^2_{0(123)} = 1 - \frac{\Delta}{\Delta_{00}} \qquad (62)$$

The pattern for generalization to cases involving more than three x-variables becomes clear.

Partial correlation coefficient.

21. Although this chapter is entitled multiple regression and correlation, it is convenient for us to have a section on partial correlation here, as these are related subjects. The concept of partial correlation may be best explained by considering three interrelated variables, say, x_1, x_2, x_3, with (pair-wise) simple correlations r_{12}, r_{13}, r_{23}. One question may be raised: what would be the correlation between x_1 and x_2 if the influence of x_3 on them were eliminated? Such a correlation will be denoted by $r_{12.3}$, the variable after the dot being the one whose influence has been <u>eliminated</u> (or being <u>held constant</u>). The influence of x_3 on x_1 and on x_2 may be assessed by the linear regression of x_1 on x_3 and of x_2 on x_3, respectively. Again, we shall use deviations from mean for brevity. The regression equation of X_1 on X_3 is then $\hat{X}_1 = b_{13} X_3$ and the full model is $X_1 = b_{13} X_3 + e_{1.3}$, where the first term $b_{13} X_3$ is the value determined by X_3 and the second term $e_{1.3}$ is the value uncorrelated with X_3. Following the notation of Yule and Kendall (1968) and Wright (1954), we write $b_{13} X_3 = X_{1(3)}$ and $e_{1.3} = X_{1.3}$. Thus,

$$X_1 = X_{1(3)} + X_{1.3}; \qquad X_{1.3} = X_1 - X_{1(3)} \qquad (63)$$

Similarly, $\qquad X_2 = X_{2(3)} + X_{2.3}; \qquad X_{2.3} = X_2 - X_{2(3)}$

Descriptively, we say that $X_{1.3}$ is the value of X_1 free of the influence of X_3, and that $X_{2.3}$ is the value of X_2 free of the influence of X_3. The partial correlation coefficient between X_1 and X_2 free of the influence of X_3 is then defined as the simple correlation between $X_{1.3}$ and $X_{2.3}$; that is,

$$r_{12.3} = r(X_{1.3}, X_{2.3}) = \frac{Cov(X_{1.3}, X_{2.3})}{\sqrt{Var\, X_{1.3}\; Var\, X_{2.3}}} \tag{64}$$

From Chapter 3 we know that $Var\, X_{1.3} = \sigma_1^2(1 - r_{13}^2)$ and $Var\, X_{2.3} = \sigma_2^2(1 - r_{23}^2)$. The numerator of (64) is

$$Cov(X_{1.3}, X_{2.3}) = E[X_1 - X_{1(3)}][X_2 - X_{2(3)}] = E(X_1 - b_{13}X_3)(X_2 - b_{23}X_3)$$

$$= (X_1 X_2 - b_{23} X_1 X_3 - b_{13} X_2 X_3 + b_{13} b_{23} X_3^2)$$

The expectations are: $E(X_1 X_2) = Cov(X_1, X_2) = \sigma_1 \sigma_2 r_{12}$, etc. and $E(X_3^2) = \sigma_3^2$. Substituting $b_{13} = r_{13}\, \sigma_1/\sigma_3$ and $b_{23} = r_{23}\, \sigma_2/\sigma_3$, and simplifying, we find

$$Cov(X_{1.3}, X_{2.3}) = \sigma_1 \sigma_2 (r_{12} - r_{13} r_{23})$$

Substitution in (64) yields the standard formula for a partial correlation:

$$r_{12.3} = \frac{r_{12} - r_{13} r_{23}}{\sqrt{1 - r_{13}^2}\, \sqrt{1 - r_{23}^2}} \tag{65}$$

This is a longhand derivation of the simplest case of partial correlation. For a general and elegant treatment of partial correlations of higher order, see Yule and Kendall (1968). Also see Ex. 5.

22. Formula (65) expresses the general relationships between the simple correlations and the partial correlation for any three variables. Returning to our problem of regression of y on x_1 and x_2, suppose that we wish to know the partial correlation between y and x_1 with the influence of x_2 eliminated. With appropriate changes of subscripts, the standard formula (65) becomes

$$r_{01.2} = \frac{r_{01} - r_{02} r_{12}}{\sqrt{1-r_{02}^2}\sqrt{1-r_{12}^2}} \tag{66}$$

This partial correlation is closely related to the corresponding partial regression coefficient of y on x_1, holding x_2 constant, as given by (35). A comparison of (35') and (66) shows that

$$b_{01.2} = r_{01.2} \frac{\sigma_0 \sqrt{1-r_{02}^2}}{\sigma_1 \sqrt{1-r_{12}^2}} = r_{01.2} \frac{\sigma_{0.2}}{\sigma_{1.2}} \tag{67}$$

which is an obvious extension of the relationship $b_{01} = r_{01} \frac{\sigma_0}{\sigma_1}$ for simple regression and correlation coefficients (Chapter 3). The relationship (67) may be extended to any number of variables being held constant; for instance,

$$b_{01.23} = r_{01.23} \frac{\sigma_{0.23}}{\sigma_{1.23}}$$

This brief review of multiple regression, multiple correlation, and partial correlation must suffice for our purpose of introducing path coefficients. We shall refer to some of the results later on.

Exercises

Ex. 1. Substituting the numerical values of the corrected sums of squares and products given by (17) and (22) into the reduced normal equations (20), we obtain

$$128 b_1 + 80 b_2 = 496$$

$$80 b_1 + 200 b_2 = 760$$

Solve for b_1 and b_2 and see if your answers agree with the values given in (9).

MULTIPLE REGRESSION AND CORRELATION

Ex. 2. From the numerical sources (1), (2), (13), (17), (22), determine that

$$\sigma_0 = \sqrt{621}, \qquad \sigma_1 = \sqrt{16} = 4, \qquad r_{12} = \frac{10}{4 \times 5} = \frac{1}{2}$$

$$r_{01} = \frac{62}{\sqrt{16 \times 621}}, \qquad r_{02} = \frac{95}{\sqrt{25 \times 621}}$$

and then verify (35')

$$b_1 = \frac{\sigma_0}{\sigma_1} \cdot \frac{r_{01} - r_{02} r_{12}}{1 - r_{12}^2} = 2.00$$

Ex. 3. The reduced normal equations (20), after being divided by n, may also be written in terms of variances and covariances:

$$b_1 \sigma_1^2 + b_2 \sigma_{12} = \sigma_{10}$$

$$b_1 \sigma_{12} + b_2 \sigma_2^2 = \sigma_{20}$$

which are equivalent to (34). Dividing the first equation by σ_1^2 and the second σ_2^2 and remembering that $\sigma_{12}/\sigma_1^2 = b_{21}$ is the simple regression coefficient of x_2 on x_1, etc., we obtain

$$b_1 + b_2 b_{21} = b_{01}$$

$$b_1 b_{12} + b_2 = b_{02}$$

Hence

$$b_1 = b_{01.2} = \frac{\begin{vmatrix} b_{01} & b_{21} \\ b_{02} & 1 \end{vmatrix}}{\begin{vmatrix} 1 & b_{21} \\ b_{12} & 1 \end{vmatrix}} = \frac{b_{01} - b_{02} b_{21}}{1 - b_{12} b_{21}}$$

which is equivalent to (35). Thus we see that a partial regression coefficient may be expressed in terms of simple regression coefficients. More generally, a partial correlation coefficient of a higher order may be expressed in terms of those of a lower order. A general treatment of the subject may be found in Yule and Kendall (1968).

Ex. 4. Given the values of x_1, x_2, and y shown in the following table, find the linear regression equation of y on x_1 and x_2.

x_1	x_2	y	$\hat{y}$	e	$y\hat{y}$	ye	$\hat{y}e$
17	3	20					
11	5	10					
18	6	17					
18	8	43					
10	11	26					
9	14	43					
20	15	74					
17	18	47					
120	80	280	280	0	12112	736	0
15	10	35	35	0			

Compare this set of data with that in Table 79 and note the similarities and differences. Compare the regression equation for this set of data with (10):

$$\hat{y} = -25 + 2x_1 + 3x_2$$

After the regression equation has been found, calculate $\hat{y}$ and e and their products as indicated in the table above. Also calculate the various corrected sums of squares and products for the variables x_1, x_2, and y. Verify

$$\text{total ssq} = \text{regression ssq} + \text{residual ssq}$$

and $\quad \Sigma y\hat{y} = \Sigma \hat{y}^2; \quad$ and $\quad \Sigma ye = \Sigma e^2$

MULTIPLE REGRESSION AND CORRELATION

Calculate $R^2_{0(12)}$ and $R^2_{0.12}$. The former is the (squared) correlation between y and $\hat{y}$; while the latter is that between y and e.

Partial answer: $\quad R^2_{0(12)} = 2312/3048 = 0.75853$

The regression equation is the same as **(10)**.

Ex. 5. *Simple, multiple, and partial correlations.* This note may be deferred until after reading the next chapter. In the simple case, write x_1 for y and x_2 for x. Then $\hat{y} = \hat{x}_1 = a + bx_2 = x_{1(2)}$ and the residual is $e = x_{1.2} = x_1 - x_{1(2)}$, so that $x_1 = x_{1(2)} + x_{1.2}$, the latter two components being uncorrelated. This situation is depicted in the left diagram below. In the previous chapter (page 60) it was shown that $r^2 + u^2 = 1$, where $r = r_{12} = r(x_1, x_2) = r(y,x)$, and

$$u = r(y,e) = r(x_1, x_{1.2}) = \sigma_{1.2}/\sigma_1$$

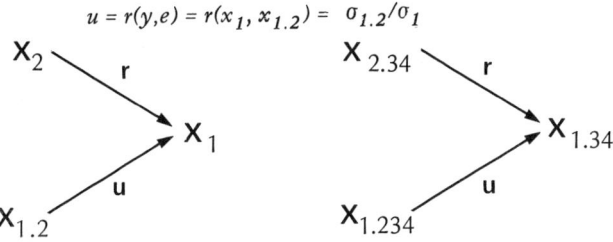

Fig. 99a. Simple correlation   Fig. 99b. Partial correlation
$r = r_{12}$ and $u = \sigma_{1.2}/\sigma_1$ $r = r_{12.34}$ and $u = \sigma_{1.234}/\sigma_{1.34}$

It may be shown that this type of relationship still holds when other variables (x_3, x_4, etc.) are added, *after the dot*, to each of the variables, as shown in the right diagram. Here, again, $r^2 + u^2 = 1$, but r is now the partial correlation

$$r = r_{12.34} = r(x_{1.34}, x_{2.34}) \text{ and } u = \sigma_{1.234}/\sigma_{1.34}$$

Thus, $\quad r^2_{12.34} = 1 - u^2 = 1 - \dfrac{\sigma^2_{1.234}}{\sigma^2_{1.34}} = 1 - \dfrac{1 - R^2_{1(234)}}{1 - R^2_{1(34)}}$

or $\quad r^2_{12.34} = \dfrac{R^2_{1(234)} - R^2_{1(34)}}{1 - R^2_{1(34)}}$

which is equivalent to an expression given by Ezekiel and Fox (1959, p. 193). Other diagrams for partial correlation are given in Fig. 127 and Fig. 173.

Chapter 5

STANDARDIZED VARIABLES; PATH COEFFICIENTS

IN THIS CHAPTER we shall deal with path coefficients for the first time. Path coefficients may be introduced in several different ways. When there is only one variable y dependent on one or several other variables $x_1, x_2, \ldots$, the path method yields the same results as linear regression. But the path method is not limited to the analysis of such a simple system, and a more general treatment of path analysis will be given in the next chapter. Since linear regression has been reviewed in the last two chapters, we shall consider path coefficients in the same context. In other words, this chapter is limited to the rewriting of the previous results of ordinary regression in terms of path coefficients. There will be some duplication between this and the next chapter. While redundancy may annoy some sophisticated statisticians, it usually helps the novice to understand the relationships better.

Standardized Simple Regression

1. Standardized Variables. We begin with the same set of data of (x, y) we used in Chapter 3 (1). They are reproduced here in Table 101 for convenience. It will be recalled that

$$\sigma_x^2 = \frac{\Sigma (x - \bar{x})^2}{n} = \frac{200}{8} = 25; \quad \sigma_x = 5 \, cm$$

$$\sigma_y^2 = \frac{\Sigma (y - \bar{y})^2}{n} = \frac{128}{8} = 16; \quad \sigma_y = 4 \, gm$$

The so-called standardized variables are those measured from their mean values in units of their own standard deviations. Denoting the standardized variables by capital letters, we have

$$X = \frac{x - \bar{x}}{\sigma_x}, \qquad Y = \frac{y - \bar{y}}{\sigma_y} \qquad (1)$$

STANDARDIZED VARIABLES; PATH COEFFICIENTS

the numerical values of which are shown in Table 101. For instance the first pair is $X_1 = (2-9)/5 = -1.4$ and $Y_1 = (3-8)/4 = -1.25$. The values of X and Y are no longer in terms of centimeters and grams; they are pure numbers without physical units.

Table 101. Original (from Chapter 3 (1)) and standardized variables.

Original		Standardized		Squares and products		
x cm	y gm	X	Y	X^2	Y^2	XY
2	3	− 1.4	− 1.25	1.96	1.5625	1.75
4	2	− 1.0	− 1.50	1.00	2.2500	1.50
5	13	− 0.8	+ 1.25	0.64	1.5625	− 1.00
7	10	− 0.4	+ 0.50	0.16	0.2500	− 0.20
10	4	+ 0.2	− 1.00	0.04	1.0000	− 0.20
13	11	+ 0.8	+ 0.75	0.64	0.5625	0.60
14	11	+ 1.0	+ 0.75	1.00	0.5625	0.75
17	10	+ 1.6	+ 0.50	2.56	0.2500	0.80
Total 72	64	0	0	8.00	8.0000	4.00
Mean 9	8	0	0	1.00	1.0000	0.50

2. Zero mean and unit variance. From (1) we see that the sum, and thus the mean, of a standardized variable is zero.

$$\overline{X} = E(X) = 0, \qquad \overline{Y} = E(Y) = 0. \tag{2}$$

The variance of a standardized variable is unity.

$$\sigma_X^2 = V(X) = E(X^2) = 1, \quad \sigma_Y^2 = V(Y) = E(Y^2) = 1 \tag{3}$$

as has been verified in Table 101. No matter what the original variables may be, the process of standardization makes them all equal in mean (zero) and all equal in variance (unity). The physical unit in which a variable is being measured is arbitrary. A variable may take various numerical appearances, depending on the unit in which it happened to be expressed, but its standardized form will remain the same. The path analysis deals with the relationships among the standardized variables. Any theory of statistics developed for concrete variables will also apply to standardized variables with the simplifications of making all means zero and all variances unity.

3. Correlation. It may be recalled from Chapter 3 **(26)** that if the original variables x and y are transformed in the following way

$$x' = k_1 + b_1 x, \qquad y' = k_2 + b_2 y \qquad (4)$$

the correlation between the new variables will remain the same: $r(x', y') = r(x, y)$, as the correlation is independent of origin and unit. Note that the transformation (1) is of the same type as (4). Hence the correlation between the standardized variables will remain the same as that between the two original variables:

$$r(X, Y) = r(x, y) \qquad (5)$$

Explicit algebraic details are as follows:

$$r(X, Y) = \frac{Cov(X, Y)}{\sigma_X \, \sigma_Y} = Cov(X, Y) \qquad (6)$$

as $\sigma_X = \sigma_Y = 1$. For standardized variables, the correlation coefficient is the same as covariance. The latter is, in turn,

$$Cov(X, Y) = E(XY) = E\left(\frac{x - \bar{x}}{\sigma_x}\right)\left(\frac{y - \bar{y}}{\sigma_y}\right) = \frac{Cov(x, y)}{\sigma_x \, \sigma_y} = r(x, y) \qquad (7)$$

Combining (7) with (6) gives (5). Briefly, we may say that a correlation coefficient is a standardized covariance. The last column of Table 101 shows that the covariance $E(XY) = \Sigma XY/n = 4/8 = 0.50$, which is precisely the value of the correlation $r(x, y)$ found in Chapter 3.

4. Linear regression of Y on X. The linear regression of Y on X may be studied in exactly the same manner as we did with that of y on x in Chapter 3. Just as $b = b_{yx}$ = regression coefficient of y on x, we let $p = p_{YX}$ = regression coefficient of Y on X. The normal equations for the original variables (x, y) are

$$a n + b \Sigma x = \Sigma y \qquad \text{Chapter 3 (10)}$$
$$a \Sigma x + b \Sigma x^2 = \Sigma xy \qquad \text{Chapter 3 (11)}$$

STANDARDIZED VARIABLES; PATH COEFFICIENTS

When the variables are standardized, $\Sigma X = \Sigma Y = 0$. The first normal equation yields $a = 0$ on substituting the standardized variables for the original ones, and the second normal equation yields (replacing b by p)

$$p = p_{YX} = \frac{\Sigma XY}{\Sigma X^2} = E(XY) = Cov(X, Y) = r(x, y) \qquad (8)$$

as $E(X^2) = \sigma_X^2 = 1$. Thus, the regression coefficient of Y on X is simply the correlation coefficient between the two original variables. (This is true only in the simple case under consideration.) Further, we have shown in Chapter 3 that

$$r_{xy} = b_{yx}\left(\frac{\sigma_x}{\sigma_y}\right) \qquad (9)$$

Hence,

$$p = p_{YX} = r_{yx} = b_{yx}\left(\frac{\sigma_x}{\sigma_y}\right) \qquad (10)$$

The regression coefficient $p = p_{YX}$ for the standardized variables is called the *path coefficient* from x to y (or from X to Y). Since $p_{YX} = b_{yx}\frac{\sigma_x}{\sigma_y}$, the path coefficient is really a standardized regression coefficient. A path coefficient carries a direction with it; we say p_{YX} is from X to Y, just as we say b_{yx} is of y on x. In Table 101 we have calculated all the individual values of X_i and Y_i. In practice this is of course not necessary, as the path coefficient may be obtained from the conventional quantities (10). For concrete variables (with physical units), the covariance, the regression coefficient, and the correlation coefficient are three different numbers. But for standardized variables, these three quantities are the same for the case of simple linear regression. When arbitrary units disappear from data, the distinction between the three statistics is no longer so clear cut. The standardized regression coefficient has been denoted by β in some statistical books; but β is also employed to denote the population parameter for the ordinary regression coefficient. So, we shall use p for path coefficients.

5. Regression equation and residual. Since the intercept $a = 0$, the equation of the regression line of Y on X is simply

$$\hat{Y} = p\,X \qquad (11)$$

where $\hat{Y} = G$ is the calculated value of Y based on the value of X. In our numerical example, $p = r(x, y) = 0.50$. The values of $G = \hat{Y}$ are given in Table 104. Then the residual is $U = Y - \hat{Y}$, also given in Table 104, so that

$$Y = G + U \qquad (12)$$

Table 104. Regression of Y on X. Equation is $G = \hat{Y} = p\,X = \tfrac{1}{2} X$ and residual $U = Y - \hat{Y}$.

X	Y	$G = \hat{Y}$	$U = Y - \hat{Y}$	YU	GU
− 1.4	− 1.25	− 0.70	− 0.55	0.6875	+ 0.385
− 1.0	− 1.50	− 0.50	− 1.00	1.5000	+ 0.500
− 0.8	+ 1.25	− 0.40	+ 1.65	2.0625	− 0.660
− 0.4	+ 0.50	− 0.20	+ 0.70	0.3500	− 0.140
+ 0.2	− 1.00	+ 0.10	− 1.10	1.1000	− 0.110
+ 0.8	+ 0.75	+ 0.40	+ 0.35	0.2625	+ 0.140
+ 1.0	+ 0.75	+ 0.50	+ 0.25	0.1875	+ 0.125
+ 1.6	+ 0.50	+ 0.80	− 0.30	− 0.1500	− 0.240
0	0	0	0	6.0000	0

From the theory of linear regression (Chapter 3), we know that G and U are uncorrelated, just as $g = \hat{y} = a + bx$ and $e = y - \hat{y}$ are uncorrelated. Actually, the residual U is the original residual e divided by σ_y. Using (10), (11), and $\hat{y} = \bar{y} + b(x - \bar{x})$, we see that

$$U = Y - \hat{Y} = \frac{y - \bar{y}}{\sigma_y} - b\,\frac{\sigma_x}{\sigma_y}\,\frac{(x - \bar{x})}{\sigma_x} = \frac{y - \hat{y}}{\sigma_y} = \frac{e}{\sigma_y} \qquad (13)$$

and equation (12) in terms of original variables, becomes

$$\frac{y - \bar{y}}{\sigma_y} = \frac{b(x - \bar{x})}{\sigma_y} + \frac{e}{\sigma_y} \qquad (14)$$

which is simply the original model equation divided by σ_y throughout.

STANDARDIZED VARIABLES; PATH COEFFICIENTS

6. Correlation and determination. Since $Y = G + U$ and the latter two components are uncorrelated, we have

$$\sigma_Y^2 = \sigma_G^2 + \sigma_U^2 = 1 \qquad (15)$$

Direct calculation from the numerical values of Table 104 shows that

$$\sigma_G^2 = \frac{\Sigma G^2}{n} = \frac{2}{8} = 0.25; \quad \sigma_U^2 = \frac{\Sigma U^2}{n} = \frac{6}{8} = 0.75 \quad (16)$$

These values remind us of the case $y = g + e$ for concrete variables and $\sigma_y^2 = \sigma_g^2 + \sigma_e^2$. Indeed, the expression (15) above is equivalent to the following equation established in Chapter 3.

$$\frac{\sigma_g^2}{\sigma_y^2} + \frac{\sigma_e^2}{\sigma_y^2} = r_{yg}^2 + r_{ye}^2 = 1 \qquad (17)$$

where

$$r_{yg}^2 = \sigma_g^2 / \sigma_y^2 \quad \text{and} \quad r_{ye}^2 = \sigma_e^2 / \sigma_y^2$$

From (13), $U = e/\sigma_y$, it follows that

$$\sigma_U^2 = \sigma_e^2 / \sigma_y^2 = r_{ye}^2 \qquad (18)$$

and from (14), $G = b(x - \bar{x})/\sigma_y$, we have

$$\sigma_G^2 = b^2 \sigma_x^2 / \sigma_y^2 = \sigma_g^2 / \sigma_y^2 = r_{yg}^2 \qquad (19)$$

These relationships establish the complete equivalence of (15) with (17). Since the correlation coefficient remains the same whether the variables are standardized or not, $r(y, x) = r(y, g) = r(Y, X) = r(Y, G) = p_{YX} = p_{YG}$. Similarly, $r(y, e) = r(Y, U) = p_{YU}$, the path coefficient from U to Y. Using the numerical values of Table 104 we obtain

$$p_{YU} = r(Y, U) = \frac{Cov(Y, U)}{\sigma_Y \sigma_U} = \frac{\sigma_U^2}{\sigma_U} = \sigma_U = \sqrt{0.75} = 0.866$$

Hence (15) and (17) may be written in terms of path coefficients:

$$p_{YG}^2 + p_{YU}^2 = p_{yg}^2 + p_{ye}^2 = 1 \qquad (20)$$

Note that as far as path coefficients are concerned, it does not matter whether we use the standardized Y and X or the concrete y and x, because it is defined as the standardized regression coefficient anyway. Nor does it matter whether we use x or $g = a + bx$; X or $G = pX$. A value such as p_{yx}^2 is called the *coefficient of determination* (for the variance of y), because it gives the fraction of σ_y^2 that can be accounted for by the variation of x.

Simple Path Diagram

7. In combining geometrical figures and algebraic expressions, there is a branch of mathematics known as *analytical geometry,* making an algebraic expression to describe a geometrical figure and making a geometrical figure to specify the algebraic equation. Somewhat analogous to the situation in analytical geometry, we may draw a diagram (according to certain rules) to represent the statistical relationships among the variables under consideration. The case of linear regression of one variable on another is the beginning point for path analysis, as it yields the simplest path diagram. Fig. 107 is the path diagram for the linear relations between x and y together with residuals e. The convention for drawing a path diagram is, however, to use capital letters for the variables involved, whether standardized or not, (saving lower case letters for other purposes) because, as pointed out before, the path coefficient always refers to their standardized forms. Hence capital letters are used in Fig. 107. Also note that we used X directly instead of using $G = pX$, because the path coefficient is the same for either case. Of course, G may be used, if the reader so wishes.

8. There are certain rules for drawing and reading a road map. So there are rules or conventions for drawing and reading a path diagram. A directed arrow indicates a path from one variable to another. In Fig. 107, there is one path from X to Y (an arrow pointing from X to Y) and one from U to Y. A path does not only have a direction but also has a quantitative

STANDARDIZED VARIABLES; PATH COEFFICIENTS 107

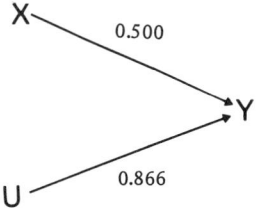

Fig. 107. Uncorrelated causes. The path coefficients are simply the correlation coefficients. Numerical calculations are based on data of Tables 101 and 104. The fact that $p_{YX}^2 + p_{YU}^2 = r_{YX}^2 + r_{YU}^2 = (0.500)^2 + (0.866)^2 = 1$ means complete determination of Y by X and U.

value to measure the importance of the path. The value assigned to the path from X to Y is the path coefficient p_{YX}. In our present case, $p_{YX} = r(Y, X) = r(y, x) = 0.50$. Similarly, the value for the path from U to Y is the path coefficient $p_{YU} = r(Y, U) = r(y, e) = 0.866$, as indicated in Fig. 107. Thus, the path diagram shows that Y is partly determined by X and partly determined by U.

We speak of *connections* between certain variables. In Fig. 107, we say that Y and X are connected by the path from X to Y. Similarly, we say that Y and U are connected by the path from U to Y. Our analytical results show that X and U are uncorrelated; so nothing has been drawn between X and U in the diagram. In tracing the connection between two variables, the first and obvious rule is that an arrow is always a connecting path between the variable at the arrowhead and the variable at the butt. This is the most direct and elementary connecting path between two variables. But how shall we express the fact that there is no connecting path between X and U? If the diagram is merely a road map, one might argue that he can travel from X to Y first and then from Y to U.

9. The mere absence of a direct connection between X and U in the diagram does not ensure that they are uncorrelated. We need rules to read a path diagram. Let us describe the movement or travelling as "forward" if it is *with* the direction of an arrow (e.g., from X to Y, or from U to Y), and call it "backward" if it is *against* the direction of an arrow. The

general rules of tracing connecting paths between two variables will be explained in the next chapter when we deal with a network of variables. For the present simple case illustrated in Fig. 107, we can set down one important rule immediately; viz.,

> the travelling route: *first forward, then backward,* is *not* permitted (21)

in reading a path diagram. This is a fundamental and universal rule in reading all path diagrams, simple or complicated. According to this rule, one cannot travel first from X to Y (forward) and then from Y to U (backward). The forbidden rule (21) *and* the absence of a direct connection between X and U ensures that X and U are unconnected in the diagram; that is, they are uncorrelated.

10. From the discussion above, the reader may see that what is absent from a path diagram is just as important as what is present. The absence of a direct connection between X and U is an essential feature of the path diagram in Fig. 107. The absence of any other arrows pointing to Y means that Y is *completely* determined by the variables shown in the diagram *(X* and *U)*. This fact is reflected by the relationship

$$p_{YX}^2 + p_{YU}^2 = (0.50)^2 + (0.866)^2 = 0.25 + 0.75 = 1 \qquad (22)$$

which means that X accounts for 25% and U accounts for 75% of the variation in Y. The expression (22) is known as the equation for complete determination. When this is true, we say that the system is *complete* or *self-contained.* A path diagram must be self-contained and consistent in order to yield meaningful analysis. The results and the diagram are derived in this chapter from the linear regression of Y on X, which is a special case of path analysis. A broader view and a more flexible terminology will be introduced in the next chapter.

Standardized multiple regression

11. We have seen how a path diagram (Fig. 107) summarizes the essential features of the relationship of the three variables: the *dependent (y),* the *independent (x),* and the *residual* (including all other variables that may affect y but not included in the data). The path diagram has in fact put the simple linear regression problem in its most succinct or canonical form, without any arbitrary units attached. We now proceed to study multiple regression by the same method. The same set of data employed in Chapter 4 will be used for the sake of comparison and checking. Table 109 gives the standardized values X_1, X_2, Y, of the original x_1, x_2, y of Table 79. The regression of Y on X_1 and X_2 introduces several additional features of path analysis. Some of them will be discussed in a more general way in the next chapter. So, for the time being, we shall limit ourselves to translating some of the regression results established in Chapter 4 to their standardized forms. A step-by-step translation may be as tedious as unnecessary, since we have already gained some experience by working through the simple regression. We shall omit certain obvious steps. Before proceeding, let us recall the various correlations between the variables:

$$r(x_1, x_2) = r_{12} = \frac{10}{\sqrt{16 \times 25}} = 0.500,000$$

$$r(y, x_1) = r_{01} = \frac{62}{\sqrt{621 \times 16}} = 0.621,994 \qquad (23)$$

$$r(y, x_2) = r_{02} = \frac{95}{\sqrt{621 \times 25}} = 0.762,444$$

Table 109. Multiple linear regression for standardized variables. $G = \hat{Y} = p_{01} X_1 + p_{02} X_2 = 0.3210 X_1 + 0.6019 X_2$. The standardized variables (X_1, X_2, Y) are derived from the concrete variables (x_1, x_2, y) of Table 79.

X_1	X_2	Y	$G = \hat{Y}$	$U = Y - \hat{Y}$	GU
− 1.25	− 1.4	− 1.0835	− 1.2440	+ 0.1605	− 0.1997
− 1.50	− 1.0	− 1.2440	− 1.0835	− 0.1605	+ 0.1739
+ 1.25	− 0.8	− 0.5618	− 0.0803	− 0.4815	+ 0.0386
+ 0.50	− 0.4	+ 0.5618	− 0.0803	+ 0.6421	− 0.0515
− 1.00	+ 0.2	− 0.3612	− 0.2006	− 0.1606	+ 0.0322
+ 0.75	+ 0.8	− 0.0803	+ 0.7223	− 0.8026	− 0.5797
+ 0.75	+ 1.0	+ 1.9663	+ 0.8427	+ 1.1236	+ 0.9469
+ 0.50	+ 1.6	+ 0.8026	+ 1.1236	− 0.3210	− 0.3607
Mean 0	0	0	0	0	0
σ^2 1	1	1	0.6586	0.3414	
σ 1	1	1	0.8115	0.5843	

These correlations are calculated directly from the data. Their values will remain so, whether we shall carry on a regression analysis or not, and whether the variables are standardized or not. In other words, these correlations are features of the given set of data.

12. Regression of Y on X_1 and X_2. Let us convert the original variables x_1, x_2, y of Table 79 into standardized variables X_1, X_2, Y in Table 109 and then regard the latter as the given set of data. Since their means are zero, the method of least squares would give us the following normal equations (Chapter 4 **(20)**):

$$b_1 \Sigma X_1^2 + b_2 \Sigma X_1 X_2 = \Sigma X_1 Y$$
$$b_1 \Sigma X_1 X_2 + b_2 \Sigma X_2^2 = \Sigma X_2 Y \qquad (24)$$

where $b_1 = b_{01.2}$ is the partial regression coefficient of y on x_1, etc. Further, these variables have been standardized; $E(X_1^2) = E(X_2^2) = 1$, and $E(X_1 X_2) = r_{12} = r_{21}$, etc. Replacing b_1 and b_2 by p_{01} and p_{02}, respectively, we obtain the standardized normal equations:

$$p_{01} + p_{02} r_{21} = r_{01}$$
$$p_{01} r_{12} + p_{02} = r_{02} \qquad (25)$$

where p_{01} is the path coefficient from X_1 to Y and p_{02} is that from X_2 to Y. Before solving **(25)**, we observe that in this case a path coefficient is no longer equal to a correlation coefficient, unless $r_{12} = 0$. Substituting the numerical values of the correlations given in **(23)** into **(25)** and solving, we obtain

$$p_{01} = 0.3210, \qquad p_{02} = 0.6019 \qquad (26)$$

approximately. Later we shall give an interpretation to the standardized normal equations **(25)** in terms of path diagrams.

STANDARDIZED VARIABLES; PATH COEFFICIENTS

13. It is seen from the previous paragraph that the path coefficients are not even slightly more difficult to find than the concrete regression coefficients, because the normal equations involved, whether standardized or concrete, are of the same form and of the same order. However, our conventional teaching and practice in statistics is to calculate the concrete regression coefficients. Hence it is desirable to give a simple formula for the conversion of the concrete b's into the standardized p's. With respect to the original variables x_1, x_2, and y, the model equation is

$$y - \bar{y} = b_1(x_1 - \bar{x}_1) + b_2(x_2 - \bar{x}_2) + e \quad (27)$$

Dividing the equation by $\sigma_y = \sigma_0$ throughout,

$$\frac{y - \bar{y}}{\sigma_0} = \frac{b_1(x_1 - \bar{x}_1)}{\sigma_0} + \frac{b_2(x_2 - \bar{x}_2)}{\sigma_0} + \frac{e}{\sigma_0}$$

Multiplying the first term on the right by σ_1/σ_1 and the second term by σ_2/σ_2,

$$\frac{y - \bar{y}}{\sigma_0} = b_1 \cdot \frac{\sigma_1}{\sigma_0}\left(\frac{x_1 - \bar{x}_1}{\sigma_1}\right) + b_2 \cdot \frac{\sigma_2}{\sigma_0}\left(\frac{x_2 - \bar{x}_2}{\sigma_2}\right) + \frac{e}{\sigma_0} \quad (28)$$

That is,

$$Y = p_{01} X_1 + p_{02} X_2 + U \quad (29)$$

where

$$p_{01} = b_1\left(\frac{\sigma_1}{\sigma_0}\right), \qquad p_{02} = b_2\left(\frac{\sigma_2}{\sigma_0}\right) \quad (30)$$

are the path coefficients (regression coefficients with respect to standardized variables). The expression (30) is the desired conversion formula for the two types of regression coefficients. As a numerical check, we recall from Chapter 4 that $\sigma_1 = 4$, $\sigma_2 = 5$, $\sigma_0 = \sqrt{621} = 24.92$, and $b_1 = 2$, $b_2 = 3$. Substituting in (30),

$$p_{01} = \frac{2 \times 4}{24.92} = 0.3210, \qquad p_{02} = \frac{3 \times 5}{24.92} = 0.6019$$

in agreement with the solutions (26) of the standardized normal equations (25). Some other algebraic details are left to the reader as exercises.

14. Multiple correlation and residual. We shall carry out some numerical calculations first. Since the path coefficients are already known, we may calculate the values of

$$G = \hat{Y} = p_{01} X_1 + p_{02} X_2 = 0.3210 X_1 + 0.6019 X_2 \quad (31)$$

and the residuals $U = Y - \hat{Y} = Y - G$, so that $Y = G + U$, just as in the case of simple regression. These values are given in Table 109. It was shown in Chapter 4 that G and U are uncorrelated; the last column of Table 109 verifies this fact. $\Sigma\, GU = 0$. Consequently,

$$\begin{aligned}
Cov(Y, G) &= E(YG) = E(G + U)G = E(G^2) = \sigma_G^2 \\
Cov(Y, U) &= E(YU) = E(G + U)U = E(U^2) = \sigma_U^2
\end{aligned} \quad (32)$$

a fact that we have also established in Chapter 4. The multiple correlation between Y and the X-variables is defined as the simple correlation between Y and G. Thus

$$R_{0(12)} = r(Y, G) = \frac{Cov(Y, G)}{\sigma_Y\, \sigma_G} = \frac{\sigma_G^2}{\sigma_G} = \sigma_G = 0.8115 \quad (33)$$

$$R_{0.12} = r(Y, U) = \frac{Cov(Y, U)}{\sigma_Y\, \sigma_U} = \frac{\sigma_U^2}{\sigma_U} = \sigma_U = 0.5843$$

and

$$\begin{aligned}
\sigma_Y^2 &= \sigma_G^2 + \sigma_U^2 = r_{YG}^2 + r_{YU}^2 = 1 \\
&= (0.8115)^2 + (0.5843)^2 = 0.6586 + 0.3414 = 1
\end{aligned} \quad (34)$$

STANDARDIZED VARIABLES; PATH COEFFICIENTS

All of these results are in agreement with those found in Chapter 4. The multiple correlation coefficient remains the same whether the variables are standardized or not.

Path diagram for multiple regression

15. We now present a path diagram to describe and summarize the situation of multiple regression. The classical diagram (Fig. 113) corresponds to the complete model (29); that is, $Y = p_{01} X_1 + p_{02} X_2 + U$. The fact that X_1 and X_2 are correlated with $r(X_1, X_2) = r(x_1, x_2) = r_{12} = 0.50$ is not shown explicity in equations (29) but is so indicated in the path diagram by a double-headed and curved "arrow." We shall always use a double-headed arrow to denote a correlation, indicating the symmetrical nature of a correlation coefficient. In tracing the paths of a diagram, a double-headed curve may be used in either direction (from X_1 to X_2, or from X_2 to X_1).

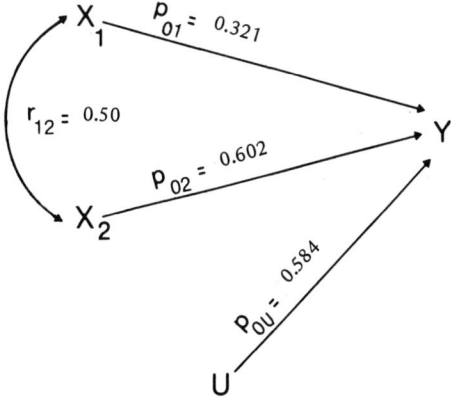

Fig. 113. Correlated causes $(X_1$ and $X_2)$ and uncorrelated residual U. The equation in terms of path coefficients is $Y = p_{01} X_1 + p_{02} X_2 + U = G + U$. The numerical calculations are based on Table 109. The degree of determination of Y by X_1 and X_2 is $R^2_{0(12)} = \sigma^2_G = p^2_{01} + p^2_{02} + 2p_{01}(r_{12})p_{02} = 0.6586$. The degree of determination of Y by U is $R^2_{0.12} = \sigma^2_U = p^2_{0U} = r^2_{0U} = 0.3414$, so that $R^2_{0(12)} + R^2_{0.12} = \sigma^2_G + \sigma^2_U = 1$.

Since this is really the first time we encounter a path diagram, other than the trivial one in Fig. 107, we shall learn the rules of reading and using it in some detail. The very first observation to be made is that there is no connecting path between U and either of the X's, because of the forbidden rule (21). This reflects the fact that U is uncorrelated with X_1 and uncorrelated with X_2, and thus uncorrelated with any linear combination of X_1 and X_2. (Review the normal equations in Chapter 4 (6) which state that $\Sigma x_1 e = 0$ and $\Sigma x_2 e = 0$. In standardized form, these conditions become $\Sigma X_1 U = 0$ and $\Sigma X_2 U = 0$.)

16. From the data (Table 79 or Table 109) we have found in (23) that

$$r(Y, X_1) \quad = \quad r(y, x_1) \quad = \quad r_{01} \quad = \quad 0.622$$

On the other hand, the standardized normal equation (25) gives

$$r_{01} \quad = \quad p_{01} + p_{02} r_{21}$$

Now turning to Fig. 113, we see that Y and X_1 are connected by two different routes. One is the direct route from X_1 to Y, with $p_{01} = 0.321$. The other is the indirect route via X_2; that is, from X_1 to X_2 and then to Y. Further, the value of the indirect route is the product of the two separate steps: $p_{02} r_{21}$. Numerically,

$$r_{01} \quad = \quad 0.321 + 0.602 \times 0.500 \quad = \quad 0.321 + 0.301 \quad = \quad 0.622$$

This simple fact illustrates a general law in path analysis: *The correlation between two variables is the sum (of the values) of all paths connecting the two variables.* Thus, a correlation may be decomposed into various components, each component corresponding to a connecting path. One of the achievements of path analysis is the decomposition of a correlation in a network of linearly related variables.

17. Partial determinations of Y. From equation (29), Y is partially determined by X_1 and X_2, and partially determined by U. The diagram Fig. 113 shows that the relationship between U and Y here is the same as that in Fig. 107. Hence, from (32), (33), and (34), the path coefficient from U to Y is

$$p_{OU} \quad = \quad r(Y, U) \quad = \quad \sigma_U \quad = \quad 0.5843 \quad (35)$$

and the coefficient of determination of Y by U is

$$p_{OU}^2 \quad = \quad r_{YU}^2 \quad = \quad \sigma_U^2 \quad = \quad 0.3414 \quad (36)$$

This says that U accounts for 34.14% of the variance of Y. In terms of tracing the paths of a diagram, this is formally obtained by starting from Y backward to U and then from U forward to Y. It is a round trip from Y to U and then from U to Y. A degree of determination of a variable is always a *round trip* for that variable via another variable(s) that determines it. Note that we have carefully stated that the round trip starts from Y. In other words, the movement of tracing the paths is *backward first and then forward*. This is the way of tracing a connecting path. The reader may recall that this tracing procedure is just the opposite of the forbidden rule (21) that no tracing may be done by going forward first and then backward.

18. The partial determination of Y by X_1 and X_2 is an important new feature in path analysis. The part of Y that is determined by X_1 and X_2 is $G = p_{01} X_1 + p_{02} X_2$. *The degree of determination is always a round trip from Y to Y via some determining variables.* The degree of determination of Y by G is, from (34), $\sigma_G^2 = 0.6586$. But $R_{0(12)}^2 = \sigma_G^2 = E(G^2) = E(p_{01} X_1 + p_{02} X_2)^2 =$

$$p_{01}^2 \quad + \quad p_{02}^2 \quad + \quad p_{01} r_{12} p_{02} \quad + \quad p_{02} r_{21} p_{01} \quad (37)$$

In terms of the path diagram Fig. 113, we see that there are four round trips from Y to Y via X_1 and/or X_2 that may be made. These four round trips are

$Y - X_1 - Y,$	p_{01}^2	=	$(0.321)^2$	= 0.1031
$Y - X_2 - Y,$	p_{02}^2	=	$(0.602)^2$	= 0.3623
$Y - X_1 - X_2 - Y,$	$p_{01} r_{12} p_{02}$	=	0.321 (0.50) 0.602	= 0.0966
$Y - X_2 - X_1 - Y,$	$p_{02} r_{21} p_{01}$	=	0.602 (0.50) 0.321	= 0.0966

Total,	$R_{0(12)}^2$	=	σ_G^2	= 0.6586 **(38)**

This, together with (36), yields $0.6586 + 0.3414 = 1.00$. It is noted that the correlation r_{12} has been used twice: once from X_1 to X_2, and once from X_2 to X_1. We describe the results (38) by saying that X_1 determines 10.31% of the variance of Y directly, X_2 determines 36.23% directly, and X_1 and X_2 *jointly* determine $2(0.0966) = 19.32\%$. The joint determination is due to the correlation between X_1 and X_2. The total fraction of determination of Y by X_1 and X_2 is precisely $R_{0(12)}^2$.

19. With the help of Fig. 113, all of these results may be read off from the path diagram directly without remembering any specific formulas. The path diagram contains all the relevant information about the multiple regression and multiple correlation, and presents the information in a logical and orderly form. It is not that the method of path coefficients will yield new mathematical results that cannot be obtained by the conventional statistical method. That could not be the case. Rather, a path diagram puts the right term in the right place of a framework so that other consequences may be read off immediately according to certain rules without resorting to longhand derivation. As an example, we have spent some effort in Chapter 4 to establish the following relationship

$$R_{0(12)}^2 = b_1^2 \left(\frac{\sigma_1}{\sigma_0}\right)^2 + 2 b_1 b_2 r_{12} \left(\frac{\sigma_1}{\sigma_0}\right)\left(\frac{\sigma_2}{\sigma_0}\right) + b_2^2 \left(\frac{\sigma_2}{\sigma_0}\right)^2 \qquad \text{Chapter 4} \quad (32)$$

but this is precisely (37), established so effortlessly and so readily understood in terms of the path diagram Fig. 113. The so-called rules of reading a diagram are, of course, based on analytical results. For instance, consider

$$R_{0(12)}^2 = b_1 r_{10} \left(\frac{\sigma_1}{\sigma_0}\right) + b_2 r_{20} \left(\frac{\sigma_2}{\sigma_0}\right) \qquad \text{Chapter 4} \quad (33)$$

STANDARDIZED VARIABLES; PATH COEFFICIENTS   117

In the standardized form, the formula becomes

$$R_{0(12)}^2 = p_{01} r_{10} + p_{02} r_{20} \tag{39}$$

which is an example of another basic law in path analysis. It says that *the degree of determination of a variable by a group of other variables* (that is, $R_{0(12)}^2$) *is the sum of such terms* $p_{0i} r_{i0}$, *where* X_i *is a determining variable.* In terms of a path diagram, the term $p_{01} r_{01}$ means that we trace the path first from Y to X_1 and then take all possible routes from X_1 back to Y. In our present example (Fig. 113) the values of r_{10} and r_{20} have been given by (25). Substituting (25) into (39) we obtain

$$R_{0(12)}^2 = p_{01}(p_{01} + p_{02} r_{21}) + p_{02}(p_{02} + p_{01} r_{12}) \tag{40}$$

which reduces to (37) or (38). It will be both a review and an exercise for the reader to go back to Chapter 4 and translate each expression into its standardized form.

20. A special case. When x_1 and x_2 are uncorrelated, we substitute $r_{12} = 0$ in all the previous expressions and delete the double-headed curve linking X_1 and X_2 from Fig. 113. Then,

$$p_{01} = r_{01}, \qquad p_{02} = r_{02} \tag{25'}$$

$$R_{0(12)}^2 = \sigma_G^2 = p_{01}^2 + p_{02}^2 = r_{01}^2 + r_{02}^2 \tag{37'}$$

and $\quad \sigma_Y^2 = \sigma_G^2 + \sigma_U^2 = r_{01}^2 + r_{02}^2 + r_{0U}^2 = 1 \tag{34'}$

The path diagram Fig. 113, without the double-headed curve between X_1 and X_2, is a simple extension of Fig. 107. A numerical example illustrating this situation is given in Exercise 4 at the end of the chapter.

Regression on three variables

21. Since the principles involved in regression on three variables are exactly the same as that on two variables, we shall merely write out some of the key expressions of Chapter 4 in standardized form without detailed comments. The standardized model equation is

$$Y = p_{01} X_1 + p_{02} X_2 + p_{03} X_3 + U \qquad (41)$$

and the corresponding path diagram is shown in Fig. 119. The correlations among the X-variables are $r(X_1, X_2) = r_{12}$, $r(X_1, X_3) = r_{13}$, $r(X_2, X_3) = r_{23}$. These are indicated by three double-headed curves in the path diagram. The mere existence of r_{12} and r_{23} does not imply that r_{13} also exists; therefore we need all three double-headed curves in the path diagram. The regression equation is

$$G = \hat{Y} = p_{01} X_1 + p_{02} X_2 + p_{03} X_3 \qquad (42)$$

which determines a part of Y. Due to the conditions specified by the normal equations, the residual U is uncorrelated with any of the X's. Thus

$$\sigma_Y^2 = \sigma_G^2 + \sigma_U^2 = 1 \qquad (43)$$

indicating the complete determination of Y by U and the group of X's. The normal equations Chapter 4 (50) or (50*), upon standardization, become

$$\begin{aligned}
p_{01} + p_{02} r_{12} + p_{03} r_{13} &= r_{10} \\
p_{01} r_{12} + p_{02} + p_{03} r_{23} &= r_{20} \\
p_{01} r_{13} + p_{02} r_{23} + p_{03} &= r_{30}
\end{aligned} \qquad (44)$$

from which the numerical values of the path coefficients p_{0i} may be solved. (The correlations among the X's, r_{ij}, and the correlations between each of the X's and Y, r_{i0}, are assumed to be known from the data.) The total fraction of determination of the variation of Y by the group of X's is

STANDARDIZED VARIABLES; PATH COEFFICIENTS

$$\begin{aligned}R^2_{0(123)} = \sigma^2_G &= E(p_{01}X_1 + p_{02}X_2 + p_{03}X_3)(p_{01}X_1 + p_{02}X_2 + p_{03}X_3) \\ &= E(p_{01}X_1)(p_{01}X_1 + p_{02}X_2 + p_{03}X_3) \\ &\quad + E(p_{02}X_2)(p_{01}X_1 + p_{02}X_2 + p_{03}X_3) \\ &\quad + E(p_{03}X_3)(p_{01}X_1 + p_{02}X_2 + p_{03}X_3) \qquad (45) \\ &= p_{01}(p_{01} + p_{02}r_{12} + p_{03}r_{13}) \\ &\quad + p_{02}(p_{01}r_{12} + p_{02} + p_{03}r_{23}) \\ &\quad + p_{03}(p_{01}r_{13} + p_{02}r_{23} + p_{03})\end{aligned}$$

Multiplying out, we obtain the nine separate components of R^2, each component corresponding to a round trip route from Y to Y via one or two of the X's in Fig. 119. These components may be condensed by using the relations (44); then (45) becomes

$$R^2_{0(123)} = p_{01}r_{10} + p_{02}r_{20} + p_{03}r_{30} \qquad (46)$$

Our (45) and (46) here are the standardized forms of Chapter 4 (52*) and (53*). These results are simple extensions of the case with two X-variables. The generalization to cases with more than three X-variables is obvious.

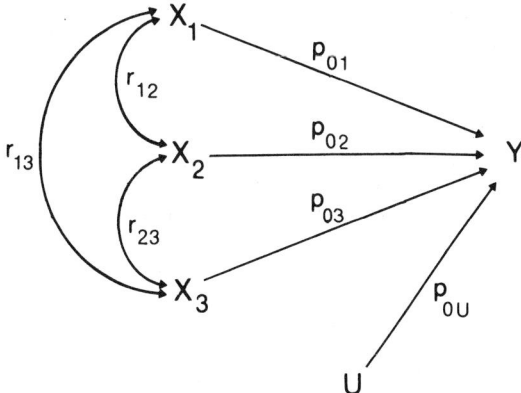

Fig. 119. Correlated causes (X_1, X_2, X_3) and an uncorrelated residual U. This is an extension of Fig. 113 and similar remarks apply. There are nine round trip routes from Y to Y; three are direct round trips: Y to X_i to Y, and six are indirect round trips through two X's: Y to X_i to X_j to Y. The sum of these nine round trips is $R^2_{0(123)}$.

22. For those who are encountering path diagrams for the first time, it will be instructive to actually trace the nine round-trip routes according to the nine terms of (45). In doing so, one will discover another general principle or rule in using a path diagram. The rule to be observed is so important that, at the risk of repetition, we list the algebraic terms of (45) and its corresponding routes in Fig. 119 as follows:

p_{01}^2	$p_{01} r_{12} p_{02}$	$p_{01} r_{13} p_{03}$
$Y \leftarrow X_1 \rightarrow Y$	$Y \leftarrow X_1 \quad X_2 \rightarrow Y$	$Y \leftarrow X_1 \quad X_3 \rightarrow Y$
p_{02}^2	$p_{02} r_{12} p_{01}$	$p_{02} r_{23} p_{03}$
$Y \leftarrow X_2 \rightarrow Y$	$Y \leftarrow X_2 \quad X_1 \rightarrow Y$	$Y \leftarrow X_2 \quad X_3 \rightarrow Y$
p_{03}^2	$p_{03} r_{13} p_{01}$	$p_{03} r_{23} p_{02}$
$Y \leftarrow X_3 \rightarrow Y$	$Y \leftarrow X_3 \quad X_1 \rightarrow Y$	$Y \leftarrow X_3 \quad X_2 \rightarrow Y$

(47)

Note that each of the correlations between the X's has been used twice, one in one direction and the other in the opposite direction. This is not new, as we have noticed it for the case of two X-variables previously. What is important to notice, though, is that none of the terms involve *two* correlation coefficients. Therefore a route of the following type is *not* permissible in tracing a path diagram

$$Y \leftarrow X_1 \quad X_2 \quad X_3 \rightarrow Y \qquad (48)$$

because there is no term like $p_{01} r_{12} r_{23} p_{03}$ in our analytical treatment. As remarked before, the existence of r_{12} and r_{23} does not imply the existence of r_{13} which links X_1 with X_3. In other words, the product $r_{12} r_{23}$ does not link X_1 with X_3, and this is not a permissible connecting route in tracing a path diagram. Hence we have arrived at another fundamental rule in using a path diagram, viz., *a connecting path cannot involve more than one correlation step, although it may involve several path steps.* This, again, has clarified the difference between a correlation coefficient and a path coefficient.

Choosing a regression equation

23. In various fields where a statistical prediction of a variable Y is desired, it is usually done in terms of many X-variables. The larger the number of X-variables, the greater the accuracy of the prediction, for $R^2_{0(123)} \geq R^2_{0(12)}$, etc. However, the cost of collecting data on a large number of X-variables dictates that only a few important factors be included in the regression equation. There is a law of diminishing return in almost everything; the inclusion of additional X-variables in reducing the residual sum of squares is certainly very much subject to the diminishing law. Hence, the practical question is not "What is the best prediction equation?" but is "What is the best compromise between the accuracy of prediction and the economy of including a few predictors?". A large amount of literature in the field of regression analysis deals precisely with such a problem: how to choose a few X-variables and yet achieve satisfactory prediction. To put the problem in another way, which of the X-variables can we exclude without jeopardizing the usefulness of the regression equation? The decision on a problem of choosing between alternatives always involves *subjective judgment;* it is not entirely a statistical decision. In the following, however, we shall discuss only the *statistical* aspect of the problem.

24. Many procedures of selecting the best compromise regression equation have been developed by theoretical and applied statisticians. Among them, we may mention: all possible regressions, backward elimination, forward selection, stepwise regression, variations on the previous methods, and stagewise regression. These procedures have been reviewed by Draper and Smith (1966, Chapter 6), an excellent book on applied regression analysis. These topics are clearly out of the scope of our present discussion. However, to show how path analysis may be used as a possible aid to choosing the variables, we shall use the same numerical example employed by

Draper and Smith (Appendix B), because the body of data is small enough for the reader to work out everything by himself and because it illustrates several points at issue in both regression and path analyses. The data are reproduced in Table 122, the bottom part of which gives all the correlations of pairs of variables. Solving the conventional normal equations, we obtain the regression coefficients and the following regression equation:

$$g = \hat{y} = 62.41 + 1.551 x_1 + 0.510 x_2 + 0.102 x_3 - 0.144 x_4 \qquad (49)$$

Table 122. Regression of y on four variables. Data are from Draper and Smith (1966, Appendix B).

	x_1	x_2	x_3	x_4	y	$\hat{y}$	e
	7	26	6	60	78.5	78.50	0
	1	29	15	52	74.3	72.79	1.51
	11	56	8	20	104.3	105.97	−1.67
	11	31	8	47	87.6	89.33	−1.73
	7	52	6	33	95.9	95.65	0.25
	11	55	9	22	109.2	105.27	3.93
	3	71	17	6	102.7	104.15	−1.45
	1	31	22	44	72.5	75.67	−3.17
	2	54	18	22	93.1	91.72	1.38
	21	47	4	26	115.9	115.62	0.28
	1	40	23	34	83.8	81.81	1.99
	11	66	9	12	113.3	112.33	0.97
	10	68	8	12	109.4	111.69	−2.29
mean	7.462	48.154	11.769	30.000	95.423	95.423	0
var.	34.60	242.16	41.03	280.17	226.31	222.32	3.99
std. dev.	5.88	15.56	6.405	16.74	15.044	14.91	2.00

Correlation coefficients

	x_2	x_3	x_4	y
x_1	$r_{12} = 0.2286$	$r_{13} = -0.8241$	$r_{14} = -0.2454$	$r_{10} = 0.7307$
x_2		$r_{23} = -0.1392$	$r_{24} = -0.9730$	$r_{20} = 0.8163$
x_3			$r_{34} = 0.0295$	$r_{30} = -0.5347$
x_4				$r_{40} = -0.8213$

STANDARDIZED VARIABLES; PATH COEFFICIENTS   123

These predicted values and residuals are also given in Table 122 which shows that

$$\sigma_y^2 = \sigma_g^2 + \sigma_e^2 = 222.32 + 3.99 = 226.31$$

and

$$R_{0(1234)}^2 = \sigma_g^2 / \sigma_y^2 = 0.98237 \quad (50)$$

25. The path coefficients p_{0i} from X_i to Y may be calculated directly from the following standardized normal equations:

$$\begin{aligned}
p_{01} + p_{02} r_{12} + p_{03} r_{13} + p_{04} r_{14} &= r_{10} \\
p_{01} r_{12} + p_{02} + p_{03} r_{23} + p_{04} r_{24} &= r_{20} \\
p_{01} r_{13} + p_{02} r_{23} + p_{03} + p_{04} r_{34} &= r_{30} \\
p_{01} r_{14} + p_{02} r_{24} + p_{03} r_{34} + p_{04} &= r_{40}
\end{aligned} \quad (51)$$

Substituting the numerical values of the correlations shown at the lower portion of Table 122 and solving, we obtain the values of the p's. Since the routine program has already given us the value of the concrete regression coefficients and the standard deviations, we would simply obtain the p's as follows using the relationship $p_{0i} = b_i \sigma_i / \sigma_0$.

concrete regression coefficients			ratio of σ_i / σ_0	path coefficients		
b_1	=	1.551	0.3910	p_{01}	=	0.6065
b_2	=	0.510	1.0344	p_{02}	=	0.5277
b_3	=	0.102	0.4258	p_{03}	=	0.0434
b_4	=	− 0.144	1.1126	p_{04}	=	− 0.1603

(52)

The path diagram for the case of four X-variables is given in Fig. 124. The path coefficient from U to Y is $p_{0U} = r(Y, U) = r(y, e) = \sigma_e/\sigma_y = \sqrt{3.99/226.31} = \sqrt{0.01763} = 0.1328$. The reader should check that the values of the p's in (52) do satisfy the equations (51) except for rounding errors. (The actual calculation involved more decimal places than indicated here.) As a final check, we calculate the total contribution of all the X-variables toward the determination of Y:

$$R^2_{0(1234)} = p_{01} r_{10} + p_{02} r_{20} + p_{03} r_{30} + p_{04} r_{40}$$

$$= 0.44318 + 0.43074 - 0.02320 + 0.13165 \quad (53)$$

$$= 0.98237$$

in agreement with our answer (50), also given by Draper and Smith (p. 165 and p. 395).

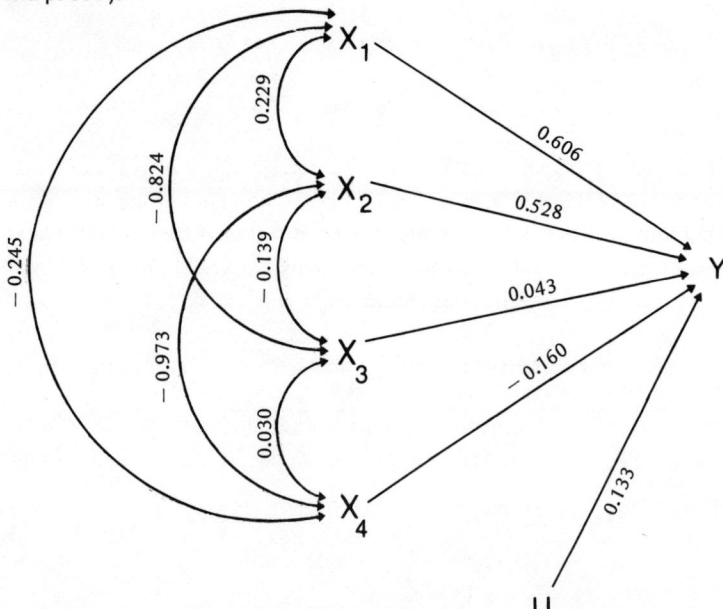

Fig. 124. Correlated causes (X_1, X_2, X_3, X_4) and an uncorrelated residual U. This is an extension of Figs. 113 and 119, and similar relationships hold. Numerical calculations are based on Table 122. It has been found that $R^2_{0(1234)} = 0.9824$ and $R^2_{0(12)} = 0.9787$. The variables X_3 and X_4 contribute very little to the determination of Y.

STANDARDIZED VARIABLES; PATH COEFFICIENTS

26. Now we come back to the original question: which variable or variables may be eliminated from the regression equation without appreciable effect on the accuracy of prediction? If we use only one x-variable for the regression, the x_4 should be used, as $r_{40} = -0.8213$ has the largest numerical value among the four r_{i0}'s. Then the degree of determination of y would be $R^2_{0(4)} = r^2_{04} = (-0.8213)^2 = 67.45\%$. However, together with other variables, Fig. 124 shows that the direct effect of X_4 to Y is comparatively small $(p_{04} = -0.160$ versus $p_{0U} = 0.133)$. Then its influence on Y must be largely through its correlation with other X-variables, chiefly the very high negative correlation (-0.973) with X_2. From (53) we see that the total contribution of X_4 to determination (via all routes to Y) is $p_{04} r_{40} = 0.1316$, considerably lower than those of X_1 and X_2.

Next, we note that the direct contribution of X_3 to Y is very low $(p_{03} = 0.0434)$ and its total contribution to determination is $p_{03} r_{30} = -0.0232$, also very low. It may be awkward to speak of a negative contribution toward determination, but this situation is not different from the elementary case in which $y = x_1 - x_2$ and $V(y) = V(x_1) + V(x_2) - 2 Cov(x_1, x_2)$, assuming the covariance itself to be positive. Although $V(y)$ must be positive, its certain components may be negative. This simply means that if x_1 and x_2 were not correlated and retained their variance, then the variance of y would be larger than it is. In a similar way, R^2 must be positive, but certain of its components may be negative. Since both p_{03} and $p_{03} r_{30}$ are low, we may eliminate x_3 from the regression analysis. Hence we decide to retain only X_1 and X_2 in the regression analysis. The new path coefficients p_{01} and p_{02} may be obtained from equations (25), substituting the numerical values of the r's (which do not change). The reader will find that

$$p_{01} = 0.574, \qquad p_{02} = 0.685 \tag{54}$$

and $\quad R^2_{0(12)} = p_{01} r_{10} + p_{02} r_{20} = 0.4195 + 0.5592 = 0.9787$

which is only slightly lower than $R^2_{0(1234)} = 0.9824$. Thus the regression on x_1 and x_2 is practically as good as the regression on all four x-variables. All of these conclusions are in agreement with Draper and Smith (1966) who have tried all possible regressions.

27. At this stage, a word of caution may be in order. In regression analysis experience shows that the addition or deletion of a variable may change the nature of the regression considerably. Path analysis is subject to the same hazard. The various selection procedures mentioned previously do not necessarily lead to the same equation for a given body of data. Path analysis may be helpful in eliminating variables but we can never be sure what would happen with the addition of new variables. That all depends on the nature of the new variable to be added. Therefore the choosing of an appropriate regression equation will necessarily involve a certain amount of trial and error.

Partial correlation and path coefficients

Toward the end of Chapter 4 we defined partial correlation coefficient and derived an expression for it in terms of simple correlations. Here we shall do essentially two things. The first is to derive the same formula by the method of path coefficients, and the second is to show the relationship between a partial correlation coefficient and a path coefficient.

28. As in Chapter 4, consider the three variables X_1, X_2, X_3, with zero means. In regressing X_1 on X_3 we obtain an estimated $\hat{X}_1$ and a residual value e_1. Similarly, the regression of X_2 on X_3 gives an estimated $\hat{X}_2$ and a residual e_2. These two uncorrelated components have been denoted by various notations under various circumstances:

$$X_1 = \hat{X}_1 + (X_1 - \hat{X}_1) = b_{13}X_3 + e_1 = X_{1(3)} + X_{1.3}$$
$$X_2 = \hat{X}_2 + (X_2 - \hat{X}_2) = b_{23}X_3 + e_2 = X_{2(3)} + X_{2.3}$$
(55)

where $X_{1(3)}$ is the value of X_1 determined by X_3, and $X_{1.3}$ is the value of X_1 free of the influence of X_3. The same applies to $X_{2(3)}$ and $X_{2.3}$. The partial correlation coefficient between X_1 and X_2 without the influence of X_3 is defined as the simple correlation between $X_{1.3}$ and $X_{2.3}$, both being free of the influence of X_3. Referring to Chapter 4 (64),

$$r_{12.3} = r(e_1, e_2) = r(X_{1.3}, X_{2.3}) \quad (56)$$

STANDARDIZED VARIABLES; PATH COEFFICIENTS

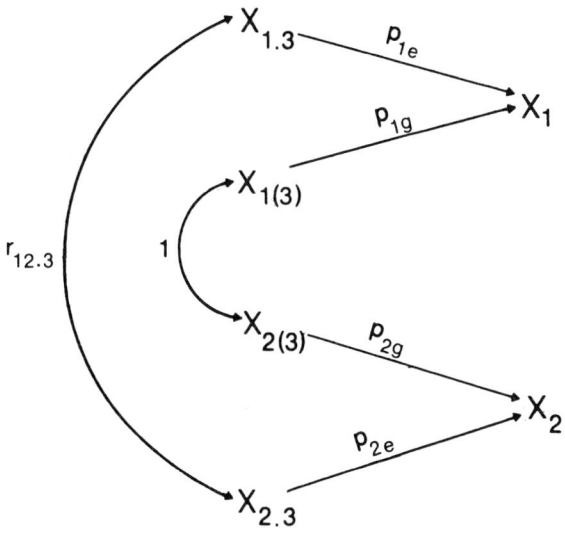

Fig. 127. Path diagram for partial correlation $r_{12.3}$. For another diagram, see Fig. 185.

Fig. 127 is the path diagram of such a situation (Wright, 1968, I, p. 313). It consists of two path diagrams linked together by correlations. The upper part represents the relationship $X_1 = X_{1(3)} + X_{1.3}$, where the latter two components are uncorrelated. Note that there is no connecting path between $X_{1(3)}$ and $X_{1.3}$, so that this part of the diagram is like Fig. 107. Exactly the same remarks apply to the lower part of the diagram depicting $X_2 = X_{2(3)} + X_{2.3}$. Then these two diagrams are linked by two correlations. One is $r_{12.3}$, the value we are trying to find; the other is

$$r(X_{1(3)}, X_{2(3)}) = r(b_{13} X_3, b_{23} X_3) = r(X_3, X_3) = 1 \quad (57)$$

Let p_{1e} be the path coefficient from $X_{1.3} = e_1$ to X_1 and p_{1g} be the path coefficient from $X_{1(3)} = g_1$ to X_1, as indicated in Fig. 127. Similarly, the two path coefficients to X_2 are p_{2e} and p_{2g}. The correlation between two variables is the sum (of the values) of all paths connecting the two variables. In Fig. 127, the variables X_1 and X_2 are connected by two paths; thus

$$r(X_1, X_2) = r_{12} = p_{1g}(1)p_{2g} + p_{1e}r_{12.3}p_{2e} \tag{58}$$

The path coefficients involved are readily found, because each part of the diagram is the same as Fig. 107 for the case of simple regression and correlation. Thus, for the determination of X_1 and X_2, respectively,

$$p_{1g}^2 + p_{1e}^2 = 1, \qquad p_{2g}^2 + p_{2e}^2 = 1$$

Since the regression is on X_3 in both cases,

$$\begin{aligned} p_{1g} &= r_{13}, & p_{1e} &= \sqrt{1 - r_{13}^2} \\ p_{2g} &= r_{23}, & p_{2e} &= \sqrt{1 - r_{23}^2} \end{aligned} \tag{59}$$

Transposing (58) and substituting (59), we obtain

$$r_{12.3} = \frac{r_{12} - p_{1g}p_{2g}}{p_{1e}p_{2e}} = \frac{r_{12} - r_{13}r_{23}}{\sqrt{1 - r_{13}^2}\sqrt{1 - r_{23}^2}} \tag{60}$$

which is the expression established in Chapter 4 (65) by the method of variance and covariance. The path method is not necessarily shorter in every problem, but the example above shows that it may be used to derive a number of statistical results in a very illuminating manner.

29. The relationships of path coefficients with regression and simple correlation coefficients have been dwelt upon in some detail. A question often raised by students is: what is the relationship of a path coefficient with a partial correlation coefficient? The answer may be most easily provided by writing out the explicit expressions of each and comparing them. Let us limit ourselves to the case of $\hat{Y} = p_{01}X_1 + p_{02}X_2$. Solving the standardized normal equations (25) we get

$$p_{01} = \frac{r_{01} - r_{02} r_{12}}{1 - r_{12}^2} \qquad (61)$$

On the other hand, by a suitable change of subscripts of (60), the corresponding partial correlation coefficient is

$$r_{01.2} = \frac{r_{01} - r_{02} r_{12}}{\sqrt{1 - r_{02}^2}\sqrt{1 - r_{12}^2}} \qquad (62)$$

Comparing these two expressions we arrive at the relationship:

$$r_{01.2} = p_{01} \cdot \frac{\sqrt{1 - r_{12}^2}}{\sqrt{1 - r_{02}^2}} \qquad (63)$$

Of course, this is merely the standardized form of the relationship between $r_{01.2}$ and $b_{01.2} = b_1$ established in Chapter 4 (67):

$$r_{01.2} = b_{01.2} \frac{\sigma_1 \sqrt{1 - r_{12}^2}}{\sigma_0 \sqrt{1 - r_{02}^2}} = b_{01.2} \frac{\sigma_{1.2}}{\sigma_{0.2}} \qquad (64)$$

where $b_{01.2}\,\sigma_1/\sigma_0 = p_{01}$. Generalizations with respect to the b's may be found in Yule and Kendall (1968), and those with respect to the p's in Wright (1968, I).

Exercises

Ex. 1. For the data in Table 101, we have found that $r(X, Y) = 0.500$ and $r(U, Y) = 0.866$, so that $(0.500)^2 + (0.866)^2 = 1$. An examination of Fig. 107 shows that the relationships of X and U to Y are symmetrical. Should a set of data yield $r(X, Y) = 0.866$, what will then be the value of $r(U, Y)$? Investigate the following set of data and see if your prediction is correct. Find the regression equation so that you can calculate $g = \hat{y} = a + bx$ and $e = y - \hat{y}$.

	x	y	$g=\hat{y}$	$e=y-\hat{y}$	e^2	ye
	12	3	5.8	-2.8	7.84	-8.4
	3	2	4.0	-2.0	4.00	-4.0
	56	13	14.6	-1.6	2.56	-20.8
	37	10	10.8	-0.8	0.64	-8.0
	1	4	3.6	$+0.4$	0.16	$+1.6$
	30	11	9.4	$+1.6$	2.56	$+17.6$
	28	11	9.0	$+2.0$	4.00	$+22.0$
	17	10	6.8	$+3.2$	10.24	$+32.0$
Total	184	64	64.0	0	32.00	32.0
mean	23	8	8	0	4.00	4.0

Partial answer: $\hat{y} = 3.4 + 0.20 x$

Ex. 2. Fig. 113 shows that U is not connected with either X_1 or X_2. This corresponds with the analytical result that U is uncorrelated with either X. Verify numerically, using the values of X_1, X_2, and U in Table 109,

$$\Sigma(X_1\ U) = 0, \qquad \Sigma(X_2\ U) = 0$$

Recognize that these are merely the conventional normal equations of Chapter 4 in standardized form:

$$\Sigma(y - a - b_1 x_1 - b_2 x_2) x_1 = 0, \quad \Sigma(y - a - b_1 x_1 - b_2 x_2) x_2 = 0$$

Ex. 3. In the text we solve the standardized normal equations (25) numerically and find that $p_{01} = 0.321$. It is desirable to show that in general the solution of (25) is $p_{01} = b_1\ \sigma_1/\sigma_0$. Let us recall that the explicit expression for b_1 is, according to Chapter 4 (35),

STANDARDIZED VARIABLES; PATH COEFFICIENTS

$$b_{01.2} = b_1 = \frac{\sigma_0}{\sigma_1} \cdot \frac{r_{01} - r_{12} r_{02}}{1 - r_{12}^2}$$

On the other hand, solving equations (25) yield (61):

$$p_{01} = \frac{r_{01} - r_{12} r_{02}}{1 - r_{12}^2}$$

Hence,

$$b_{01.2} = p_{01} \frac{\sigma_0}{\sigma_1}, \qquad p_{01} = b_{01.2} \frac{\sigma_1}{\sigma_0}$$

In a similar way, show that $p_{02} = b_{02.1} \sigma_2/\sigma_0$.

Ex. 4. Consider the following two sets of data as if you have never seen them before.

	Set I			Set II		
	x_1	x_2	y	x_1	x_2	y
	10	3	8	17	3	20
	9	5	4	11	5	10
	20	6	21	18	6	17
	17	8	49	18	8	43
	11	11	26	10	11	26
	18	14	33	9	14	43
	18	15	84	20	15	74
	17	18	55	17	18	47
Total	120	80	280	120	80	280
mean	15	10	35	15	10	35

(i) Find the various sums of squares and products and the correlation coefficients for each set of data.

(ii) Find the concrete multiple regression equation of y on x_1 and x_2 for each set of data. What do you notice?

(iii) Find the path coefficients p_{01} and p_{02} and draw a path diagram for each set of data. Are the two path diagrams different?

(iv) Calculate $R^2_{0(12)}$ and check its relationship with path coefficients.

This exercise summarizes nearly all of the material covered in the last two chapters. The reader should work out the details regardless of his possible previous knowledge of the two sets of data. The partial answers provided in the following are meant to ensure that your understanding of the problem and your numerical calculations are correct.

(i) Let the capital letters stand for deviations from mean. The various sums of squares and products are required by the normal equations.

	Set I				Set II		
	X_1	X_2	Y		X_1	X_2	Y
X_1	128	80	496	X_1	128	0	256
X_2		200	760	X_2		200	600
Y			4968	Y			3048

The correlations for the two sets of data are, respectively,

$r_{10} = 0.62199$ | $r_{10} = 0.40985$
$r_{12} = 0.500$ | $r_{12} = 0$
$r_{20} = 0.76244$ | $r_{20} = 0.76847$

(ii) Solving the normal equations, you will find that $b_1 = 2$ and $b_2 = 3$ for both sets of data and the intercept is also the same, so that the regression equation is, for both sets,

$$\hat{y} = 2x_1 + 3x_2 - 25$$

STANDARDIZED VARIABLES; PATH COEFFICIENTS

(iii) The path coefficients may be obtained independently from the standardized normal equations, or from the concrete regression coefficients already obtained by multiplying them by the ratio $\sigma_i/\sigma_0 = \sqrt{\Sigma X_i^2}\big/\sqrt{\Sigma Y^2}$. The values of the path coefficients and the corresponding path diagrams for the two sets of data are as follows:

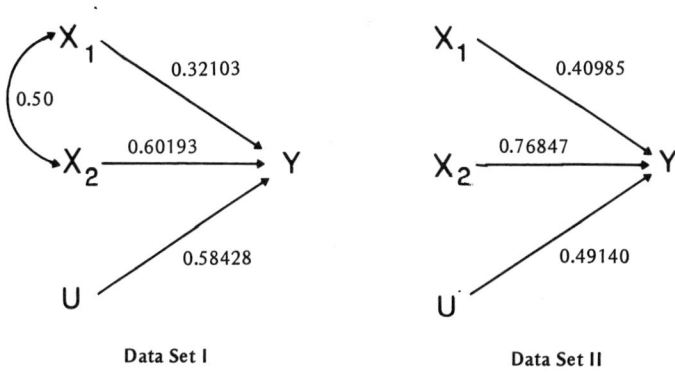

Data Set I   Data Set II

Note that the ratios of the two path coefficients in the two diagrams are the same: *60193/32103 = 76847/40985 = 1.8750.* See Ex. 5 for details. For Set II, where the *x*'s are uncorrelated, the path coefficients are identical with the correlation coefficients. But for Set I, check that

$$r_{10} = 0.32103 + (0.50)(0.60193) = 0.62199$$

$$r_{20} = 0.60193 + (0.50)(0.32103) = 0.76244$$

Here we learn the lesson that a path diagram gives a more complete description of the relationships among the three variables than the conventional regression equation.

(iv) To obtain $R_{0(12)}^2$, it is easiest to calculate first the sum of squares due to regression:

STANDARDIZED VARIABLES; PATH COEFFICIENTS

	Set I			Set II	
				ssq	R^2
Regression ssq	= 3272	$R^2_{0(12)}$	= 0.6586	2312	0.7585
Residual ssq	= 1696	p^2_{OU}	= 0.3414	736	0.2415
Total ssq	= 4968		1.0000	3048	1.0000

To relate $R^2_{0(12)}$ with the path coefficients shown in the diagram, we have the total fraction of determination

$$R^2_{0(12)} = p^2_{01} + p^2_{02} + 2 p_{01} r_{12} p_{02}$$

and $\quad R^2_{0(12)} + p^2_{OU} = 1$

Check that these are true for both sets of data, using the path values shown in the respective diagrams.

Ex. 5. With respect to the diagrams in the previous page, let x_1, x_2, and u be the path coefficients of the left one (Set I) and let p_1, p_2, and v be the path coefficients of the right one (Set II). Then we may write, with $r = r(X_1, X_2)$,

$$x_1^2 + x_2^2 + 2x_1 x_2 r + u^2 = 1, \qquad p_1^2 + p_2^2 + v^2 = 1$$

$$x_1^2 + x_2^2 = 1 - 2x_1 x_2 r - u^2 = C, \qquad p_1^2 + p_2^2 = 1 - v^2 = D$$

$$\frac{x_1^2}{C} + \frac{x_2^2}{C} = 1, \qquad \frac{p_1^2}{D} + \frac{p_2^2}{D} = 1$$

Hence, $x_1^2/x_2^2 = p_1^2/p_2^2$ and $x_1/x_2 = p_1/p_2$. The path coefficient in one diagram may be converted into one in the other diagram by using the relationships shown above; for instance, $p_1 = x_1 \sqrt{D/C}$. The reader should verify this numerically.

Chapter 6

PATH ANALYSIS OF CAUSAL SYSTEMS

PATH COEFFICIENTS were introduced in the last chapter entirely as standardized partial regression coefficients. The basic statistical model employed is the same as that in conventional multiple regression situations, except that the variables involved have been standardized first. The few properties of path coefficients mentioned in the last chapter are merely consequences of standardization of the well established results of conventional statistics. If the path method stopped there, we would have not much to gain in carrying out such an analysis. The usefulness of path analysis is in extending the single-multiple-regression-equation treatment to a network of variables involving more than one equation. This chapter will introduce some basic network patterns amenable to path analysis, but applications in biology and social sciences are left to later chapters. We continue to assume that the variables involved are linearly related, or, at least, essentially so.

1. Multiplicity of patterns of relationship. Consider, for example, five variables. They may be related in various ways. As a rough method of classifying the patterns of relationship, we may consider the following fifteen ways of partitioning the number 5.

(1, 1, 1, 1, 1)	*(1, 1, 3)*	*(1, 2, 2)*
(1, 1, 1, 2)	*(1, 3, 1)*	*(2, 1, 2)*
(1, 1, 2, 1)	*(3, 1, 1)*	*(2, 2, 1)*
(1, 2, 1, 1)	*(2, 3)*	*(1, 4)*
(2, 1, 1, 1)	*(3, 2)*	*(4, 1)*

If we limit ourselves to drawing arrows from left to right between two adjacent variables, we will obtain 15 possible patterns of relationship as shown in Fig. 136. This, however, does not exhaust all possible patterns of relationship. The direction of an arrow in some cases (e.g., *2, 2, 1)* may

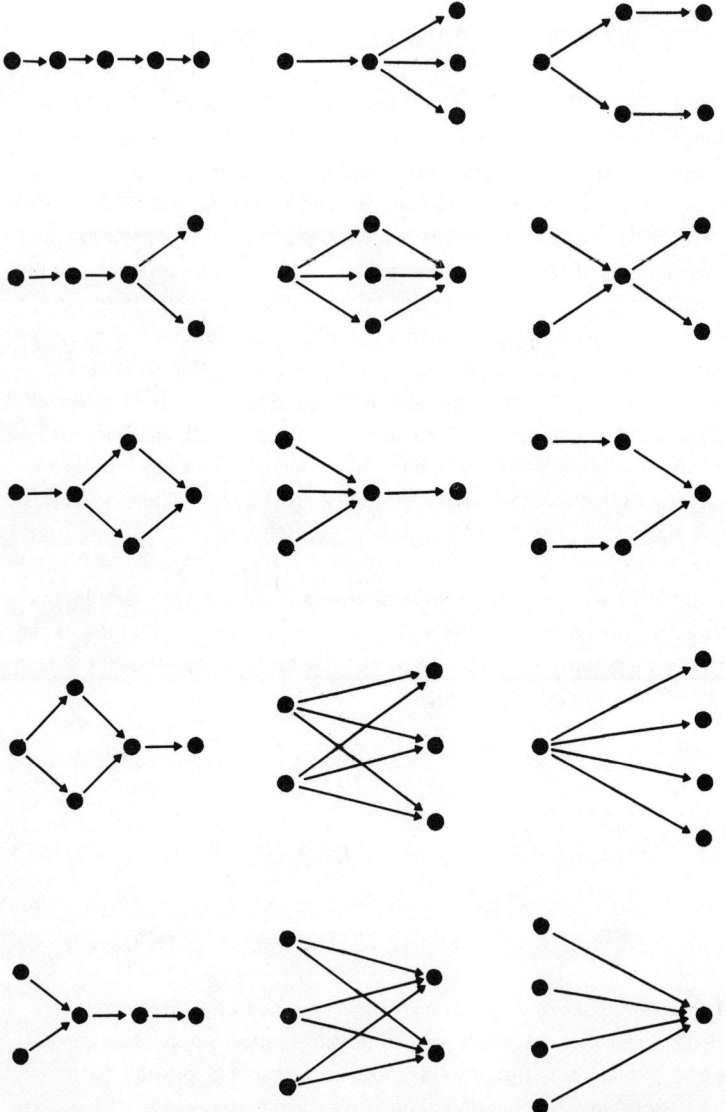

Fig. 136. **Possible arrangements (by partition) of five variables.** Direction of influence is indicated by arrows. The arrows are limited to adjacent variables. The variables on the left of each diagram may be correlated (though not actually shown). By shifting the direction of some of the arrows, more path diagrams may be obtained.

PATH ANALYSIS OF CAUSAL SYSTEMS

be shifted 45° to yield another pattern. Moreover, arrows may be added between nonadjacent variables, and double-headed arrows (correlations) may be added to the independent variable on the left. Furthermore, each variable has been represented merely by a solid dot in the diagram without specifying which dot is which one of the five variables (a, b, c, d, e, say). If each variable is specified, then for the pattern *(1, 1, 1, 1, 1)* alone there will be 5! = 120 different ways of arranging them. Thus, the reader sees that even with as few as five variables under consideration, there will be many possible types of relationship among them. Of these, only the *(4, 1)* pattern at the lower right corner of Fig. 136 is equivalent to a multiple regression model (with correlations added between the independent variables on the left). Path analysis deals with all possible patterns of relationship.

Chain of uncorrelated causes

2. We begin with the simplest type of causal systems, viz., a chain of uncorrelated causes as shown in Fig. 137 (c). Here, W and V are uncorrelated and both influence X. In turn, X and U are uncorrelated and both influence Y. In addition, U is uncorrelated with W and V. The chain under consideration is $W \to X \to Y$. The chain may obviously be extended to any number of steps. Incidentally, the diagram (c) in Fig. 137 is not covered by the patterns of Fig. 136, but may be obtained from the *(2, 2, 1)* pattern by changing the direction of one arrow (making it from V to X instead of from V to U).

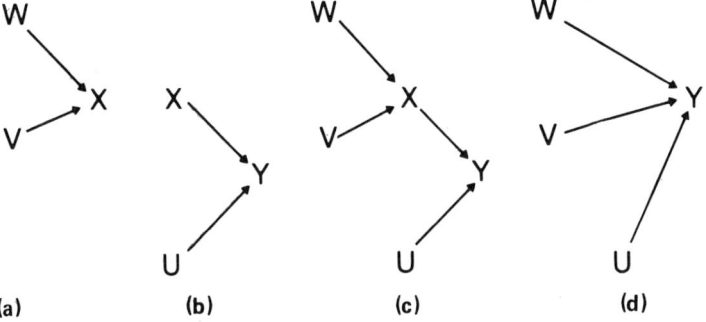

Fig. 137. Chain of uncorrelated causes. (a) W and V are uncorrelated. (b) X and U are uncorrelated. (c) is a combination of (a) and (b); the crucial assumption is that U and V are uncorrelated. (d) is an equivalent form of (c) without studying X explicitly.

3. One of the characteristics of a chain of causes (correlated or uncorrelated) is that each step of the chain is a complete (self-contained) path diagram by itself and can be analyzed separately without affecting the remaining parts of the chain. Thus, the causal diagram (c) of Fig. 137 may be taken apart to form the other two diagrams, (a) and (b), of the same figure, each being complete by itself. Using the theorems developed in the last chapter, and regarding U as residual from regression of Y on X, and V as residual from regression of X on W, we have

$$\text{from Fig. 137 (a),} \quad p_{XW}^2 + p_{XV}^2 = r_{XW}^2 + r_{XV}^2 = 1$$
$$\text{from Fig. 137 (b),} \quad p_{YX}^2 + p_{YU}^2 = r_{YX}^2 + r_{YU}^2 = 1 \quad (1)$$

In dealing with a network of variables there will be many path coefficients, and keeping track of the subscripts of the p's will be wearisome. To facilitate rapid writing and reading, it is conventional to use a lower case letter to denote the path coefficient from the variable denoted by the corresponding capital letter. Thus,

$$w = p_{XW}, \qquad v = p_{XV}, \qquad x = p_{YX}, \qquad u = p_{YU} \quad (2)$$

It is advisable for the reader to insert these lower case letters on the respective paths of Fig. 137 to facilitate subsequent reading. Such notation, though simple, is incomplete by itself. For instance, x denotes a path coefficient from the variable X but it does not tell which variable it goes to. In our present case, $x = p_{YX}$, meaning from X to Y. Thus, the simplified notations must always be read with respect to a specified diagram. With this much understood, the expression (1) becomes

$$w^2 + v^2 = 1, \qquad x^2 + u^2 = 1 \quad (3)$$

Needless to say, these lower case letters are not to be confused with the conventional notation for variables.

4. Abstract diagrams such as those in Fig. 137 are very much like mathematical expressions, the concrete physical meaning of which must be assigned to the symbols. In the paragraph above it was mentioned that U may be

regarded as the residual for the linear regression of Y on X, and V as the residual for the linear regression of X on W. However, it is not the only meaning that the diagrams of Fig. 137 can take. As a matter of fact, they can take many different meanings as long as the relationships among the variables conform to the patterns shown in the diagrams. For example, V and W may represent two grandparents who are unrelated, and X their offspring. Then X and U are the two parents who are unrelated, and Y their offspring. Further, U is also unrelated to the two grandparents. Then the relationships among the five individuals would be exactly like those shown in Fig. 137. As a second example, let us imagine that human stature (standing height) consists of three major segments: the first, top of head to neck; the second, neck to buttock; and the third, buttock to heel; and let us call these segments W = head length, V = body length, and U = leg length, respectively, for brevity. Then $W + V = X$ is the sitting height, and $X + U = Y$ is the standing height. Physical anthropologists measure these segments (and many others) separately because each segment has its own biological characteristics. If the lengths of these three segments are uncorrelated, then Fig. 137 (c) would be a representation of the relationships among the five measurements of heights. (Cases in which the causes, that is, the segments, are correlated, will be considered later.)

5. Now we proceed to find the path coefficient from W directly to Y without the intermediate variable X. A new symbol is needed for such a compound path, as our previous notation (2) does not cover this case. Let $\overline{w}$ be the path coefficient from W to Y without considering the intermediate variable X. Let the linear functions be

$$Y = X + U \qquad \text{and} \qquad X = W + V \qquad (4)$$

where X, U are uncorrelated and so are W, V. This model will be readily understood in the light of the example on human height described in the last paragraph, assuming that the separate segments of human stature are uncorrelated. If we think of U and V as residuals from linear regression, then X would stand for $\hat{Y}$ and W would stand for $\hat{X}$, where the "hat" means the calculated collinear values of that variable. So, the expressions (4) may be considered as the general model. By substitution, model (4) may also be written (Fig. 137d)

$$Y = W + V + U \tag{5}$$

where W, V, U are uncorrelated components. In terms of our example on human stature, it amounts to saying that standing height is the sum of head length, body length, and leg length, without considering the intermediate variable, sitting height. For uncorrelated causes, the path coefficient from X to Y is the correlation coefficient between X and Y. In order to obtain the latter, we need the covariance:

$$\begin{aligned} Cov(Y, X) &= E(Y - \bar{Y})(X - \bar{X}) = E(X - \bar{X} + U - \bar{U})(X - \bar{X}) \\ &= E(X - \bar{X})^2 + E(U - \bar{U})(X - \bar{X}) = \sigma_X^2 \end{aligned} \tag{6}$$

as $Cov(U, X) = E(U - \bar{U})(X - \bar{X}) = 0$. Similarly, from (4) and (5),

$$Cov(X, W) = \sigma_W^2 \quad \text{and} \quad Cov(Y, W) = \sigma_W^2 \tag{7}$$

Hence the path coefficients (correlation coefficients in this case) are

$$\bar{w} = \frac{Cov(Y, W)}{\sigma_Y \sigma_W} = \frac{\sigma_W}{\sigma_Y}, \quad x = \frac{Cov(Y, X)}{\sigma_Y \sigma_X} = \frac{\sigma_X}{\sigma_Y}, \quad w = \frac{Cov(X, W)}{\sigma_X \sigma_W} = \frac{\sigma_W}{\sigma_X} \tag{8}$$

and

$$\bar{w} = \frac{\sigma_W}{\sigma_Y} = \frac{\sigma_X}{\sigma_Y} \cdot \frac{\sigma_W}{\sigma_X} = x w \tag{9}$$

Thus we have reached one of the most elementary and fundamental theorems of path analysis, viz., the value of the path from W directly to Y is equal to the product of the two component paths: that from W to X and that from X to Y. This of course implies that $r(Y, W) = r(Y, X) r(X, W)$. To write (9) in a more general way,

$$p_{02} = p_{01} p_{12} \tag{10}$$

As long as the causes are uncorrelated, this chain may be extended to any number of steps. Thus,

$$p_{03} = p_{01} p_{12} p_{23} \tag{11}$$

for the chain $X_3 \to X_2 \to X_1 \to X_0$ when the other causes of the X's are uncorrelated.

PATH ANALYSIS OF CAUSAL SYSTEMS

6. We see from Fig. 137c that the path coefficient from V directly to Y, without considering the intermediate variable X, is $\bar{v} = x\,v$. The fractions of determination with respect to the separate parts, (a) and (b) of Fig. 137, have already been given by (3). With respect to (d) of Fig. 137, we have

$$\bar{w}^2 + \bar{v}^2 + u^2 = 1 \qquad (12)$$

Comparing this with (3), we conclude that x^2 must be equal to $\bar{w}^2 + \bar{v}^2$. More formally,

$$x^2 = x^2(w^2 + v^2) = (xw)^2 + (xv)^2 = \bar{w}^2 + \bar{v}^2 \qquad (13)$$

To put it verbally, the fraction of determination of Y's variance by X is the sum of those determined by W and V without considering X. Therefore the diagrams (c) and (d) of Fig. 137 are entirely equivalent with respect to Y. The choice between them depends on the availability and our interest in X.

7. In view of the results obtained above, let us consider yet another example for the interpretation of Fig. 137c. This time let V and W be two unrelated parents and X be the "midparent," viz., $X = (½)(V + W)$. Then Y is the offspring of the couple V and W, and U represents all non-genetical factors, the environmental effects in the broadest sense. From (3), (12), and (13) it follows that we can analyze the data by either using the midparent X (without using V and W) or using parents V and W (without using X). In the former case we obtain a joint fraction of determination by both parents, and in the latter, separate fractions of determination by the two parents. In either case, the residual fraction $p_{0U}^2 = u^2$ remains the same. For numerical analysis it is unnecessary to calculate the midvalue X; it is more convenient to use the parental total $V + W$ directly, as path coefficients refer to standardized variables, and the fractions of determination remain the same whether we use the midvalue or the total.

Common causes and correlation

8. Consider two variables, X and Y. We have studied their relationship of the type $X \to Y$ in the previous section. But not all relationships between

X and Y are of that type. It is true that X and Y may be highly correlated, but that does not imply that one variable causes or influences the other in any direct sense. Two variables may be correlated because both are influenced by some common factors. It is these common influences of the common factors that contribute to the correlation between X and Y. In such a case, while a simple linear regression of Y on X still describes the data, it has no interpretative value and does not increase our understanding of the relationships between X and Y. In this section we shall examine the contributions of the common factors to the correlation between X and Y, still assuming all causes to be uncorrelated.

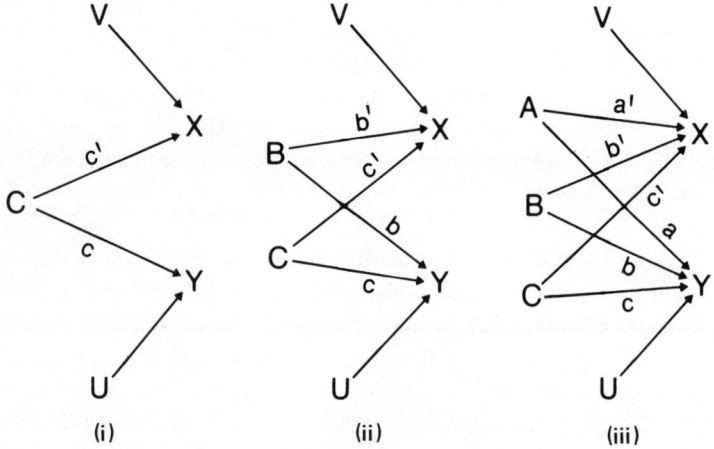

Fig. 142. Uncorrelated common cause(s) of two variables. (i) one common factor; C is the common factor of X and Y. (ii) two common factors B and C. (iii) three common factors, A, B, C. The causes (variables on the left in all diagrams) are uncorrelated.

9. Fig. 142 gives three path diagrams showing X and Y linked by common factors: (i) one common factor, (ii) two common factors, (iii) three common factors. These factors (A, B, C) and others (V, U) are all uncorrelated. We shall first study situation (i) with one common cause C. The pattern of relationship among the five variables in Fig. 142 (i) is quite different from that among the five variables in Fig. 137 (c), although each

PATH ANALYSIS OF CAUSAL SYSTEMS

diagram involves four arrows. Let the variables be expressed in terms of deviations from the mean, so that $E(VC) = E(UC) = E(UV) = 0$, and $E(C^2) = \sigma_C^2$, etc. Note that Fig. 142 (i), like Fig. 137 (c), can also be taken apart to form two self-contained path diagrams, the upper one involving V, C, and X, and the lower one involving U, C, and Y. Again, we use a lower case letter to denote path coefficients. Those with primes are from a cause to effect X, and those without primes are from a cause to effect Y. Since the causes are uncorrelated, the path coefficients are the corresponding correlation coefficients. Under the model

$$X = C + V \quad \text{and} \quad Y = C + U \quad (14)$$

we have by the usual method of expectation

$$Cov(X, C) = \sigma_C^2; \qquad Cov(Y, C) = \sigma_C^2$$

and the path coefficients (correlation coefficients in this case) are:

$$c' = p_{XC} = r(X, C) = \frac{\sigma_C}{\sigma_X}$$

$$c = p_{YC} = r(Y, C) = \frac{\sigma_C}{\sigma_Y} \quad (15)$$

The covariance of X and Y is

$$Cov(X, Y) = E(XY) = E(C + V)(C + U) = E(C^2) = \sigma_C^2 \quad (16)$$

so that

$$r_{XY} = \frac{Cov(X, Y)}{\sigma_X \sigma_Y} = \frac{\sigma_C}{\sigma_X} \cdot \frac{\sigma_C}{\sigma_Y} = c'c \quad (17)$$

that is, $r_{XY} = r(X, C) \cdot r(C, Y)$

Reading (17) in connection with Fig. 142 (i), we see that the correlation between X and Y is given by the (value of the) connecting path $c'c$. The uncorrelated variables V and U do not contribute to the correlation $r(X, Y)$.

10. As before, we note that the model (14) is a general expression, the specific meaning of which is to be assigned by the concrete problem under investigation. If V is the head length, C is the body length, and U is the leg length, then $X = C + V$ will be the sitting height, and $Y = C + U$ will be the shoulder standing height. The latter two measurements will be correlated through the common component C. In a conventional regression problem, the model is $X = \hat{X} + V$, where $\hat{X} = b_{XC} \, C$ is the calculated value of X, and V is the residual. In such a case, the symbol C in the expression $X = C + V$ would stand for $\hat{X}$. Similarly, the C in the expression $Y = C + U$ would stand for $\hat{Y} = b_{YC} \, C$. The reader may wonder how one symbol (C) could represent two different quantities: $b_{XC} \, C$ and $b_{YC} \, C$. The reason is, as we have noted in the last chapter dealing with partial correlations, that these two quantities have correlation unity and, after standardization, they will reduce to the same variable. This is, in fact, one of the advantages of using path coefficients with respect to standardized variables.

11. Now we proceed to the case with two common causes shown in Fig. 142 (ii). Since the general procedure is the same, we shall only indicate the essential algebraic steps. The model is

$$X = V + B + C \qquad \text{and} \qquad Y = B + C + U \qquad (18)$$

where B and C are the two common causes contributing to the correlation between X and Y. All the causes (V, B, C, U) are uncorrelated with each other, so that $E(VB) = 0$, etc., and

$$Cov(X, B) = Cov(Y, B) = \sigma_B^2, \qquad Cov(X, C) = Cov(Y, C) = \sigma_C^2$$

$$Cov(X, Y) = E(V + B + C)(B + C + U) = E(B^2 + C^2) = \sigma_B^2 + \sigma_C^2 \qquad (19)$$

The path coefficients from B and C to X and Y are:

$$b' = r(B, X), \qquad b = r(B, Y), \qquad c' = r(C, X), \qquad c = r(C, Y)$$

where all the correlations take the form $r(B, X) = \sigma_B/\sigma_X$, etc. Finally, the correlation between X and Y is, by (19),

PATH ANALYSIS OF CAUSAL SYSTEMS   145

$$r_{XY} = \frac{Cov(X, Y)}{\sigma_X \; \sigma_Y} = \frac{\sigma_B}{\sigma_X} \cdot \frac{\sigma_B}{\sigma_Y} + \frac{\sigma_C}{\sigma_X} \cdot \frac{\sigma_C}{\sigma_Y}$$

$$= r(B, X) \, r(B, Y) + r(C, X) \, r(C, Y) = b'b + c'c \qquad (20)$$

Reading this expression in connection with Fig. 142 (ii), we see that the total correlation between X and Y is the sum (of the values) of the two connecting paths between X and Y, one through B and one through C. The value of r_{XY} has been decomposed into two components; the component $b'b$ is due to the common cause B and the component $c'c$ is due to the common cause C. It is through analysis of this type that we may be able to evaluate the relative importance of the causes contributing to an observed correlation between two variables.

12. The case with three common causes shown in Fig. 142 (iii) hardly needs any explanation. It is a good exercise for review and for following the procedure adapted above to show that

$$r_{XY} = a'a + b'b + c'c \qquad (21)$$

when all the causes are uncorrelated. The generalization to cases involving more than three common causes is obvious. To summarize, the correlation between two variables is the sum of all connecting paths between them.

Correlated causes

13. The preceding two sections, wherein we have assumed that the causes are uncorrelated, are of course special cases. Now we proceed to the more general cases in which the causes themselves are correlated. In doing so, however, we still need the preliminary theorems established in the two preceding sections. In dealing with correlated causes we should always remember that the path coefficients are no longer the same as correlation coefficients. We shall begin with the three path diagrams, (a), (b), (c), of Fig. 146. First, consider the model

$$Y = W + X \qquad (22)$$

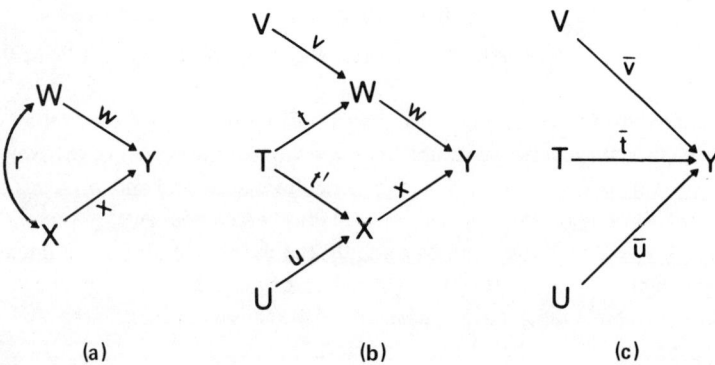

Fig. 146. Correlated causes. (a) The causes W and X are correlated. (b) The correlation between the causes W and X is due to a remote common factor T. The remote variables V, T, U are uncorrelated. (c) is an equivalent form of (b) without studying the intermediate variables W and X explicitly, but directly in terms of V, T, U.

where W and X are correlated with coefficient $r = r_{WX}$ as indicated in (a) of Fig. 146. This diagram is the same as Fig. 113 for multiple regression, except for the absence of the residual factor. Hence the only modification to be made of our previous results of Chapter 5 is to set the residual factor equal to zero, and consequently $R^2_{0(12)} = 1$. However, we shall write out the essential results of the model $Y = W + X$ here anyway for two reasons. One is to provide continuous reading without constantly referring to the last chapter, assuming that a certain amount of redundancy does not irritate the reader. The second reason is that path diagram (a) of Fig. 146 itself has important applications in analysis and understanding of data involving human inbreeding (as we shall see in later chapters) and deserves an independent description.

14. With respect to model (22) and its corresponding Fig. 146 (a),

$$\sigma_Y^2 = \sigma_W^2 + \sigma_X^2 + 2\,Cov(W, X) \tag{23}$$

PATH ANALYSIS OF CAUSAL SYSTEMS 147

Dividing both sides by σ_Y^2 and writing $Cov(W, X) = \sigma_W r_{WX} \sigma_X$, we obtain

$$1 = \left(\frac{\sigma_W}{\sigma_Y}\right)^2 + \left(\frac{\sigma_X}{\sigma_Y}\right)^2 + 2\left(\frac{\sigma_W}{\sigma_Y}\right) r_{WX} \left(\frac{\sigma_X}{\sigma_Y}\right) \quad (24)$$

The path coefficients from W to Y and from X to Y are respectively

$$w = p_{YW} = \frac{\sigma_W}{\sigma_Y}, \qquad x = p_{YX} = \frac{\sigma_X}{\sigma_Y} \quad (25)$$

remembering that these are *not* correlation coefficients between Y and W or between Y and X. Substituting the notation (25), the expression (24) becomes

$$w^2 + x^2 + 2wrx = 1 \quad (26)$$

where $r = r_{WX}$. Note that this result may be interpreted as self-correlation, $r(Y, Y) = 1$. As explained in the last chapter, (26) indicates that there are four ways of making round trips from Y to Y in Fig. 146 (a). But the correlation between two variables is the sum of all connecting paths. Therefore the sum of the four round trips from Y to Y is the correlation $r(Y, Y)$, which is unity. The three terms (actually four) of (26) give the fractions of determination of the variance of Y by W, X, and their joint effects. The correlations between Y and its causes are

$$r_{YW} = w + rx, \qquad r_{YX} = x + rw \quad (27)$$

each being the sum of two connecting paths, one direct and one indirect.

15. If we have information on W, X, and Y only (Fig. 146, a), the correlation $r = r_{WX}$ between the two causes is called "unanalyzed." In the multiple regression situation (Chapter 5), the r_{ij}'s among the X-variables are all unanalyzed. Such a correlation is simply accepted as a given fact which cannot be further decomposed under the given causal scheme. This topic will be discussed in more detail in paragraph 26.

16. However, if we have information on other variables which we believe are behind the immediate causes W and X, the path diagram may then be extended to these remote variables to account for the correlation r_{WX}. Fig. 146 (b) represents such an attempt. For simplicity, only one common cause T has been shown to contribute to the correlation between W and X. From the previous section the reader realizes that the analysis remains the same when there are two or more common causes as long as these remote variables are uncorrelated among themselves. The middle portion of Fig. 146 (b), involving only W, X, and T, is a self-contained path diagram by itself, from which we see that

$$r_{WX} = t't \tag{28}$$

where t' is the path coefficient from T to X and t is that from T to W. Since the remote variables V, T, U are uncorrelated, these t's are correlation coefficients. By introducing the variable T into consideration, we have "explained" the correlation r_{WX}. Comparing the two diagrams, (a) and (b), it is important to note that a diagram *either* has the double-headed arrow $W \longleftrightarrow X$ to indicate the unanalyzed correlation r_{WX}, *or* shows the connecting path $W \leftarrow T \rightarrow X$, which has the value $t't$, but never both. Since $r_{WX} = t't$, we may use either one of them but one only. A double-headed arrow between W and X in Fig. 146 (b) would be a redundancy which should be avoided in all path diagrams.

17. Having introduced the remote and uncorrelated variables V, T, U into Fig. 146 (b), we see that it is equivalent to saying that Y is determined by them, while W and X are merely intermediate observations. So the relationship may also be represented by Fig. 146 (c) without W or X. Let $\bar{v}, \bar{t}, \bar{u}$ be the direct path coefficients from V, T, U to Y, respectively, without considering the intermediate variables W and X. From the results established previously about chains of causes, we have

$$\bar{v} = v\,w, \qquad \bar{u} = u\,x. \tag{29}$$

We wish to find an expression for $\bar{t}$ in terms of the single-step path coefficients of Fig. 146 (b). The variable T influences Y through two routes:

PATH ANALYSIS OF CAUSAL SYSTEMS 149

one through W and one through X. Substituting (28) in (26), we obtain

$$w^2 + x^2 + 2wtt'x = 1 \qquad (30)$$

From the self-contained path diagram involving V, T, W only, and that involving T, U, X only, in Fig. 146 (b), we have

$$v^2 + t^2 = 1, \qquad\qquad t'^2 + u^2 = 1 \qquad (31)$$

and finally, from Fig. 146 (c),

$$\bar{v}^2 + \bar{t}^2 + \bar{u}^2 = 1 \qquad (32)$$

Using the last four expressions (29 - 32), we arrive at

$$\begin{aligned}
\bar{t}^2 &= 1 - \bar{v}^2 - \bar{u}^2 \\
&= (w^2 + x^2 + 2wtt'x) - (vw)^2 - (ux)^2 \\
&= w^2(1-v^2) + x^2(1-u^2) + 2wtt'x \\
&= w^2 t^2 + x^2 t'^2 + 2wtt'x = (wt + xt')^2
\end{aligned} \qquad (33)$$

giving the total fraction of determination of Y's variance by T. The direct path coefficient from T to Y without the intermediate variables W and X is thus

$$\bar{t} = wt + xt' \qquad (34)$$

Reading this expression in connection with diagrams (b) and (c) of Fig. 146 we see that the direct path $\bar{t}$ is the sum of the two paths, tw and $t'x$, linking T with Y. The direct path $\bar{t}$ is of course the correlation $r(T, Y)$. The result (34) confirms once again the basic theorem that the correlation between two variables is the sum of all connecting paths between the two variables. The correlations (27) may also be written in terms of Fig. 146 (b):

$$r_{YW} = w + tt'x, \qquad\qquad r_{YX} = x + t'tw \qquad (35)$$

Multiple correlation as a path

18. From the foregoing account of path diagrams and their analysis, we see that the diagrammatic representation of the relationships among the variables is quite flexible. A given relationship may be represented in more than one way, depending on the number of variables to be included in the diagram. Just as we may omit certain intermediate variables, or cut off some remote variables by unanalyzed correlations, we may also insert intermediate variables or extend the diagram to remote variables. As an illustration of the flexibility of the path method, it will be shown that a multiple correlation coefficient may also be used as a path coefficient by inserting a variable as an intermediate. Consider Fig. 119 which shows the regression of Y on X_1, X_2, X_3 and the uncorrelated residual U. The multiple correlation between Y and the group of X's is defined as the simple correlation between the actual Y and the calculated $\hat{Y} = G$. If the intermediate variable G is inserted into Fig. 119, it becomes Fig. 151 of this chapter. Note carefully that when an intermediate variable is inserted, it cuts the original path into two segments, each with its own (new) path coefficients. For example, the inserted G in Fig. 151 cuts the original path from X_1 to Y in Fig. 119 into two segments: one from X_1 to G with value p_{G1}, and one from G to Y with value p_{OG}. Explicitly, there are no longer p_{01}, p_{02}, p_{03} as such in Fig. 151. The old and new path coefficients are, by the theorem on chains, related in the following way:

$$p_{01} = p_{OG}\, p_{G1}, \qquad p_{02} = p_{OG}\, p_{G2}, \qquad p_{03} = p_{OG}\, p_{G3} \quad (36)$$

where p_{OG} is a common multiplier (factor of proportionality). The old and new correlation coefficients are related in a similar manner:

$$r_{01} = p_{OG}\, r_{G1}, \qquad r_{02} = p_{OG}\, r_{G2}, \qquad r_{03} = p_{OG}\, r_{G3} \quad (37)$$

PATH ANALYSIS OF CAUSAL SYSTEMS

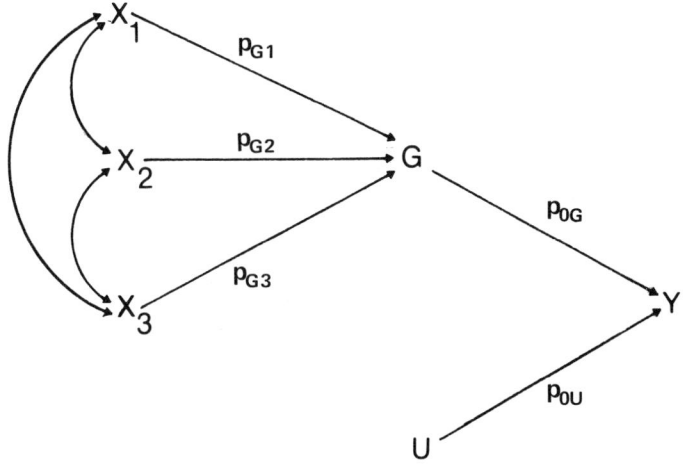

Fig. 151. Multiple correlation coefficient as a path coefficient. The present diagram should be compared with Fig. 119 to which it is equivalent. Thus, $p_{0G} p_{G1}$ here = p_{01} of Fig. 119. Now, the right portion of the diagram involving G, U, and Y is complete by itself, the three X-variables being reduced to one variable G. The path p_{0G} is the multiple correlation $R_{0(123)}$.

where, again, p_{0G} is the common multiplier. From the two relations above, we see that $p_{01} r_{01} = p_{0G}^2 \cdot p_{G1} r_{G1}$, etc.

19. Fig. 151 consists of two self-contained path diagrams, and each may be analyzed separately. The lower right portion involving G, U, and Y only, is a complete diagram by itself, equivalent to Fig. 107, showing the complete determination of Y by G and U. Hence

$$p_{0G}^2 + p_{0U}^2 = r_{0G}^2 + r_{0U}^2 = 1 \qquad (38)$$

where $\qquad p_{0G} = r_{0G} = r(Y, \hat{Y}) = R_{0(123)} \qquad (39)$

is the path coefficient from G to Y. Thus it has been shown that a multiple correlation coefficient may also be used as a path coefficient when we choose to ignore the detailed separate contributions from each of the X's and treat the group of X's as one variable.

20. The upper left portion of Fig. 151, involving the X's and G only, is also a self-contained diagram depicting the complete determination of G by X_1, X_2, X_3. In this respect, it is a generalization of Fig. 146 (a) with only two determining factors. When we analyze this upper left portion of Fig. 151 as a separate path diagram, we will be considering the fractions of determination of the variance of G, not of Y. The path coefficients involved are p_{G1}, p_{G2}, p_{G3} not p_{01}, p_{02}, p_{03}. From the account of Chapter 5, we know that there are nine round trips that may be made from G to G. The nine terms corresponding to the nine round trip routes may be condensed into three terms by using correlations. We shall use the latter form for brevity. Thus, for the complete determination of G,

$$p_{G1} r_{1G} + p_{G2} r_{2G} + p_{G3} r_{3G} = 1 \qquad (40)$$

Now we may put the two sub-diagrams together and read the double diagram, Fig. 151, as a whole. Multiplying the expression (40) by p_{0G}^2 will give us the fraction of determination of Y's variance by G. That is, from the relations (36) and (37),

$$\begin{aligned} p_{0G}^2 &= p_{0G}^2 \left(p_{G1} r_{1G} + p_{G2} r_{2G} + p_{G3} r_{3G} \right) \\ &= p_{01} r_{10} + p_{02} r_{20} + p_{03} r_{30} = R_{0(123)}^2 \end{aligned} \qquad (41)$$

which is identical with (46) in Chapter 5. The insertion of an intermediate variable into a path diagram does not change the consistency of the path system. The final results are the same, although the intermediate variable introduces some intermediate steps in the analysis.

PATH ANALYSIS OF CAUSAL SYSTEMS

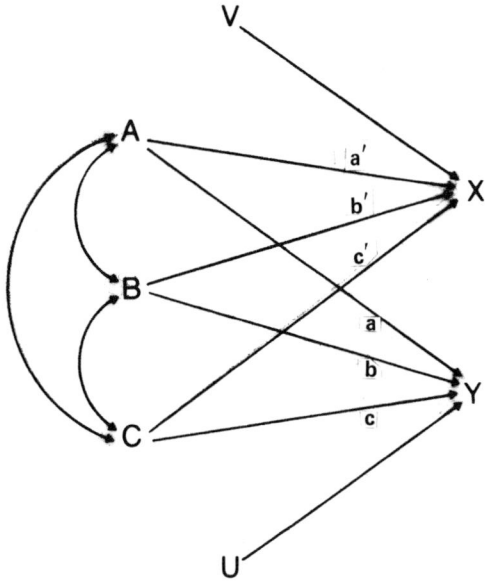

Fig. 153. Correlation between two linear functions. The diagram is the same as Fig. 142 (iii), except that now the common causes A, B, C are correlated. The diagram may also be regarded as the linking of two multiple regression diagrams, each being that of Fig. 119. There are nine connecting chains between X and Y.

Correlation between linear functions

21. Speaking of correlated common causes, we still have to generalize diagram (iii) of Fig. 142, in which the common factors A, B, C are uncorrelated. Now, three double-headed arrows should be added to Fig. 142 (iii) for the correlations r_{AB}, r_{BC}, r_{AC}. The result of doing so is Fig. 153. Although it may be regarded as a fairly general case for correlated causes contributing to the correlation between two variables, it is still a special case of the correlation between two linear functions in general. However, we shall study the simpler case first, as it is a direct extension of Fig. 142 (iii).

The notation in Figs. 142 and 153 is the same. However, the three common causes could be designated by, say, A_1, A_2, A_3 in lieu of A, B, C. Then the notation for the three correlations and the six path coefficients may be converted as follows:

$$(r_{AB}, r_{BC}, r_{AC}) = (r_{12}, r_{23}, r_{13})$$

$$(a', b', c'; a, b, c) = (p'_1, p'_2, p'_3; p_1, p_2, p_3)$$

As different books use different notations, one should be able to switch freely from one notation system to another. For a more general treatment than that indicated in Fig. 153, see Ex. 7 at the end of this chapter.

22. In studying the system shown in Fig. 142 (iii), it was noted that the uncorrelated variables V and U do not contribute to the correlation r_{XY} since they are not common causes and they are not correlated with each other. For the same reason, the variables V and U in Fig. 153 do not contribute to correlation r_{XY} either. They may be ignored in the process of finding the correlation between X and Y. At this stage, it is useful to point out the similarities and differences between Fig. 153 and the diagram for regression on three variables, Fig. 119 of Chapter 5. If X, and consequently all arrows leading to X, were deleted from Fig. 153, the resultant diagram would be identical with Fig. 119, the multiple regression diagram in Chapter 5. Similarly, if Y and all arrows leading to Y are deleted, the result is another diagram for multiple regression. Therefore, Fig. 153 is the linking of two multiple regression diagrams into one. Now, if we employ the method of Fig. 151 to reduce A, B, C into one variable, and let the multiple correlation coefficient between X and (ABC) be $R_{X(ABC)}$, it follows immediately that the correlation between X and Y is simply

$$r_{XY} = R_{X(ABC)} \cdot r(\hat{X}, \hat{Y}) \cdot R_{Y(ABC)} \tag{42}$$

In the following few paragraphs, however, the total correlation r_{XY} will be decomposed into finer additive components, so that we can evaluate the relative contributions of the common factors A, B, C.

PATH ANALYSIS OF CAUSAL SYSTEMS

23. For each multiple regression diagram like the one in Fig. 119, the reader will remember that there are nine round trips that may be made from Y to Y, not counting the one $Y - U - Y$. Now, turning to our present Fig. 153, we see that there are nine connecting paths between X and Y; we shall write out the nine terms later. In both cases there are nine terms, similar in appearance too, but their meaning is very different. The difference between these two sets of terms should be clear in view of what we have developed about connecting paths and fractions of determination. To crystallize the difference, however, we shall repeat once more the principle involved by the simplest example:

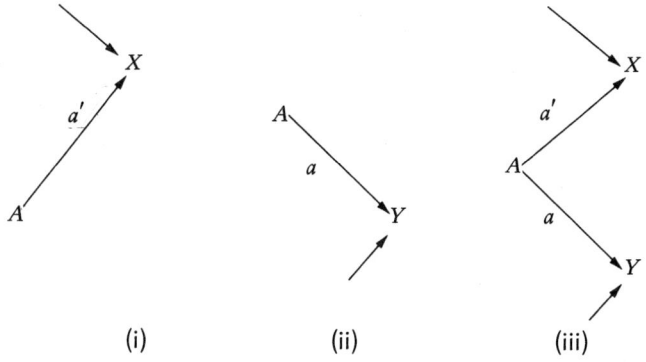

(i)　　　　　　　　(ii)　　　　　　　(iii)

From (i), $a' = p_{XA} = r_{XA}$, determination $d_{XA} = r_{XA}^2 = a'^2$

From (ii), $a = p_{YA} = r_{YA}$, determination $d_{YA} = r_{YA}^2 = a^2$

In (i), we are dealing with the relationship between A and X. In (ii), we are dealing with the relationship between A and Y. The coefficient of determination of X by A is obtained by making the round trip from X to A and then from A to X. Following the rule of tracing connecting paths, first backward and then forward, the coefficient of determination of Y by A is obtained in the same way. The figure (iii) is obtained by linking (i) and (ii) together to form a more comprehensive system. From (iii), we may find, in addition to the values of a' and a and their squares, the correlation coefficient $r_{XY} = a'a$. Hence, the regression diagram of Fig. 119 corresponds to our present (ii) and permits us to calculate $R_{0(123)}^2$; and the diagram in Fig. 153 corresponds to our present (iii) and permits us to calculate the correlation r_{XY}.

24. Now we proceed to find the correlation between X and Y in Fig. 153. To reduce the algebra to its simplest form, we let all the variables be standardized from the beginning, so that $E(A^2) = \sigma_A^2 = 1$ and $E(AB) = \sigma_{AB} = r_{AB}$, etc. Then

$$X = a'A + b'B + c'C + V$$
$$Y = aA + bB + cC + U \tag{43}$$

where a', b', c', and a, b, c are path coefficients as indicated in the diagram. The correlation between the two linear functions X and Y is

$$r_{XY} = E(XY) = E(a'A + b'B + c'C + V)(aA + bB + cC + U) \tag{44}$$

Since all cross-products involving V or U have zero expectation, the correlation is simply

$$\begin{aligned} r_{XY} &= E(a'A + b'B + c'C)(aA + bB + cC) \\ &= a'a \quad\quad + \quad a'r_{AB}b \quad + \quad a'r_{AC}c \\ &\quad + b'r_{BA}a \quad + \quad b'b \quad\quad + \quad b'r_{BC}c \\ &\quad + c'r_{CA}a \quad + \quad c'r_{CB}b \quad + \quad c'c \end{aligned} \tag{45}$$

Reading these nine terms in the light of Fig. 153, we see that these are the nine connecting paths between X and Y, completing the proof that a correlation between two variables is the sum of all connecting paths between them. Comparing these nine terms with those of Chapter 5 (**45** or **47**) for $R_{0(123)}^2$, we see that one of the path coefficients in each term here has acquired a prime, indicating that the path goes to X, instead of going back to Y. Hence, if the primes are deleted from a', b', c' in (45), we will obtain $R_{Y(ABC)}^2$. Similarly, if primes are added to a, b, c in (45), we will obtain $R_{X(ABC)}^2$. To summarize (Fig. 153):

PATH ANALYSIS OF CAUSAL SYSTEMS

$$r_{XY} = a'a + a' r_{AB} b + \ldots$$

$$R^2_{Y(ABC)} = a a + a r_{AB} b + \ldots \quad (46)$$

$$R^2_{X(ABC)} = a'a' + a' r_{AB} b' + \ldots$$

Those who are familiar with matrix algebra will recognize that the correlation (45) is the bilinear form:

$$r_{XY} = (a', b', c') \begin{pmatrix} 1 & r_{AB} & r_{AC} \\ r_{BA} & 1 & r_{BC} \\ r_{CA} & r_{CB} & 1 \end{pmatrix} \begin{pmatrix} a \\ b \\ c \end{pmatrix}$$

while $R^2_{Y(ABC)}$ and $R^2_{X(ABC)}$ are the two corresponding quadratic forms. Matrices will not be used in the sequel. We mention them here in the hope of bringing the method of path coefficients closer to traditional statistics.

25. The nine components of r_{XY} in (45) are arranged in systematic order, so that they may be condensed into three terms by taking the three row-totals or the three column-totals. For instance, the sum of the three in the first row of (45) is

$$a' (a + r_{AB} b + r_{AC} c) = a' r_{AY} = p_{XA} r_{AY} \quad (47)$$

as the three terms within the brackets of (47) represent the three connections between A and Y in Fig. 153. Similarly, the sum of the three terms of the first column of (45) is

$$a (a' + b' r_{BA} + c' r_{CA}) = a r_{AX} = p_{YA} r_{AX} \quad (48)$$

Thus, we have arrived at the more condensed formulas; they are based on row and column totals of (45), respectively:

$$r_{XY} = p_{XA} r_{AY} + p_{XB} r_{BY} + p_{XC} r_{CY} = \Sigma p_{Xi} r_{iY} \qquad (49)$$

$$r_{YX} = p_{YA} r_{AX} + p_{YB} r_{BX} + p_{YC} r_{CX} = \Sigma p_{Yi} r_{iX} \qquad (50)$$

where i stands for a common factor. These are the basic formulas for the correlation between two effects due to common causes. Note that the expressions (49) and (50) are symmetrical (being of the same form) but they are not equal term by term; that is, the first term $p_{XA} r_{AY}$ of (49) is not necessarily equal to the first term $p_{YA} r_{AX}$ of (50). In other words, (49) and (50) represent two different partitions of the same total correlation $r(X, Y)$. Hence, we cannot say that $p_{XA} r_{AY}$ or $p_{YA} r_{AX}$ is the total contribution from variable A to correlation $r(X, Y)$. The specific contributions to $r(X, Y)$ are given by the nine terms of (45). If the variable A is eliminated from Fig. 153, then the five (not three) terms involving either $a' = p_{XA}$ or $a = p_{YA}$ in (45) will be dropped out and only four terms will be left. A clear-cut subdivision of the contributions of the common causes is possible on two occasions. One is when X and Y are similarly affected by the causes A, B, C, in which case $(a', b', c') = (a, b, c)$ and the row and column totals of (45) will be the same. Another occasion is that when all the causes are uncorrelated, in which case $r(X, Y) = a'a + b'b + c'c$, identical with (21).

26. As noted before, the correlations r_{AB}, r_{BC}, and r_{AC} in Fig. 153 are unanalyzed ones. These correlations are probably due to remote common factors behind them (not shown in the diagram). In fact, there must be an endless string of factors, one set behind another. We will never know all the factors. In any given analysis there must be a limit as to how far back we can or should go. The unanalyzed correlations mark the limit of the analysis; they summarize all the effects of the remote and unknown causes. Even when some of the remote causes are known, we may choose to use the unanalyzed correlations as cut-off points to simplify the diagram. The important thing to remember is that when we do this, it will not affect the consistency and validity of the analysis of the remaining diagram. It is this property that makes the method of path analysis highly flexible and applicable in a wide range of situations.

27. The two linear functions (43) we have been studying are special cases in that they are functions of the same variables, except V and U which do not contribute to the correlation between the two functions X and Y. More generally, the number of variables involved in the two linear functions need not be the same, nor need the variables appear in both functions. To illustrate, consider the following two linear functions:

$$X = p'_1 A_1 + p'_2 A_2 + p'_3 A_3$$
$$Y = p_2 A_2 + p_3 A_3 + p_4 A_4 + p_5 A_5 \qquad (51)$$

where the variables have been standardized and the p's are path coefficients. There are three variables in X and four variables in Y. Two of these variables appear in both X and Y; they are the common factors. The correlation between the two linear functions is

$$r_{XY} = E(XY) = E(p'_1 A_1 + p'_2 A_2 + p'_3 A_3)(p_2 A_2 + p_3 A_3 + p_4 A_4 + p_5 A_5) \qquad (52)$$

The reader should write out the 12 terms and draw the corresponding path diagram as an exercise. In doing so, he will notice that two of the terms are $p'_2 p_2$ and $p'_3 p_3$; the other ten terms are $p'_1 r_{12} p_2$, etc. where $r_{ij} = r(A_i, A_j)$. The contribution to r_{XY} of these ten terms depends solely on the correlations r_{ij} among themselves. The terms of the type $p'_i p_i$ are due to the fact that A_i is a common factor. A common factor always contributes to r_{XY} regardless of the values of r_{ij}. It follows that if all $r_{ij} = 0$, the correlation between X and Y depends on their common factors only.

28. Correlation between two sums of independent variables of equal variance. A direct application of the considerations in the previous paragraph is to determine the correlation between two special linear functions —two sums of variables. Let

$$X = A_1 + \ldots + A_k + \ldots + A_m; \qquad Y = A_1 + \ldots + A_k + \ldots + A_n \qquad (53)$$

where X is the sum of m factors and Y is the sum of n factors. Further, let us assume that these factors are all uncorrelated, $r_{ij} = r(A_i, A_j) = 0$, and all have the same variance, $\sigma_i^2 = \sigma^2$. Let k of these factors appear in both X and Y. Without loss of generality, we may take the first k as the common factors of X and Y. The remaining $m-k$ factors in X and the remaining $n-k$ factors in Y are different A's. We wish to find the correlation $r(X, Y)$ under these circumstances. From (53),

$$\sigma_X^2 = \sigma_1^2 + \ldots + \sigma_m^2 = m\sigma^2; \qquad \sigma_Y^2 = \sigma_1^2 + \ldots + \sigma_n^2 = n\sigma^2$$
$$\mathrm{Cov}\,(X, Y) = \sigma_1^2 + \ldots + \sigma_k^2 = k\sigma^2 \qquad (54)$$

so that (Wright, 1968, I:309)

$$r_{XY} = \frac{\mathrm{Cov}\,(X, Y)}{\sigma_X \, \sigma_Y} = \frac{k}{\sqrt{m}\,\sqrt{n}} \qquad (55)$$

which is the general formula for the correlation between two sums of independent variables of equal variance. Several interesting special cases may be noted. If X and Y have the same number of factors, we have

$$m = n, \qquad r_{XY} = \frac{k}{n} \qquad (56)$$

which is the proportion of the factors that are common to X and Y. For instance, if half of the factors are common to both sums,

$$m = n = 2k, \qquad r_{XY} = \frac{1}{2} \qquad (57)$$

If X has twice as many factors as Y does, and all Y's factors are in X,

$$m = 2n = 2k, \qquad r_{XY} = \frac{n}{\sqrt{2n}\,\sqrt{n}} \frac{1}{\sqrt{2}} = 0.7071 \qquad (58)$$

The last two results are particularly useful in understanding the correlations between relatives due to additive and independent common genetic factors of equal importance.

PATH ANALYSIS OF CAUSAL SYSTEMS

29. Since **(50)**, derived from Fig. 153, is a general theorem in the theory of path coefficients, other formulas may be derived from it by special substitutions. Replacing the common causes A, B, C, by the more general notation A_1, A_2, A_3, and using subscript 0 for Y, we may rewrite **(50)** in the following form:

$$r_{OX} = p_{01} r_{1X} + p_{02} r_{2X} + p_{03} r_{3X} \quad (50')$$

where X is a variable correlated with the A's. Consider Fig. 153 once more. If we move X toward $A = A_1$ until it coincides with A, then Fig. 153 becomes identical with that for multiple regression, Fig. 119. Analytically, this is equivalent to putting $X = A = A_1$ in **(50')**. Then $r_{1X} = r_{11} = 1$. Similarly, if we put $X = B = A_2$, a new expression results with $r_{2X} = r_{22} = 1$. Thus, by putting $X = A_1, A_2, A_3$, successively, the general expression **(50')** becomes, respectively,

$$\left. \begin{aligned} r_{01} &= p_{01} + p_{02} r_{21} + p_{03} r_{31} \\ r_{02} &= p_{01} r_{12} + p_{02} + p_{03} r_{32} \\ r_{03} &= p_{01} r_{13} + p_{02} r_{23} + p_{03} \end{aligned} \right\} \quad (59)$$

which, the reader will recognize, are precisely the standardized normal equations of Chapter 5 **(44)**. Thus, multiple correlation is a special case of the correlation between two linear functions.

Review of rules for reading path diagrams

30. We have in one place or another covered all the rules for tracing connecting paths of a causal diagram. Identification of connecting paths is a prerequisite of obtaining a correlation coefficient, as the latter is the sum of all connecting paths between two variables. This section is necessarily redundant but it summarizes all the rules in one place by examining one diagram (Fig. 162) which has the various features to illustrate the various rules. In Fig. 162, we dispense with the letters, for ease of reading, and

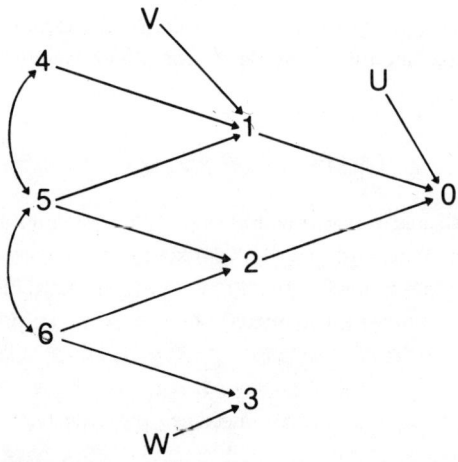

Fig. 162. A fairly complicated network of relations to illustrate the various rules of reading a path diagram. The variables denoted by numerals (0, 1, 2, 3, 4, 5, 6) are observed ones, while U, V, W are residuals. (Wright, 1934)

use the numerals for the variables, so that the path coefficient from variable 1 to variable 0 is p_{01}. The residuals are still denoted by letters, U, V, W. Let us proceed from the simple to the more complicated cases. All statements and formulas refer to Fig. 162.

(i) First, we note that $r_{46} = 0$. This illustrates the rule that no connecting path should use more than one correlation coefficient. The existence of r_{45} and r_{56} does not connect 4 with 6. Since we must begin tracing by going backward along an arrow, we see that there are no other paths in the diagram that connect 4 with 6.

(ii) The next simple thing to observe is that $p_{36} = r_{36}$. All the other variables determining 3 are uncorrelated with 6, as shown by the arrow from W to 3. The path coefficient from an uncorrelated cause to an effect is the respective correlation coefficient. Similarly, $r_{0U} = p_{0U}$.

(iii) But p_{14} is not equal to r_{14}, since cause 4 is correlated with other cause(s) of 1. There are two connecting paths between 1 and 4. Thus,

$$r_{14} = p_{14} + p_{15} r_{54}$$

(iv) A compound path through an intermediate variable is the product of the two elementary paths along the route of connection. Thus, the path coefficient from *4* to *0* is

$$p_{04} = p_{014} = p_{01} p_{14}$$

where the middle subscript *1* reminds us that the intermediate variable is *1*. Note carefully that it is the path coefficient p_{04}, not correlation r_{04}, as cause *4* is correlated with other causes leading to *0*. The new notation p_{014} is introduced to help us to read the diagram.

(v) The chain property exemplified above may be extended to any number of variables, permitting one change in direction. For instance,

$$p_{015} = p_{01} p_{15}$$

$$p_{0152} = p_{01} p_{15} p_{25}$$

The chain may include one correlation, whether it appears at the end or in the middle of a chain. Examples:

$$p_{0145} = p_{01} p_{14} r_{45}$$

$$p_{01452} = p_{01} p_{14} r_{45} p_{25}$$

Wright introduced more elaborate symbols to distinguish a change of direction in the arrows from a two-headed correlation. Here the subscripts *01452* are used only to guide the reader in following the path. No matter how long a connecting path may be, it cannot pass through the same variable more than once. That is, the subscripts *01452*... should not have duplicate numerals.

(vi) The path coefficient from *5* to *0* involves two intermediate variables, *1* and *2*. The total path from *5* to *0* is then the sum of the two connecting paths:

$$p_{05} = p_{015} + p_{025} = p_{01} p_{15} + p_{02} p_{25}$$

Again, it is the path, not the correlation coefficient, as variable 5 is correlated with other causes leading to 0.

(vii) We have to be careful in tracing the connection between 1 and 3. The zig-zag line $1-5-2-6-3$ is not permissible because it violates the rule that there shall be no "first forward and then backward movement." All movements are first backward and then forward. The part $1-5-2$ is all right by itself, and so is the part $2-6-3$ by itself. But we cannot connect these two parts, because it would involve the movement $5-2-6$, which is the forbidden (forward-backward) step. The proper connection between 1 and 3 is through the correlation $r(5, 6)$. This is the only connection between 1 and 3, as all other factors leading to 1 and 3 are uncorrelated (no connection between W and V). Hence the connecting path is also the correlation coefficient:

$$r_{13} = p_{15}\, r_{56}\, p_{36}$$

(viii) Being familiar by now with the methods of tracing connections, the reader should have no difficulty in seeing the following results, each correlation being the sum of two, three, or four connecting paths.

$$r_{23} = p_{26}\, p_{36} + p_{25}\, r_{56}\, p_{36}$$

$$r_{12} = p_{15}\, p_{25} + p_{14}\, r_{45}\, p_{25} + p_{15}\, r_{56}\, p_{26}$$

$$r_{02} = p_{02} + p_{0152} + p_{01452} + p_{01562}$$

$$r_{05} = p_{015} + p_{025} + p_{0145} + p_{0265}$$

(ix) For every self-contained portion of the path diagram (Fig. 162) we may calculate the fractions of determinations of the effect under consideration; these fractions add up to unity, expressing complete determination of that variable. The following expressions are for the determinations of variables 3, 2, 1, and 0, respectively.

$$p_{3W}^{2} + p_{36}^{2} = 1$$

$$p_{25}^{2} + p_{26}^{2} + 2p_{25}\, p_{26}\, r_{56} = 1$$

$$p_{1V}^2 + p_{14}^2 + p_{15}^2 + 2p_{14}p_{15}r_{45} = 1$$

$$p_{0U}^2 + p_{01}^2 + p_{02}^2 + 2p_{01}p_{02}r_{12} = 1$$

where r_{12} is as given in (viii) above. In practice we must have enough observations on the various correlation coefficients so that the values of the path coefficients may be found from these equations.

Viewpoints and interpretations

31. It should be emphasized that the formulas involving path coefficients, unlike mathematical expressions in physics, have no absolute meaning, as they do not describe the relationships among the variables in any absolute sense. They describe the relationships only from a particular point of view taken by the investigator who thinks that his viewpoint makes sense to him. All of the expressions in the last section are true only with respect to the viewpoint expressed by Fig. 162. Without the accompanying diagram, those expressions would become unreadable, and, in fact, meaningless. Should the seven variables $(0, 1, \ldots, 6)$ be rearranged in some other way, the absolute measurements (the correlations among the variables) would remain the same, but the expressions involving path coefficients would be no longer true. For instance, the value of r_{36} would remain the same after the rearrangement of the variables but it would no longer be equal to p_{36}, which depends on the new arrangement. The relationship $r_{36} = p_{36}$ is a consequence of the particular arrangement (Fig. 162) of the seven variables and is not an intrinsic property of the variables *3* and *6*. Therefore all expressions involving path coefficients must always be read with respect to a specified path diagram.

32. The interpretation of the results of analysis is, of course, also based on the viewpoint expressed by the diagram. A diagram represents a set of assumptions or tentative hypotheses relating the direction and strength of influences among the variables. Since many path diagrams may be constructed with a given number of variables, it is the responsibility of the investigator to choose one or more (preferably more) to serve as starting points for analysis and interpretation. The investigator should always be ready for new diagrams

before he obtains a consistent and reasonable interpretation of the causal system within the limit of his data. There are many expensive instruments for scientific research, but trial and error remains the most important single tool at the disposal of a scientist.

33. Mathematically speaking, any viewpoint that might be taken by an investigator is entirely arbitrary. When properly analyzed, each path diagram will yield a set of consistent conclusions for that particular causal system. Since beginners frequently have difficulty in grasping this seemingly abstract point, it is desirable to supply a very simple concrete example to clarify the situation. Of the several examples I can think of, none can match Wright's original example in simplicity and clarity. Besides, it has other features which bring out additional properties of path coefficients. Let A and B be two uncorrelated variables of equal variance; $\sigma_A^2 = \sigma_B^2 = \sigma^2$, and $r(A, B) = 0$. Then let S be the sum of the two variables, $S = A + B$ with variance $\sigma_S^2 = \sigma_A^2 + \sigma_B^2 = 2\sigma^2$. The covariance of S and A (or B) is $Cov(S, A) = Cov(S, B) = E(SA) = E(A+B)A = \sigma_A^2 = \sigma^2$, so that the correlation between S and A (or B) is

$$r_{SA} = r_{SB} = \frac{Cov(S, A)}{\sigma_S \, \sigma_A} = \frac{\sigma^2}{\sqrt{2}\, \sigma \cdot \sigma} = \frac{1}{\sqrt{2}} = 0.7071 \quad (60)$$

All of these should be thoroughly familiar to the reader by now. So far, only the data without any path analysis have been described. The values of the three correlations, $r_{SA} = r_{SB} = 1/\sqrt{2}$ and $r_{AB} = 0$, will not change, no matter what path analysis we might make.

34. Let us carry out a path analysis of the data on A, B, and S. First, consider the model $S = A + B$ as shown in Fig. 167 (i). Since A and B are uncorrelated, we have

$$\left. \begin{array}{rcccccc} a & = & p_{SA} & = & r_{SA} & = & 1/\sqrt{2} \\ b & = & p_{SB} & = & r_{SB} & = & 1/\sqrt{2} \end{array} \right\} a = b \quad (61)$$

and
$$a^2 + b^2 = 2a^2 = 1$$

PATH ANALYSIS OF CAUSAL SYSTEMS

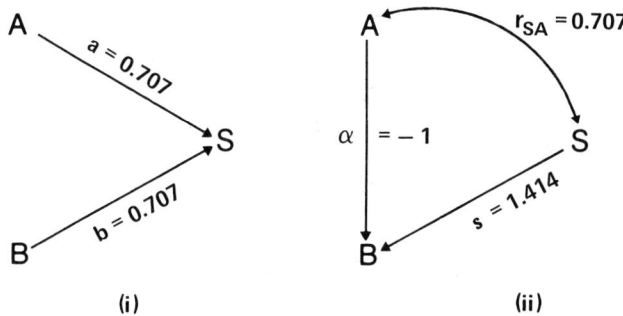

Fig. 167. Two different viewpoints on the same three variables. Left: $S = A + B$, where the causes A and B are uncorrelated. Right: $B = S - A$, where the causes S and A are correlated. Note that the path from S to B is $s > 1$, and the path from A to B is $\alpha < 0$.

Now, let us carry out the path analysis of an alternative model. Mathematically, it is equally sound to regard B as an effect determined by the causes S and A. Then the path diagram would be that shown in Fig. 167 (ii) under the model $B = S - A$. A rearrangement of the variables is equivalent to changing the position and direction of the arrows. Let α be the path coefficient from A to B and s be that from S to B, as shown in Fig. 167 (ii). Then B is determined now by two correlated causes (S and A) with $r(S, A) = 1/\sqrt{2}$. To find the values of these path coefficients, we recall that $p_{ji} = b_{ji} \sigma_i/\sigma_j$, implying that the p and the corresponding b always have the same sign. Under the model $B = S - A$, the regression coefficients are $b_{BS} = +1$ and $b_{BA} = -1$. Hence the path coefficients are

$$s = p_{BS} = (+1) \frac{\sigma_S}{\sigma_B} = \frac{\sigma\sqrt{2}}{\sigma} = \sqrt{2} = 1.4142$$

$$\alpha = p_{BA} = (-1) \frac{\sigma_A}{\sigma_B} = -1 \qquad (62)$$

For the complete determination of B, we have

$$\alpha^2 \quad + \quad s^2 \quad + \quad 2\alpha s\, r_{SA} \quad = \quad 1 \tag{63}$$

i.e., $\quad\quad 1 \quad + \quad 2 \quad + \quad 2(-1)\sqrt{2}\left(\dfrac{1}{\sqrt{2}}\right) \;=\; 1$

correctly. As a final check for the correctness of the results (62) and (63), we read from Fig. 167 (ii)

$$r_{AB} = \alpha + s\, r_{SA} = -1 + \sqrt{2}\left(\dfrac{1}{\sqrt{2}}\right) = 0$$

$$r_{SB} = s + \alpha r_{SA} = \sqrt{2} - \dfrac{1}{\sqrt{2}} = \dfrac{1}{\sqrt{2}} \tag{64}$$

which are both correct. Thus, the alternative model $B = S - A$ yields perfectly consistent results. Note carefully that $b = p_{SB}$ in model (i) is not equal to $s = p_{BS}$ in model (ii). In this particular example, one is the reciprocal of the other. Also note that $r(A, B) = 0$ in model (ii) is due to the cancelling off of two equally important forces along the two connecting paths. Although $r(A, B) = 0$, the path coefficient $\alpha = p_{BA}$ still exists. (For those who need reinforcement on the various points we have discussed but dislike irrational numbers, an exercise is provided at the end of the chapter in which all correlation and path coefficients are rational numbers.)

35. The important lesson we learn from the preceding example (Fig. 167) is that when one (no matter which one of several possibilities) path diagram is chosen, all subsequent analyses must be done according to that diagram and the path coefficients must be attached to the arrows of that diagram only. When another path diagram is chosen, a new set of path coefficients will be obtained which belong to the new diagram only. Path coefficients vary from diagram to diagram and are not transferable from one to another, as they do not describe any absolute relationships among the variables. Be sure that the two diagrams in Fig. 167 are based on the same set of data: $r(A, B) = 0$, and $r(S, A) = r(S, B) = 0.707$, but analyzed

from different viewpoints. We cannot say that one of the diagrams in Fig. 167 is right and the other is wrong. It is not a matter of right and wrong here. They are both correct, in the sense of being consistent, once the viewpoint is specified.

36. The results of path analysis of model (ii), $B = S - A$, as shown in Fig. 167 (ii) give us additional insight into the properties and meaning of path coefficients. It should be noted that the path coefficient from S to B is $p_{BS} = s = \sqrt{2} = 1.414$. This shows that a path coefficient, unlike a correlation or a partial correlation coefficient, may take values greater than unity. My experience shows that some students regard this property as a defect of path coefficients because it does not have a "scale" ranging from 0 to 1. Hence we should clarify this point here so that the meaning of a path coefficient greater than unity may be properly appreciated. First, as a regression coefficient of one standardized variable on another, there is nothing in the mathematics to limit its numerical values to a certain range. Concrete regression coefficients have no numerical limits either. Only the symmetrical and absolute measurements such as correlation coefficients have a numerical limitation. Second, the meaning of $p > 1$ will be clear if we examine the expression (63) for complete determination of an effect. In general, if p_1 and p_2 are the two path coefficients from causes 1 and 2 to an effect O, then

$$p_1^2 + p_2^2 + 2p_1p_2r_{12} = 1 \qquad (63')$$

The terms p_1^2 and p_2^2 are always positive. If one of the p's is greater than unity, then the term $p_1p_2r_{12}$ has to be negative. In our example, the reader will recall (Fig. 167) that the path from A to B is $p_{BA} = \alpha = -1$. The expression (63) or (63') is nothing more than the standardized form of the familiar formula for the variance of $y = x_1 - x_2$:

$$\sigma_y^2 = \sigma_1^2 + \sigma_2^2 - 2\sigma_1\sigma_2r_{12} \qquad (63'')$$

Dividing both sides by σ_y^2, we obtain the three fractions of determinations of the variance of y through the three different routes of influence toward y. The negative fraction of determination simply means that if the two variables x_1 and x_2 exert their influences on y independently and maintain their observed magnitude of variance (above), then the variance of y would be larger than the observed. This was mentioned in discussing multiple correlation in the last chapter. Consequently, whenever there is one $p_{ji} > 1$ in a causal system, there must exist a compensatory variable in the system to diminish the variance of y. In other words, the presence of a path coefficient greater than unity in a causal system is a sure signal for the existence of a compensatory mechanism (negative influence) in that system. When viewed this way, the property that a path coefficient may take values greater than unity is not only meaningful but of practical importance.

Partial correlation and path coefficients

37. Consider the three intercorrelated variables A, B, C. A common practice for "testing" their relationships is to calculate the partial correlation coefficient between two of them, holding the third constant. Suppose that $r_{AC \cdot B}$ vanishes or becomes small, the conclusion is that factor B must have caused or is responsible for the correlation between A and C. The reasoning for this testing procedure is apparently borrowed from the methodology of experimentation. In most experimental designs, the investigator will "control" by holding constant all sources of variation except the variable(s) he is studying. In this way, the experimenter will be able to assess the effects of these variables under specified conditions. But the analogy is not exact. A study of simultaneously intercorrelated variables is not the same as a study by designed experiments; nor is the calculation of a partial correlation equivalent to controlling an extraneous source of variation. The interpretative value of a partial correlation coefficient (a symmetrical measurement) is limited (see below). Historically, the unsatisfactory nature of interpretation on the basis of partial correlation is one of the reasons that led Wright to develop the directional path coefficient.

38. The remarks above are not meant to imply that a partial correlation coefficient is useless and should not be used. It may be useful in certain circumstances, especially when used in conjunction with some structural model. What we have said above is that when $r_{AC \cdot B}$ is much smaller than r_{AC}, the usual assertion that B is an intermediary between A and C is not always true. A path analysis will quickly exhaust all the possibilities,

PATH ANALYSIS OF CAUSAL SYSTEMS

including the case when $r_{AC \cdot B}$ does not differ much from r_{AC}. Suppose that the data on $A, B, C,$ yield the following correlations (Wright, 1934):

$$r_{AB} = 0.50, \qquad r_{BC} = 0.50, \qquad r_{AC} = 0.25 \qquad (65)$$

The partial correlation between A and C, holding B constant, is (Chapter 4)

$$r_{AC \cdot B} = \frac{r_{AC} - r_{AB}\, r_{BC}}{\sqrt{1 - r_{AB}^2}\, \sqrt{1 - r_{BC}^2}} = \frac{0.25 - 0.50(0.50)}{\sqrt{0.75}\, \sqrt{0.75}} = 0 \qquad (66)$$

indicating that B is somehow playing a role in correlating A with C. The path analysis, on the other hand, is primarily concerned with a causal scheme that is consistent with the *actually observed* results (65) themselves.

39. When there are three intercorrelated variables, the possible partitions of 3 are *(1, 1, 1), (1, 2), (2, 1),* and *(3, 0),* if we count the latter as a partition. In order to be consistent with the observed correlations (65), the path diagrams shown in Fig. 171 may be constructed.

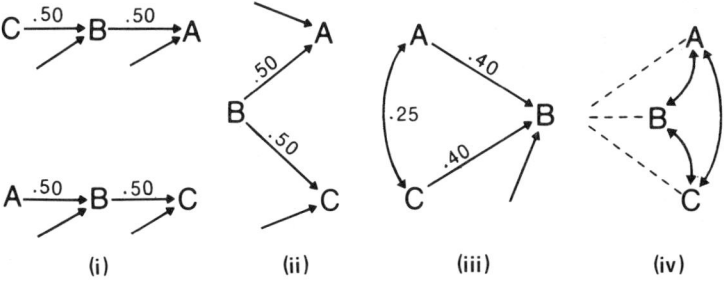

Fig. 171. Four possible interpretations of the observed correlations: $r_{AB} = 0.50$, $r_{BC} = 0.50$, $r_{AC} = 0.25$. (i) B is intermediate between A and C, in two possible sequences. (ii) B is a common factor of A and C. (iii) B is a joint product of A and C. (iv) The correlations are due to remote unknown common factors. (Wright, 1934)

(i) For the *(1, 1, 1)* chain model, Fig. 171 (i), there are two possible sequences, both being consistent with the observed correlations **(65)**, as $p_{AC} = p_{AB}\, p_{BC} = r_{AB}\, r_{BC} = 0.25 = r_{AC}$. The calculated result $r_{AC \cdot B} = 0$ in **(66)** is compatible with this path diagram.

(ii) For the *(1, 2)* partition to be consistent with observed correlations **(65)**, the path diagram in Fig. 171 (ii) will suffice. Since B is a common cause, we have $r_{AC} = p_{AB}\, p_{CB} = 0.25$. Again, the calculated value $r_{AC \cdot B} = 0$ is compatible with the common cause interpretation. The analytical derivation of the formula for partial correlation is actually based on the common cause model (**Ex. 4**).

(iii) For the *(2, 1)* partition pattern to be consistent with the observed facts **(65)**, the path diagram in Fig. 171 (iii) is correct. It is shown that A and C are correlated causes with $r_{AC} = 0.25$; the path coefficients from A and C to B are $a = 0.40$ and $c = 0.40$, respectively. This path diagram is perfectly consistent with the observed correlations **(65)**, as

$$r_{AB} = a + c\, r_{AC} = 0.40 + 0.40(0.25) = 0.50$$

Then the calculated result $r_{AC \cdot B} = 0$ **(66)** makes no sense. If Fig. 171 (iii) is in fact the true state of affairs yielding the observed correlations **(65)**, a calculation based on the partial correlation formula would have led us to the wrong conclusion that B is a cause for the correlation or an intermediary between A and C. It is not; it is the joint effect of A and C. The correlation r_{AC} exists whether B is observed or not. Thus we see that $r_{AC \cdot B} = 0$ does not necessarily imply that B plays any role in the correlation r_{AC} after all.

(iv) With a set of observed correlations, we may not be able to construct any causal scheme at all, either because of our insufficient knowledge of the variables or simply because all correlations are due to some remote common factors that have not been identified yet. In such a case it seems that the reasonable thing to do at the moment is to limit ourselves to description and presentation of these correlations. Fig. 171 (iv) is supposed to represent such a state of affairs. It would be helpful if we can suggest some causal relationship but we should never be pushed into it. To construct a nonsensical path diagram is certainly no service to our effort at understanding the relationships among the variables. Scientific research usually begins with description; explanation may not come until much later.

PATH ANALYSIS OF CAUSAL SYSTEMS

40. Diagram for partial correlation. We have already given (Chapter 5) a path diagram for partial correlation (Fig. 127) in terms of the linear regressions

$$X_1 = X_{1(3)} + X_{1.3} \quad \text{and} \quad X_2 = X_{2(3)} + X_{2.3}$$

where $X_{1(3)}$ is the value of X_1 determined by regression on X_3 and $X_{1.3}$ is the value of X_1 when the influence of X_3 is eliminated; that is, $X_{1.3}$ is the residue after fitting the regression line. In view of the causal diagrams developed in this chapter and in order to generalize it to partial correlations of higher order, we now simplify the previous Fig. 127 to the present Fig. 173 (i), showing X_3 as a common cause of X_1 and X_2. Then the partial correlation between X_1 and X_2, eliminating the common influence of X_3, is simply the correlation between the two residuals $U_1 = X_{1.3}$ and $U_2 = X_{2.3}$.

$$r(U_1, U_2) = r(X_{1.3}, X_{2.3}) = r_{12.3} \qquad (67)$$

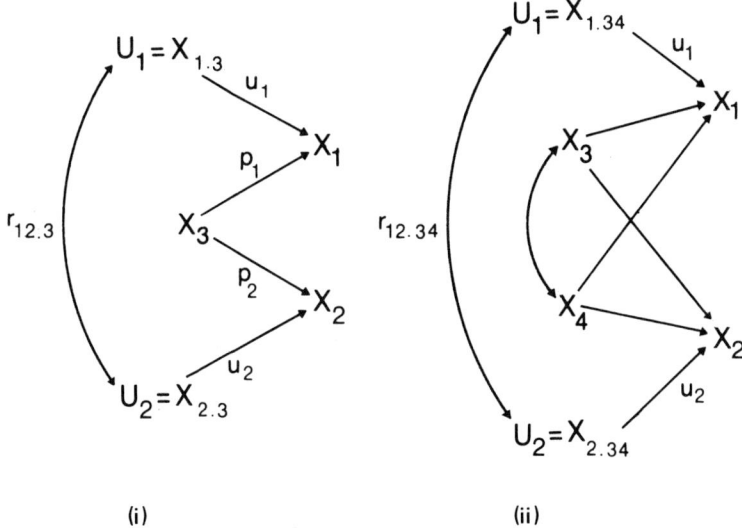

(i)   (ii)

Fig. 173. Diagrammatic representation of partial correlations. (i) first order partial correlation $r_{12.3} = r(U_1, U_2) = r(X_{1.3}, X_{2.3})$. (ii) second order partial correlation $r_{12.34} = r(U_1, U_2) = (X_{1.34}, X_{2.34})$.

The path coefficients from these residuals to X_1 and X_2 are denoted by the lower case letters u_1 and u_2 in Fig. 173 (i). Also, for brevity, let the path coefficients from X_3 to X_1 and X_2 be $p_1 = p_{13}$ and $p_2 = p_{23}$, as shown in Fig. 173 (i). By the theorem of complete determination,

$$u_1^2 + p_1^2 = 1, \qquad u_1 = \sqrt{1 - p_1^2}$$
$$u_2^2 + p_2^2 = 1, \qquad u_2 = \sqrt{1 - p_2^2} \qquad (68)$$

The total correlation between X_1 and X_2 is the sum of two connecting chains:

$$r(X_1, X_2) = r_{12} = p_1 p_2 + u_1 (r_{12.3}) u_2 \qquad (69)$$

Substituting $p_1 = p_{13} = r_{13}$, $p_2 = p_{23} = r_{23}$, and the u's of (68) we obtain

$$r_{12.3} = \frac{r_{12} - p_1 p_2}{u_1 u_2} = \frac{r_{12} - r_{13} r_{23}}{\sqrt{1 - r_{13}^2} \sqrt{1 - r_{23}^2}} \qquad (70)$$

which is the desired expression (Chapter 5).

41. In a similar way, Fig. 173 (ii) depicts the situation for the second order partial correlation $r_{12.34}$, a straightforward extension of $r_{12.3}$. In Fig. 173 (ii), $U_1 = X_{1.34}$ is the residual of X_1 after fitting the regression of X_1 on X_3 and X_4, and the path coefficient from U_1 to X_1 is denoted by u_1 as before. Similar meaning for U_2 and u_2. Thus,

$$r(U_1, U_2) = r(X_{1.34}, X_{2.34}) = r_{12.34} \qquad (67')$$

It has been shown (Fig. 151) that a multiple correlation coefficient is a path coefficient from the group of independent variables to the dependent. Let the path coefficient from X_3 and X_4 to X_1 be $R_1 = R_{1(34)}$ and that to X_2 be $R_2 = R_{2(34)}$, analogous to p_1 and p_2 in Fig. 173 (i). Although this is not shown as such explicitly in Fig. 173 (ii), we may make it so by using Fig. 151. Complete determination of X_1 and X_2 yields

$$R_1^2 + u_1^2 = 1, \qquad u_1 = \sqrt{1 - R_1^2}$$
$$R_2^2 + u_2^2 = 1, \qquad u_2 = \sqrt{1 - R_2^2} \qquad (68')$$

Then the total correlation between X_1 and X_2 is the sum of two connecting chains:

$$r_{12} = R_1 \cdot r(G_1, G_2) \cdot R_2 + u_1 (r_{12.34}) u_2 \qquad (69')$$

$$r_{12.34} = \frac{r_{12} - R_1 \cdot r(G_1, G_2) \cdot R_2}{\sqrt{1 - R_1^2} \quad \sqrt{1 - R_2^2}} \qquad (70')$$

where $R_1 = R_{1(34)}$ and $R_2 = R_{2(34)}$ are the multiple correlation coefficients; and $G_1 = \hat{X}_1$ and $G_2 = \hat{X}_2$ are the calculated or regression values of X_1 and X_2. It is noted that $(67') - (70')$ are of the same form as $(67) - (70)$, except that multiple correlation coefficients take the place of simple correlation coefficients. The partial correlation of a higher order, $r_{12.345}$, may be obtained by adding X_5 to Fig. 173 (ii); the expression will remain the same as $(70')$ with $R_1 = R_{1(345)}$ and $R_2 = R_{2(345)}$. Thus, $(70')$ is a general expression for all partial correlations (Li, Mazumdar, and Rao, 1975).

Unobserved variable

42. The example of Fig. 171 (iv) leads us to consider yet another property of path analysis. A path diagram may include a variable (or variables) that is not observable or unobserved. This may sound very strange to those who have been trained to process "data" (cards and tapes). Of course, in conventional statistical analysis, it is impossible to include any variable that is not observed, with mean and variance unknown. In path analysis, however, a relevant factor (zero mean and unit variance) should be included whether it is actually observed or not, if the formal completeness of the path diagram and the interpretation of the relationships require the presence of such a factor. Should this sound abstract, a concrete example in biology will help at this stage. Applications of path method in genetic

studies will be given in subsequent chapters. The reader will find that gametes (reproductive cells, male or female) are included as variables in path diagrams, although gametes have no observable traits that can be measured (e.g., weight, height, musical talent, mathematical ability, etc.). But they are there; and they are treated as if they have certain intrinsic quantitative values, so that we can speak of correlations and path coefficients relating to such gametes. In fact, it is the inclusion of such gamete variables in path analysis that enabled Wright to arrive at many genetic conclusions.

43. As an example of including an unobservable variable in path analysis, let us consider the three measurable variables *1, 2, 3*. Suppose that the observed correlations among the three variables are

$$r_{12} = 0.50, \qquad r_{13} = 0.40, \qquad r_{23} = 0.45 \qquad (71)$$

If there is reason to believe that an underlying common factor is responsible for these correlations, then a path diagram as shown in Fig. 177 may be constructed, where G represents the general factor and the U's are the residuals uncorrelated with G or among themselves. To simplify the notation we write p_1, p_2, p_3, for p_{1G}, p_{2G}, p_{3G}, the path coefficients from G to variables *1, 2, 3*, respectively. In a similar way, the path coefficients from the residuals to the measured variables are denoted by u_1, u_2, u_3, as shown in Fig. 177. Then the path diagram represents our concept of the correlated variables, even though the general factor G is not observable. Its influence, however, may be analyzed from the viewpoint specified by the diagram. There are six paths in the diagram and there are six equations:

$$\left.\begin{array}{ll} \text{(i)} \quad r_{12} = p_1 p_2 = 0.50 & \text{(iv)} \quad p_1^2 + u_1^2 = 1 \\[4pt] \text{(ii)} \quad r_{13} = p_1 p_3 = 0.40 & \text{(v)} \quad p_2^2 + u_2^2 = 1 \\[4pt] \text{(iii)} \quad r_{23} = p_2 p_3 = 0.45 & \text{(vi)} \quad p_3^2 + u_3^2 = 1 \end{array}\right\} \quad (72)$$

The set of equations (72) are particularly easy to solve. The first three equations constitute a set by themselves from which the three p's may be determined. Then the three u's may be found from the remaining three equations.

PATH ANALYSIS OF CAUSAL SYSTEMS

We have already solved the first three equations in Chapter 2, Ex. 4. Thus, the complete solutions are:

$$p_1 = 2/3, \qquad u_1 = \sqrt{1-p_1^2} = 0.745$$
$$p_2 = 3/4, \qquad u_2 = \sqrt{1-p_2^2} = 0.661 \qquad (73)$$
$$p_3 = 3/5, \qquad u_3 = \sqrt{1-p_3^2} = 0.800$$

In passing, we may mention that Fig. 177 and the corresponding analysis of the observed correlations (71) is the beginning of a branch of statistics known as "factor analysis" involving unobservable common factors behind a set of observed correlations. A review of the technique of factor analysis is clearly out of the scope of this book.

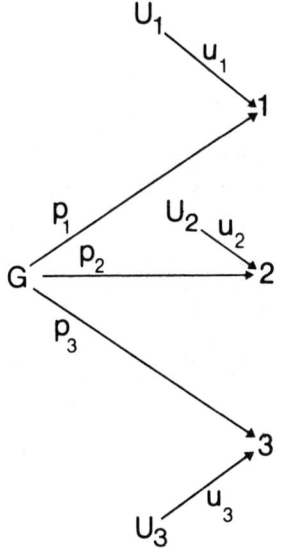

Fig. 177. The assumption of the existence of an unobserved variable G to account for the observed correlations among the measured variables $1, 2, 3$. The U's are residuals.

44. In the example above there are six path coefficients to be found from six equations. In general the number of equations formulated from observed correlations and complete determination must match the number of unknown path coefficients. If there are more observed values than unknowns, it is said to be overdetermined. If there are more unknowns than observed, it is said to be underdetermined. When the total number of equations and the total number of unknowns are equal, it is still not sufficient to guarantee solutions, because it may be *overdetermined* in one part of the path diagram and *underdetermined* in another part. When it is overdetermined, some type of adjustments (e.g., by the method of least squares) may be made so that the solutions will satisfy the observed values approximately, or the diagram may be revised by introducing more paths. When it is underdetermined, it implies that we simply do not have enough observations to accommodate the type of interpretation specified by the path diagram. In formulating a causal scheme, the investigator should always question whether the available observations are adequate for the interpretation he has in mind.

Summary: path analysis in a nutshell

When studying path theory for the first time, one needs a couple of nails (key points), on which to hang various things, so as to avoid being confused by the details and the endless modifications of the main theme. In many problems, the application of path analysis is based on only two main theorems, one expressing the complete determination of a variable and one decomposing a correlation into components. Eliminating the common cause C from Fig. 153 for brevity and letting $u = p_{OU}$, and $a' = p_{XA}$, $a = p_{YA}$, etc., the two main "laws" are:

$$u^2 + a^2 + b^2 + 2ab\, r_{AB} = 1$$

$$r_{XY} = a'a + a'r_{AB}b + b'b + b'r_{AB}a$$

Almost all other properties may be derived from these two. When remote factors behind A and B are introduced, the correlation $r(A, B)$ itself

PATH ANALYSIS OF CAUSAL SYSTEMS

may be expanded into components according to the same formula for $r(X, Y)$. Thus, path analysis involves the repeated use of a few fundamental relations. The formulation of hypotheses, and consequently the construction of a path diagram, is arbitrary in the mathematical sense; it depends on the knowledge of the investigator about the variables and the particular viewpoint taken in interpreting the data. The investigator may have to try a number of causal schemes before achieving consistent and plausible results.

Exercises

Ex. 1. In practical applications it is not uncommon for a path diagram to involve ten or more variables. Before there can be hope of tackling complicated situations, one must learn how to trace simple connections. In Fig. 179 the lower case letters associated with the arrows are path coefficients. Calculate the correlations between X and Y in all three diagrams; and calculate the correlation between P and Q in two of them. Can we analyze $r(A, B)$?

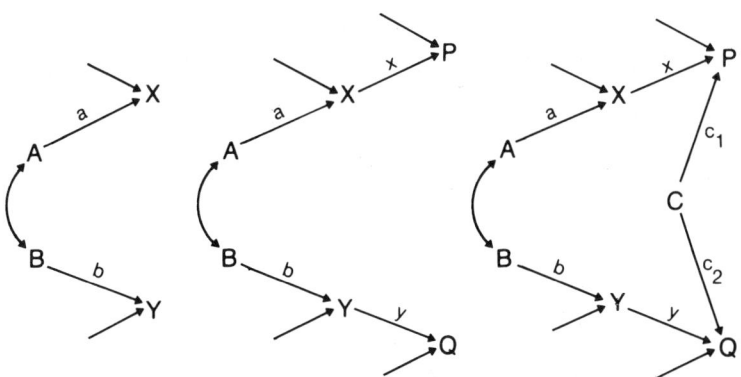

Fig. 179. See Exercise 1.

Ans.: $r_{XY} = a(r_{AB})b$ for all three diagrams (Fig. 179).

$r_{PQ} = x(r_{XY})y = x\, a(r_{AB})by$ for the middle diagram.

$r_{PQ} = x(r_{XY})y + c_1 c_2,$ where C is a common factor.

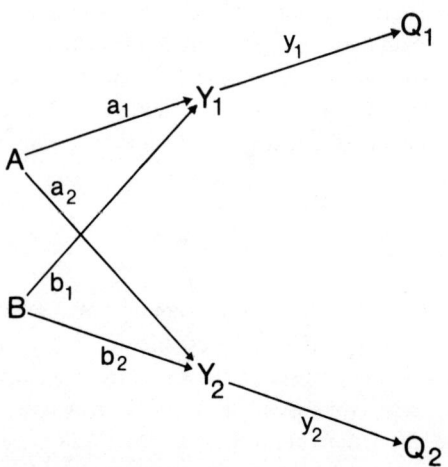

Fig. 180. See Exercise 2.

Ex. 2. Study Fig. 180. Are A and B correlated? Is there any connecting path between A and B in the diagram? Calculate the correlations between Y_1 and Y_2 and between Q_1 and Q_2.

Ans.: $r(A, B) = 0.$ $r(Y_1, Y_2) = a_1 a_2 + b_1 b_2.$

$r(Q_1, Q_2) = y_1 \cdot r(Y_1, Y_2) \cdot y_2 = y_1 a_1 a_2 y_2 + y_1 b_1 b_2 y_2.$

Ex. 3. This numerical exercise summarizes much of what we have learned about correlation due to common causes, and you are urged to carry out the arithmetic as a means of review. The first four columns of numbers of Table 181 (A, B, X, Y) are the "observed data;" the rest are calculations based on them. To simplify the arithmetic, observations are given in terms of deviations from mean. Only five sets of observations are given, so that the entire arithmetic may be done even without a desk calculator. The purpose is to verify certain relationships predicted by theory.

PATH ANALYSIS OF CAUSAL SYSTEMS 181

Table 181 (Exercise 3). The variables *A* and *B* are assumed to be the common causes of the variables *X* and *Y*. Draw a path diagram to depict the situation. The table gives the sum of squares, sum of products, variances, and covariances.

	A	*B*	*X*	*Y*	*AX*	*BX*	*AY*	*BY*	*AB*	*XY*
	−3	−8	−10	−19	30	80	57	152	24	190
	−2	4	−2	6	4	−8	−12	24	−8	−12
	0	−2	−1	−4	0	2	0	8	0	4
	1	0	2	1	2	0	1	0	0	2
	4	6	11	16	44	66	64	96	24	176
	0	0	0	0						
(1*)	30	120	330	670	80	140	110	280	40	360
(2*)	6	24	66	134	16	28	22	56	8	72

(1*) sum of squares and sum of products; e.g., $\Sigma Y^2 = 670$; $\Sigma XY = 360$.

(2*) variances and covariances; obtained by dividing the corresponding sum by $n = 5$. $Cov(A, Y) = 110/5 = 22$. The results on regressions, correlations, and path coefficients will remain the same if the divisor is $n − 1 = 4$, because only their ratios are relevant. Hence we may use the totals in (1*) to accomplish the same purpose.

From either the sums (1*) or the averages (2*) calculate the six correlation coefficients. It is convenient to arrange them in some kind of order.

$$r_{AX} = \frac{16}{\sqrt{6 \times 66}} = 0.80403; \quad r_{BX} = \frac{28}{\sqrt{24 \times 66}} = 0.70353;$$

$$r_{AY} = \frac{22}{\sqrt{6 \times 134}} = 0.77588; \quad r_{BY} = \frac{56}{\sqrt{24 \times 134}} = 0.98748;$$

$$r_{AB} = \frac{8}{\sqrt{6 \times 24}} = 2/3; \quad r_{XY} = \frac{72}{\sqrt{66 \times 134}} = 0.76561$$

In the above we have shown both the exact fractional form with radicals and their approximate values in decimals. You may use the exact form in the following calculations, as it avoids arithmetic and many of the common factors will cancel off. However, for ease of printing, I will use the approximate decimals as symbols for the exact expressions. Organize the calculations in five steps as follows:

I. Find the concrete regression coefficients for (i) X on A, B and (ii) Y on A, B. Use the ordinary normal equations: for regression of X on A, B

$$b_{XA}\Sigma A^2 + b_{XB}\Sigma AB = \Sigma AX$$

$$b_{XA}\Sigma AB + b_{XB}\Sigma B^2 = \Sigma BX$$

For regression of Y on A, B, replace X by Y in the equations above.

Ans.: $b_{XA} = 2$, $b_{XB} = 1/2$; and $b_{YA} = 1$, $b_{YB} = 2$.

II. Find (i) the path coefficients from A, B to X and (ii) those from A, B to Y. For part (i), use the standardized normal equations:

$$p_{XA} + p_{XB}r_{AB} = r_{AX}$$

$$p_{XA}r_{AB} + p_{XB} = r_{BX}$$

For part (ii), replace X by Y in the equations above.

Ans.: $p_{XA} = 0.60302$, $p_{XB} = 0.30151$; and $p_{YA} = 0.21160$, $p_{YB} = 0.84641$

III. Check your answers by the relationships $p_{0i} = b_{0i}\dfrac{\sigma_i}{\sigma_0}$ for each of the four coefficients ($0 = X, y$; and $i = A, B$).

IV. See if the following decompositions are true.

$$r_{XY} = p_{XA}r_{AY} + p_{XB}r_{BY}$$

$$r_{YX} = p_{YA}r_{AX} + p_{YB}r_{BX}$$

Ans.: r_{XY} = 0.60302(0.77588) + 0.30151(0.98748)

= 0.46787 + 0.29774 = 0.76561

r_{YX} = 0.21160(0.80403) + 0.84641(0.70353)

= 0.17013 + 0.59547 = 0.76560

V. Write out completely the four components of the correlation $r(X, Y)$.

r_{XY} = $p_{XA} p_{YA}$ + $p_{XA} r_{AB} p_{YB}$

+ $p_{XB} r_{AB} p_{YA}$ + $p_{XB} p_{YB}$

= 0.12760 + 0.34027

+ 0.04253 + 0.25520

Find the row and column totals of the four components as arranged above. Are these totals related to the answers found in IV previously?

Ex. 4. This exercise is to be done in connection with the two diagrams in Fig. 167. The best way to proceed is to copy down the two diagrams on a separate sheet, using the same notation for the variables but without the numerical values for the path coefficients. As in the text, let A and B be two uncorrelated variables with equal variance σ^2. The problem is to find the path coefficients under the following two models:

(i) $S = 3A + 4B$;   (ii) $4B = S - 3A$

In the diagram, then, the symbol A actually denotes $(3A)$ and B denotes $(4B)$. In algebraic manipulations $(3A)$ and $(4B)$ may be regarded as single letters. The variance of S is then

$$\sigma_S^2 = \sigma_{3A}^2 + \sigma_{4B}^2 = 9\sigma^2 + 16\sigma^2 = 25\sigma^2$$

Under model (i), you should find:

$$a = \sqrt{\frac{9}{25}} = \frac{3}{5}; \quad b = \sqrt{\frac{16}{25}} = \frac{4}{5}; \quad a^2 + b^2 = 1$$

Under model (ii), the correlation $r(A, S) = 0.60$ remains the same, but

$$s = \frac{\sigma_S}{\sigma_{4B}} = \frac{5}{4}; \quad \alpha = (-1)\frac{\sigma_{3A}}{\sigma_{4B}} = \frac{-3}{4}$$

You may now insert these numerical values onto your own diagrams. Check your answers by seeing that

$$\alpha^2 + s^2 + 2\alpha s\, r_{AS} = 1$$

and $\quad r_{AB} = \alpha + r_{AS}\, s = 0, \quad r_{BS} = s + r_{AS}\, \alpha = 0.80$

Ex. 5. This exercise is to be done in connection with Fig. 171, where the symmetrical relationship is purely for arithmetic convenience. The argument applies for any situation. Again, one should copy down the diagrams of Fig. 171, leaving the numerical values to be inserted later. Suppose that our observations on variables A, B, C, yield the following correlations:

$$r_{AB} = \frac{1}{2}, \quad r_{BC} = \frac{2}{3}, \quad r_{AC} = \frac{1}{3}$$

Calculate the partial correlation coefficient $r_{AC \cdot B}$. Insert the observed correlation values on the first two diagrams and see that they are consistent with the calculated $r_{AC \cdot B}$. Find the path coefficients a and c, assuming that the causal relations are those shown in Fig. 171 (iii). Check your results by seeing that

$$r_{AB} = a + \frac{1}{3}c = \frac{1}{2}$$

$$r_{BC} = \frac{1}{3}a + c = \frac{2}{3}$$

PATH ANALYSIS OF CAUSAL SYSTEMS

If the causal system of Fig. 171 (iii) is indeed correct (it could be, because they are consistent with observed results), what is the use of calculating the partial correlation coefficient $r_{AC \cdot B}$?

Ans.: $a = 5/16 = 0.3125$ and $c = 9/16 = 0.5625$.

The result $r_{AC \cdot B} = 0$ is misleading. Under the system Fig. 171 (iii), B is not responsible for the correlation between A and C.

Ex. 6. Equivalent diagrams. In Fig. 137 we see that X and U determine Y, where X itself is determined by W and V. If our purpose is to study Y, then diagram (b) is sufficient. The linked diagram (c) is produced to show the chain property of path coefficients. In the (c) diagram, the intermediate variable X plays no particular role if W and V are brought into consideration, as the (d) diagram is equivalent to (c) and shows that Y is determined by W, V, U. Similarly, in Fig. 146, the intermediate variables W and X in the (b) diagram may be eliminated, yielding the equivalent diagram (c) with new path coefficients. Along the same vein, you may examine the two diagrams of Fig. 185 and establish their equivalence. Note that in studying $r(X, Y)$ in Fig. 185 (i), the variables S and T play no role and may be deleted if the variable B has been observed. Alternatively, we may delete B and retain S and T as the common factors of X and Y. Then the path coefficient from S to X will be sb', and that from S to Y will be sb, etc. Analytically, we have from Fig. 185

$$s^2 + t^2 = 1$$
$$r_{XY} = b'b = b'(s^2 + t^2)b = sb' \cdot sb + tb' \cdot tb.$$

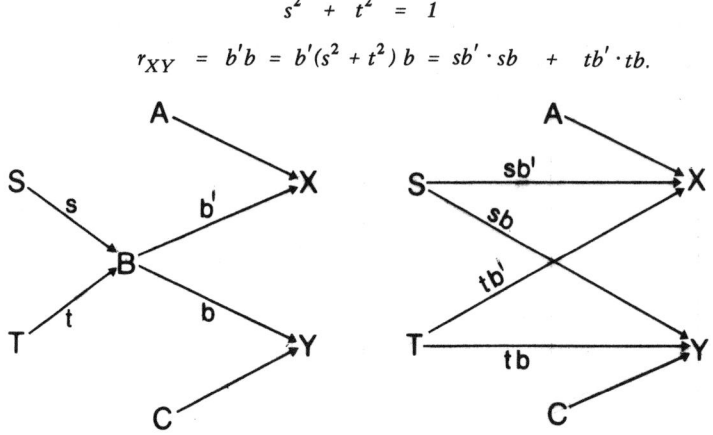

Fig. 185. See Exercise 6.

Ex. 7. The general theorem relating to a path diagram was developed in this chapter in connection with Fig. 153 for ease of understanding. Wright (1934, 1968) derived it in a more general manner. Let Y be a variable to (be represented by subscript 0) completely determined by $X_1, \ldots, X_m$ which are intercorrelated with $r_{ij} = r(X_i, X_j)$ and an uncorrelated variable (residual) U. Before proceeding further, you should draw a path diagram up to this stage, consulting Fig. 153 if necessary. In standardized form,

$$Y = p_{01} X_1 + \ldots + p_{0m} X_m + p_{0u} U$$

Then consider a variable T which is correlated with every term of Y above, including U. Add this variable to your diagram. Then the correlation between Y and T will be

$$r_{0T} = E(YT) = E(p_{01} X_1 + \ldots + p_{0m} X_m + p_{0U} U) T$$

$$= p_{01} r_{1T} + \ldots + p_{0m} r_{mT} + p_{0U} r_{UT} \quad \text{(W)}$$

This expression may then be accepted as the basic equation in path theory. From this the other theorems may be derived by appropriate substitutions. For instance, putting $T = X_1$ so that $r_{1T} = r_{11} = 1$ and $r_{UT} = r_{U1} = 0$, we obtain from **(W)**

$$r_{01} = p_{01} + p_{02} r_{21} + \ldots + p_{0m} r_{m1}$$

which is the first of the standardized normal equations. The remaining ones are obtained by putting $T = X_2, \ldots, X_m$, successively. If we put $T = Y$ itself, the general expression **(W)** becomes

$$r_{00} = p_{01} r_{10} + \ldots + p_{0m} r_{m0} + p_{0U}^2 = 1$$

as $r_{UT} = r_{UY} = p_{0U}$, U being an uncorrelated cause. The equation above expresses the complete determination of Y's variance. The sum of the terms $p_{0i} r_{i0}$ is of course $R^2 = R_{0(1 \ldots m)}^2$ so that we have $R^2 + p_{0U}^2 = 1$, or $R^2 = 1 - p_{0U}^2$, as shown in Chapter 4.

Chapter 7

ELEMENTS OF POPULATION GENETICS

THE MOST clearcut and successful applications of the path method are in the field of population genetics, especially in problems concerning the systems of mating. The major reason for the success of the path method in population genetics is that the "causes" and "effects" are known through the Mendelian laws of heredity. From the expository viewpoint, it is best to begin with genetic applications. The author well realizes that many readers are not biologists. For this reason we shall first cover the elements of Mendelian genetics and population genetics in this chapter, only to the extent that these materials make the subsequent chapters on genetic applications readable without a background in biology. In order to cover the elements of Mendelian genetics in a few paragraphs, it is necessary to commit the "crime" of presenting the subject as a statistical model, without describing what is probably one of the greatest biological experiments and discoveries in the history of science.

Mendelian genetics

   1. Particulate, not blending. In spite of the widespread teaching of biology in high schools, most laymen still think of heredity as a blending process. There is nothing wrong in describing a person as half Spanish and half French, for it merely means that one of the parents is Spanish and the other French. But we should not visualize the hybridization process as a blending of one cup of vinegar with one cup of water. The hybridization is rather like mixing a bag of blue marbles with a bag of yellow marbles. Although the hybrid bag contains two kinds of marbles, each marble still maintains its own identity; and the blue and yellow ones may be sorted out by an appropriate process. The important feature here is that the blue and yellow marbles do not blend together to form (new) green ones. There will be no green marbles in the hybrid bag. Genes, the hereditary units in man and in all other organisms, are chemical entities and maintain their identity from parent to offspring, just as a blue marble remains a blue marble no matter in which bag it happens to be. The analogy between marbles and genes is an extremely crude one. The only purpose it serves at this moment is to establish the concept that the genes are particulate in nature and maintain their own biochemical properties through the generations.

2. Chromosomes in pairs. In each nucleus of the body cells of animals or the vegetative cells of plants there are a certain fixed number of *pairs* of chromosomes. Corn has ten pairs; the fruit fly has four pairs; man has 23 pairs. The organisms with pairs of chromosomes are called *diploids* (with two sets of chromosomes). The genes are located in the chromosomes; in fact, they are arranged in linear order along the length of each chromosome. Again, we may use an old and crude analogy: if the genes are like beads, a chromosome is like a string of beads. Since chromosomes exist in pairs, so do genes. Each pair of chromosomes has many pairs of genes, depicted as follows:

$$\tag{1}$$

Each pair of genes occupies a definite and fixed position in the chromosome. The two chromosomes that form a pair are said to be *homologous*. Similarly, the two genes occupying the same position in a pair of homologous chromosomes are also said to be homologous. Thus, the gene at position 1 in one chromosome and the gene at position 1 in the other chromosome of the pair are homologous (to each other). Of the 23 pairs of chromosome of man, 22 are the same for males and females. These chromosomes are called *autosomes* and the genes they carry are said to be autosomal. The remaining pair is very special and its two members are called the *sex chromosomes*, as they determine the sex of the organism. In this book we consider only the autosomal genes. The path coefficients for sex-linked genes are different from those for autosomal ones. The reader who wishes to know more than covered here may consult a textbook on population genetics (e.g., Li, 1975).

3. One pair of genes. Since each pair of genes occupies a definite position in the chromosome, such positions identify the genes. These positions are called the *loci*. Each pair of genes has definite and highly specific biochemical properties which in turn control or influence the development of a certain trait of an organism. For instance, the genes at locus *1* may control the formation of red pigment of the flowers of a plant; the genes at another locus (either in the same pair or in another pair of chromosomes) control the development of hair on the leaves; and still another pair of genes determines whether the coat of the seed be smooth or wrinkled; and so on. Since there are many thousands of pairs of genes in an organism, it is impossible to study the joint effects of all of them simultaneously. In fact, should we try to study the heredity "as a whole," the results would be more confusing than enlightening. The crucial feature of the methodology that has led to the discovery of *laws of heredity* is to concentrate attention on one locus at a time. Most of the following sections deal with only *one pair* of autosomal genes.

4. Reproductive cells. We have seen that chromosomes and hence genes exist in pairs in the body or vegetative cells of a diploid organism. The reproductive cells, or *gametes* (sperm, egg) of such an organism, however, contain only one chromosome of each pair. In other words, a reproductive cell has only *one set* of chromosomes. In a sense, it is a "half cell." Let us consider a pair of autosomal genes at a certain locus of a pair of chromosomes. Call this pair of genes (A_1, A_2). As far as this pair of genes is concerned, we say that the *genotype* of the organism is $A_1 A_2$, which is the gene content of the organism. Its reproductive cells will have either A_1 or A_2 with equal chance. That is, 50% of its reproductive cells will carry A_1 only, and 50% will carry A_2 only. To summarize:

$$\text{genotype} \qquad \qquad \text{gametes}$$
$$A_1 A_2 \quad \longrightarrow \quad \tfrac{1}{2} A_1 + \tfrac{1}{2} A_2 \qquad (2)$$

The expression (2) is true for both males and females. We also see that expression (2) is true for all pairs of genes *when taken separately;* that is, if we concentrate our attention on one fixed pair at a time. If we consider many pairs of genes simultaneously, there will be many different kinds of gametes. That a genotype produces many different kinds of gametes is a key feature of genetics.

5. Fertilization and zygote. A new individual (child) is formed when two reproductive cells—one from the mother and one from the father—unite, at fertilization of the egg by the sperm, to form a new diploid zygote. A fruit fly sperm with four chromosomes and an egg with four chromosomes will produce at fertilization a zygote with four pairs of chromosomes. Let us concentrate on one locus again. Suppose the genotype of the father is A_1A_2, and the genotype of the mother is A_3A_4 at the same locus. Then the possible children to be produced by the mating $A_1A_2 \times A_3A_4$ are calculated according to the probabilities for the various types of fertilization as follows:

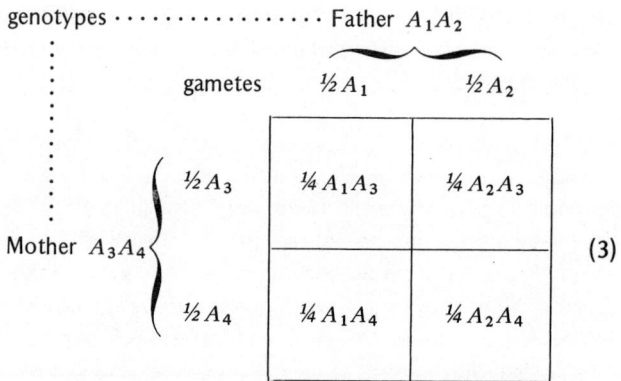

(3)

There will be four possible *recombinations* of genes, or four *genotypes* for the children, viz., A_1A_3, A_1A_4, A_2A_3, A_2A_4, each with a probability of ¼. These basic biological facts constitute the foundation of Mendelian inheritance.

6. Alleles at a locus. If all the genes at a certain locus are the same, having the same biochemical property and producing the same trait (color, hair, wrinkle, etc.), then we have no way of knowing their presence and therefore we cannot study their effects. In order to be susceptible to being studied, some of the genes at a locus must have a different biochemical property and influence the trait in a different way from the other genes at the same locus. For example, if the genes A_1, A_2, A_3, A_4 in (3) are all

ELEMENTS OF POPULATION GENETICS   191

the same, then the father and mother have exactly the same trait and so do all their children. There is then nothing to study. Hence we shall here assume that at a certain locus, the genes are of two kinds. Some of them, say, can produce a red pigment in flowers; some cannot. The former will be designated by A and the latter by a. The genes A and a are all located at the same locus in a chromosome. Only their functions and capabilities are different. We may also say that the genes at this locus exist in two forms or phases, A and a. The genes A and a are called *alleles*. It simply means that they are members of the same locus. When a locus has two alleles (A and a), there will be three possibilities for an organism's genotype, viz., AA, Aa, aa. The genotypes AA and aa are called *homozygotes*, because the two genes are of the same phase. The genotype Aa is called a *heterozygote* as the two genes are of different phases.

Table 191. Mendelian law of inheritance for various types of families.

Parents' mating type			Children's genotype		
			AA	Aa	aa
AA	X	AA	1	0	0
AA	X	Aa	½	½	0
Aa	X	Aa	¼	½	¼
AA	X	aa	0	1	0
Aa	X	aa	0	½	½
aa	X	aa	0	0	1

7. Mendelian law of inheritance. When there are two alleles (A and a) at an autosomal locus, there are three possible genotypes (AA, Aa, aa) for that locus and six possible types of mating as shown in Table 191, together with their possible children. The type of mating includes reciprocal matings. The type AA X Aa, for instance, refers to the type (mother AA) X (father Aa) or the type (father AA) X (mother Aa), as it makes no difference for autosomal genes. The proportions of the genotypes of the children are calculated according to scheme (3) for each type of mating, viz., by uniting the gametes produced by the two parents. The results of Table 191 are in essence the Mendelian laws of heredity for families, as far as any given locus is concerned. The mating $AA \times AA$ produces all AA children; the mating $aa \times aa$ produces all aa children. These two types of families are said to *breed true,* as *all* their children are exactly like their parents. The matings $AA \times Aa$, $Aa \times Aa$, and $Aa \times aa$ produce children of different genotypes; they are said to be *segregating families.* The ratios $\frac{1}{2} : \frac{1}{2} = 1 : 1$, and $\frac{1}{4} : \frac{1}{2} : \frac{1}{4} = 1 : 2 : 1$, are called Mendelian ratios (due to segregation of genes). Finally, the mating $AA \times aa$ produces all Aa children. It is not segregating, as all children are of the same genotype, and it is not breeding true either, as the children are different from their parents.

8. Genotype and phenotype. In Table 191 only the genotypes of parents and offspring are indicated, primarily to show the underlying Mendelian mode of inheritance, without mentioning the actual appearance (or the trait) of the organism. The appearance of an organism of a given genotype is quite something else. For ease of understanding, let us consider again the pair of genes controlling the flower color of certain plant species. Suppose that the allele A produces the red pigment and the allele a does not. The flower color of the three genotypes may be as follows:

Genotype (gene content)	Phenotype (flower color)			
	case I	case II	case III	
AA	red	red	red	
Aa	pink	red	red —	(4)
aa	white	white	white	

The actual appearance (or expression of the trait) of an organism is called the phenotype of the organism. In case I above, the phenotype of the AA genotype is red flower; the phenotype of aa is white; the phenotype of the heterozygote Aa is pink, a color intermediate between red and white. In such a case it is said there is *no dominance* of one allele over the other; the expression of the heterozygote is intermediate between the two homozygotes. In case II, however, the heterozygote Aa is also red, indistinguishable from the phenotype of AA. Apparently, the flower will be red whether

ELEMENTS OF POPULATION GENETICS

it has two A genes or only one A gene. In such a case, we say that the allele A is *dominant* over allele a, or, alternatively, allele a is *recessive* to allele A. There is no *a priori* reason for the phenotypes to be like those in case I or those in case II; it is a matter of fact to be observed. In many instances, upon closer examination (by whatever means, e.g., biochemical analysis) we discover that the color of Aa is almost but not exactly as red as AA, as depicted in case III above. They may look alike to the naked eye, but there is a difference in the amount of red pigment between the AA and Aa flowers. We say that allele A is *incompletely dominant* over a as far as biochemical analysis is concerned, but completely dominant as judged by the naked eye. Now, we see that the term *dominance* describes partly the color of the heterozygote and partly the ability or inability of the observer to distinguish the various genotypes.

9. Phenotype and environment. Since a genotype develops into an organism with a certain phenotype, the latter is bound to be influenced by environmental factors. The study of phenotypes is a study of the joint effects of heredity *and* environment. A phenotype is the end result of development of a genotype. If the environmental factors have not been explicitly mentioned in our example (4) above, it is understood that the colors of the various genotypes are true under normal conditions. If an organism is raised in a very different environment, then the expression of the trait may be different from its normal phenotype under normal conditions. For instance, in a breed of poultry, the genotypes AA and Aa will develop yellow legs while genotype aa has white legs with normal diet. However, if the diet is completely lacking in corn, the AA and Aa birds will no longer be able to develop yellow pigments. Then all birds will have white legs. This, however, does not change the fact that three genotypes (AA, Aa, aa) still exist. The development of a genotype into a phenotype is a major field of study of geneticists; thus, there are developmental genetics, physiological genetics, biochemical genetics, etc. So-called "genetic studies" include the study of environmental effects, among other things.

10. Quantitative traits. The previous examples of genetic control of flower color are chosen for ease of comprehension. Quantitative traits, though less obvious than qualitative ones, are affected by genetic factors in a similar way. One of the seven traits of a pea species originally studied by Gregor Mendel was the height of the pea plant. It was discovered through many crosses that the genotypes *AA* and *Aa* develop into tall plants (six feet or more) while the genotype *aa* develops into dwarf plants (18 inches or less) when grown in the same garden. In this particular case the allele *A* for tallness is dominant to the allele *a*. Intercross of hybrid plants *(Aa* X *Aa)* yield offspring, of which ¾ are tall and ¼ dwarf, confirming the familiar Mendelian 3 : 1 ratio for such families (see Table 191). In general, of course, there may or may not be dominance in alleles affecting quantitative traits, just as in the case of qualitative traits. For many quantitative traits experimentally studied, the heterozygotes *(Aa)* frequently assume a value more or less intermediate between those of the homozygotes *(AA* and *aa)*. There may be no dominance, there may be mild dominance, and rarely there may be complete dominance. In addition to these possibilities, the heterozygote *Aa* may assume a value greater than those of either homozygote. For example, while the heights of *AA* and *aa* plants are 6 ft. and 1½ ft., respectively, the heterozygote *Aa* may be 8 ft. high, under the same cultural conditions, of course. In such a case, we say there is *overdominance* or *heterosis*, with respect to the trait (tallness in our example) under consideration. In the applications to be described subsequently, we assume most of the time that there is no dominance, or the dominance is so mild that it does not affect the statistical results appreciably. The important point of this paragraph for non-biologists is that quantitative traits may be genetically controlled in the same way as qualitative traits. We must realize that the terms "qualitative" and "quantitative" are merely for convenience of the investigator; they do not refer to any intrinsic properties of a trait. The color (qualitative) of a flower may be studied in terms of the amount (quantitative) of red pigment in the tissue. The height (quantitative) of a plant may be simply classified as tall and short (qualitative). There is no fundamental difference between a qualitative and a quantitative trait.

11. Multiple interrelationships. We have so far presented a pleasantly simple picture of the correspondence of one locus with one trait. For many traits (including flower color), however, the correspondence is not so simple. A gene may affect two or more traits, and the multiple effects of genes are called *pleiotropism*. It does not contradict the "one gene, one polypeptide"

ELEMENTS OF POPULATION GENETICS

theory because the resultant polypeptide may enter more than one pathway leading to further reactions. Conversely, a trait may be jointly influenced by several or even many loci. This is not surprising, since what we call a trait is usually a mixture of many elementary traits; the standing height is the sum of many segments, each of which may be influenced by a certain number of loci. The connections between the genes and the traits thus form a complicated network, crudely depicted as follows:

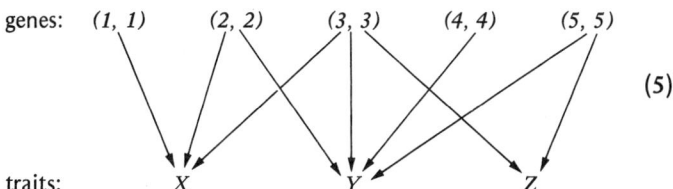

(5)

It is a many-to-many relationship, not one-to-one. The genes indicated above are not necessarily located in the same pair of chromosomes. The intermediate steps between the primary gene action and the final phenotype have been omitted. In this book we shall study only one trait at a time.

Population genetics

12. In Table 191 are listed the various types of parental combinations and their children's genotype. If we ask: "what is the frequency of the AA genotype in a population?" or "what is the frequency of the parental combination $AA \times Aa$ in a population?" these questions cannot be answered by referring to the Mendelian laws of segregation listed in Table 191. The Mendelian laws are familial laws, applicable to given types of families. Problems concerning the various frequencies in a population belong to population genetics, which studies the consequences of Mendelism at the population level. The path method applies to the analysis of hereditary and environmental effects on a trait at the population level. Hence the following few sections are devoted to the elements of population genetics. Again, these sections have to be outlined as a statistical model without discussion of the biological details.

13. Genotype and gene frequencies. Let us begin with an arbitrary population, in which there are 70 AA, 20 Aa, 10 aa, totaling 100 individuals. Then 0.70 is the frequency of the genotype AA in this group of individuals, etc. If we disregard the pairing of the genes and concentrate on the genes themselves, we see that these 100 individuals have 200 genes. Each AA individual has two A genes; each Aa individual has one A gene and one a gene; and each aa individual has two a genes. Thus the total number of A genes in this group is $2(70) + 20 = 160$; and the total number of a genes is $20 + 2(10) = 40$. The frequencies of these genes are

$$p = freq(A) = \frac{160}{200} = 0.80, \qquad q = freq(a) = \frac{40}{200} = 0.20 \tag{6}$$

and $\qquad p + q = freq(A) + freq(a) = 1$

Since we are studying only one autosomal locus, the gene frequency is also the gamete frequency. In our example we may say that 80% of the gametes produced by the group (0.70, 0.20, 0.10) as a whole are A-gametes and 20% are a-gametes. Thus, $p(A)$ and $q(a)$ are called the *gametic output* of the group. In general, if (D, H, R) are the frequencies of (AA, Aa, aa) in a population, where $D + H + R = 1$, then the gene frequencies are

$$p = D + \tfrac{1}{2}H, \qquad\qquad q = R + \tfrac{1}{2}H \tag{7}$$

e.g., $\quad p = 0.70 + \tfrac{1}{2}(0.20) = 0.80, \quad q = 0.10 + \tfrac{1}{2}(0.20) = 0.20$

From the numerical calculations (6) and the general formulation (7), it is seen that once the genotype frequencies (D, H, R) are known, the gene frequencies (p, q) will also become known. The reverse is, however, not true in general. Suppose that we know $(p, q) = (0.80, 0.20)$. There are many populations that could have produced the same gametes, e.g., (0.80, 0, 0.20), (0.75, 0.10, 0.15), (0.60, 0.40, 0), (0.64, 0.32, 0.04), etc. The mechanism by which a relationship is established between the gene frequencies (p, q) and the genotype frequencies (D, H, R) is the *mating system* of the population. The gene frequencies tell us how much there is for each allele; the mating system tells us how these genes are associated in pairs to form the genotypes.

ELEMENTS OF POPULATION GENETICS   197

14. Random mating. The simplest mating system is random mating (panmixia) with respect to the pair of genes under consideration. For example, human blood may be classified into three types, viz., M, MN, N, and these are controlled by a pair of autosomal genes, say, A_M and A_N. The three genotypes $A_M A_M$, $A_M A_N$, $A_N A_N$, correspond to the three blood types (phenotypes) M, MN, N, respectively. These blood types do not affect blood transfusion and hence, are not typed in the usual blood typing procedure. For all practical purposes, they are unknown to the layman. There are ample empirical data to show that matings between the various blood types are entirely at random. The technical definition of random mating is that the parental combination (with respect to a pair of genes) is a chance event. If the frequency of the three blood types (M, MN, N) is (0.64, 0.32, 0.04) in both sexes, then the frequency of the mating $M \times M$ is $0.64 \times 0.64 = 0.4096$, and the frequency of the matings $MN \times N$ and its reciprocal is $2(0.32)(0.04) = 0.0256$, etc. If we shift our attention from the MN blood system to the stature of individuals of the same population, we will discover that the matings with respect to stature are no longer at random. The correlation between mates with respect to height is found to be approximately $+0.30$. Thus we see that random mating is not an absolute term but *always* relative to some specified trait. The question "Is there random mating in human populations?" would be meaningless without specifying the trait involved in the matings. In genetic literature we sometimes simply say "random mating" for brevity, but it should always be understood that the geneticist means "random mating with respect to a certain trait or locus under consideration."

15. The stationary state. Given any initial population (D, H, R), we can readily calculate the gametic output from the population, viz., the values of $p(A)$ and $q(a)$, by formula (7). Random mating among the individuals of the population is equivalent to random union of their gametes. We ask the reader to accept this principle for the time being; it will be made more acceptable in a later section. Now, if the gene frequencies (p, q) are the same for both sexes, random union of their gametes will produce the next generation as follows:

male gametes

	$p(A)$	$q(a)$
$p(A)$	p^2 (AA)	pq (Aa)
$q(a)$	pq (Aa)	q^2 (aa)

female gametes (on the left)

(8)

In other words, the frequencies of the genotypes in the offspring generation will be:

genotype:	AA	Aa	aa
frequency:	p^2	$2pq$	q^2

(9)

We may continue the process of calculating the gametic output of a population and then unite them at random. The gametic output (p', q') of the offspring generation (9) is calculated by the same formula (7), using $(D, H, R) = (p^2, 2pq, q^2)$. Thus,

$$p' = D + \tfrac{1}{2}H = p^2 + pq = p, \qquad q' = R + \tfrac{1}{2}H = q^2 + pq = q \qquad (10)$$

The gametic output of the offspring generation remains the same: $(p', q') = (p, q)$. Random union of these gametes in the fashion of (8) will yield the next generation exactly like (9) again. The process goes on and on, but the genetic composition of the population will remain $(p^2, 2pq, q^2)$ from generation to generation, assuming no disturbing forces at work. The condition (9), achieved after only one generation of random mating, is called the "stationary" or "equilibrium state" of the population. The condition (9) is usually known as the "Hardy-Weinberg law" (1908) for random mating populations. Li (1967) notes that Castle (1903) discovered the same law, expressed in proportional numbers instead of algebraic symbols, and stressed the generality of the equilibrium condition whatever the values of (p, q). Keeler (1968), in "Some oddities in the delayed appreciation of Castle's law," gives a fascinating story about the historical details. The stationary

ELEMENTS OF POPULATION GENETICS

state (9), viz., the *Castle-Hardy-Weinberg law,* is the cornerstone of population genetics. The fact that it is achieved in merely one generation of random mating, whatever the initial population, enables us to assume the equilibrium state (9) to be true in empirical studies of the particular locus. A mathematical law is never fully realized in the real world, especially in the biological world. Nevertheless, the binomial distribution (9) seems robust enough to serve as a starting point for genetic studies of populations.

16. Frequency of matings and offspring. Suppose that the frequencies of genotypes in both sexes are given by the equilibrium state (9). Then under the system of random mating, the frequency of the various types of matings will be as follows:

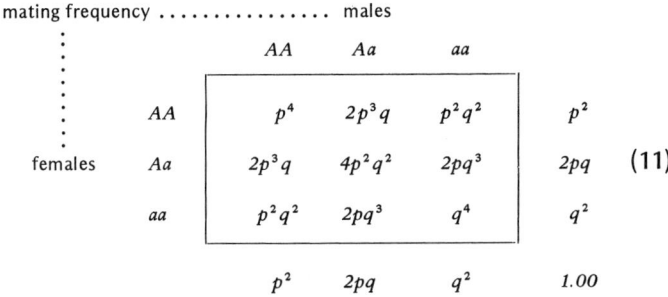

The entries in the table above are products of the corresponding marginal frequencies. This may be accepted as the *definition of random mating.* The table is symmetrical with respect to the principal diagonal (upper left to lower right). This means the reciprocal crosses are equally frequent; for example, the frequency of female AA × male Aa is $p^2 \times 2pq = 2p^3q$, and the frequency of male AA × female Aa is also $p^2 \times 2pq = 2p^3q$, so that the total frequency of $AA \times Aa$, without specifying the sex of the mates, is $4p^3q$. Table 191 lists the various types of families as given conditions (without frequencies). Now, we insert the frequencies (11) to each type of mating; the result is shown in Table 200, which is essentially the same as the original table of Weinberg (Stern, 1943). As to the construction of the table, consider the matings $Aa \times Aa$ with frequency $4p^2q^2$. These parents produce offspring of genotypes AA, Aa, aa, with

Table 200. Frequency of mating and offspring in a random-mating population. Reciprocal crosses have been combined as one type of mating.

Mating type (including reciprocals)	mating frequency from (11)	frequency of offspring		
		AA	Aa	aa
AA × AA	p^4	p^4	0	0
AA × Aa	$4p^3q$	$2p^3q$	$2p^3q$	0
Aa × Aa	$4p^2q^2$	p^2q^2	$2p^2q^2$	p^2q^2
AA × aa	$2p^2q^2$	0	$2p^2q^2$	0
Aa × aa	$4pq^3$	0	$2pq^3$	$2pq^3$
aa × a	q^4	0	0	q^4
Total	1.00	p^2	$2pq$	q^2

frequencies ¼, ½, ¼, respectively. Multiplying these fractions by $4p^2q^2$, we obtain p^2q^2, $2p^2q^2$, p^2q^2, respectively, for the various genotypes of offspring. The column totals of Table 200 give the genetic composition of the next generation. Thus, the total frequency of heterozygotes in the offspring generation is the sum of the column under Aa offspring:

$$2pq(p^2 + 2pq + q^2) = 2pq$$

Thus we see that the parent population $(p^2, 2pq, q^2)$, under a random mating system (11), will produce an offspring generation also of the composition $(p^2, 2pq, q^2)$. It follows that all subsequent generations will also be $(p^2, 2pq, q^2)$ under the random mating system, barring other disturbing forces. There are many genetic properties in Table 200, most of which we have to ignore in this text. Further details may be found in Li (1961). However, let us note one of the properties here. The matings Aa × Aa are twice as frequent as those of AA × aa, reciprocals included. We shall return to Table 200 again when we study quantitative traits.

17. Parent-child combinations. In both biological and sociological research we are interested in the joint distribution and correlation between parent and child. In a random-mating population the parent-child joint distribution with respect to one pair of autosomal genes may be easily derived. It will be recalled that in Table 200 the reciprocal crosses have been

ELEMENTS OF POPULATION GENETICS

pooled as one type of mating. If the sex of parents is distinguished, there will be nine, instead of six, types of mating as originally indicated by (11). The upper portion of Table 201 is an expansion of Table 200, with mothers and fathers separately listed. The method of constructing the table, however, remains the same as before and needs no comment. Now, let us condense the table by summing up (telescoping) the fathers according to each genotype of mother. The result is a mother-child joint distribution table shown in the lower portion of Table 201. For example, the frequency of AA-Aa mother-child combinations is $p^3q + p^2q^2 = p^2q(p+q) = p^2q$, ignoring the genotype of the father. The mother-child combination table is useful for legal-medical applications, where the father is not definitely known. For autosomal genes, the mother and father may be interchanged. In fact, the four types of parent-child combinations: father-son, father-daughter, mother-son, and mother-daughter, have the same joint distribution for autosomal loci. (This is not so for sex-linked genes.)

Table 201. Parent-child combinations in a random-mating population.

Mother		Father		Offspring		
				AA	Aa	aa
p^2	AA	p^2	AA	p^4	0	0
		$2pq$	Aa	p^3q	p^3q	0
		q^2	aa	0	p^2q^2	0
$2pq$	Aa	p^2	AA	p^3q	p^3q	0
		$2pq$	Aa	p^2q^2	$2p^2q^2$	p^2q^2
		q^2	aa	0	pq^3	pq^3
q^2	aa	p^2	AA	0	p^2q^2	0
		$2pq$	Aa	0	pq^3	pq^3
		q^2	aa	0	0	q^4

Condensation of the table above by summing up the fathers

		Offspring			
		AA	Aa	aa	
Mother	AA	p^3	p^2q	0	p^2
	Aa	p^2q	pq	pq^2	$2pq$
	aa	0	pq^2	q^3	q^2
	Total	p^2	$2pq$	q^2	1.00

18. Sib-pair distribution. The frequencies of the various types of sib-sib combinations in a random-mating population may be obtained by enumerating the possible types of sib-pairs produced by each type of parental combination. This is not the shortest way of obtaining the distribution of sib-pairs, but it is readily understandable, as it requires no additional knowledge of genetic properties. There are six types of parental combinations, as listed in Table 200. The $AA \times AA$ parents can produce AA children only and therefore all sib-pairs from these families are of the type AA-AA. Similar remarks apply to $aa \times aa$ parents. Also, the children of $AA \times aa$ parents are all heterozygotes and therefore all the sib-pairs produced from such families are of the type Aa-Aa. Before proceeding further, the reader should locate these three entries in the upper part of Table 203, where the frequencies of sib-pairs as well as their familial sources are indicated. As to the $4p^3q$ ($AA \times Aa$) families, their children will form pairs according to the four combinations $(\frac{1}{2}AA + \frac{1}{2}Aa)(\frac{1}{2}AA + \frac{1}{2}Aa)$; the first factor gives the probabilities for the first child and the second factor gives the probabilities for the second child. These four types of sib-pairs, each with frequency p^3q, are entered in the upper left four cells of Table 203. The reader should identify these four entries in the table before he proceeds further. Similarly, the four types of sib-pairs, each with frequency pq^3, from the $4pq^3$ ($Aa \times aa$) families, are entered in the lower right four cells of Table 203. Finally, the $4p^2q^2$ ($Aa \times Aa$) families will produce all nine types of sib-pairs according to the expansion of $(\frac{1}{4}AA + \frac{1}{2}Aa + \frac{1}{4}aa)(\frac{1}{4}AA + \frac{1}{2}Aa + \frac{1}{4}aa)$; again, the first factor gives the probabilities for the first child and the second factor gives the probabilities for the second child. Thus, these families contribute $4p^2q^2$ ($1/16$) = $\frac{1}{4}p^2q^2$ AA-AA sib-pairs, etc. The reader should identify the nine entries contributed by the $Aa \times Aa$ families in the upper portion of Table 203. Now the table is complete. Summing the entries in each cell, we obtain the table of joint distribution of sib-pairs shown in the lower portion of Table 203. In summary, Table 191 gives the basic Mendelian laws of heredity; Tables 200, 201, and 203 give the genetic relationships among immediate family members in a random-mating population. In subsequent sections when we refer to Tables 201 and 203, we mean the lower portion of those tables; the upper part is to show the procedure by which the lower part is obtained. The genetic relationships among more distant relatives are also well known, but we shall not deal with them in this book.

ELEMENTS OF POPULATION GENETICS

Table 203. Sib-sib combinations in a random-mating population. (The familial source of the sib-pairs is indicated in parenthesis.)

	AA	Aa	aa
AA	(AA × AA) p^4 (AA × Aa) p^3q (Aa × Aa) $\frac{1}{4}p^2q^2$	(AA × Aa) p^3q (Aa × Aa) $\frac{1}{2}p^2q^2$	(Aa × Aa) $\frac{1}{4}p^2q^2$
Aa	(AA × Aa) p^3q (Aa × Aa) $\frac{1}{2}p^2q^2$	(AA × Aa) p^3q (Aa × Aa) p^2q^2 (AA × aa) $2p^2q^2$ (Aa × aa) pq^3	(Aa × Aa) $\frac{1}{2}p^2q^2$ (Aa × aa) pq^3
aa	(Aa × Aa) $\frac{1}{4}p^2q^2$	(Aa × Aa) $\frac{1}{2}p^2q^2$ (Aa × aa) pq^3	(Aa × Aa) $\frac{1}{4}p^2q^2$ (Aa × aa) pq^3 (aa × aa) q^4

Condensation of the entries above

sibs	AA	Aa	aa	Total
AA	$\frac{1}{4}p^2(1+p)^2$	$\frac{1}{2}p^2q(1+p)$	$\frac{1}{4}p^2q^2$	p^2
Aa	$\frac{1}{2}p^2q(1+p)$	$pq(1+pq)$	$\frac{1}{2}pq^2(1+q)$	$2pq$
aa	$\frac{1}{4}p^2q^2$	$\frac{1}{2}pq^2(1+q)$	$\frac{1}{4}q^2(1+q)^2$	q^2
Total	p^2	$2pq$	q^2	1.00

The identity of genes by descent

19. The lower portion of Tables 201 and 203 may be regarded as correlation tables for parent-child and sib-sib, respectively, when some numerical values are assigned to the genotypes AA, Aa, aa. The variance and covariance may all be calculated in the usual manner, purely algebraic. It will be discovered that the parent-child correlation and the sib-sib correlation have a certain relationship, again expressed purely as an algebraic result. In order to

understand the difference between these two correlations on biological grounds, we must consider the concept of the identity of genes by descent, as briefly as possible. This concept will enable us to view the correlation tables in a new way that exhibits the difference between the parent-child relationship and the sib-sib relationship. It not only shortens the calculation but also facilitates our understanding and interpretation of the algebraic results.

20. Let us consider the fertilization scheme (3) once more. In that we merely say the "genotype" of father is $A_1 A_2$, a pair of genes at a certain locus, without specifying whether he is AA, Aa, or aa. The gene A_1 may happen to be A or a; likewise, the gene A_2 may happen to be A or a. No matter the phase (A or a) in which a gene may exist, we are now primarily interested in that particular gene itself and give it a license number (or a label) to identify it so that we can trace it from one generation to the next. We merely wish to trace the two genes (A_1 and A_2) of the father, and are not yet interested in the phases (AA, Aa, aa) in which the genotype happens to be. Similarly, the two genes of the mother may be labeled as A_3 and A_4, without caring about their phases for the time being. To facilitate writing, we may omit the symbol A and merely write the license number of the genes. Thus, $A_1 A_2 \times A_3 A_4$ may be simply written as (1, 2) × (3, 4), remembering that these digits (1, 2, 3, 4) are labels, not quantities. The reader will soon appreciate the simplified notation. The fertilization scheme (3) may now be rewritten as follows:

Parents	Children				
(1, 2) × (3, 4) ⟶	(1, 3),	(1, 4),	(2, 3),	(2, 4)	(12)

It is understood that the four types of children are equally likely. We also see that the symbolic expression (12) is universally true, no matter what genotypes these individuals may happen to be. It covers all the possibilities listed in Table 191.

21. Consider the parent (1, 2) and child (1, 3). The gene 1 of the child is the same gene 1 of the parent, duplicated and transmitted through the reproductive cell (gamete). We say that these two genes are *identical by*

descent. An examination of scheme (12) shows that a parent-child pair (any parent and any child) *always* has one identical gene in common, and there are no exceptions. This is the characteristic of the genetic relationship for parent-child pairs. The relationship has nothing to do with gene frequencies. It is a law of transmission of genes from parent to offspring. In other words, it is an abstract form of the basic Mendelian mode of inheritance.

22. Now we come to the genetic relationships for sib-pairs, assuming again the parents to be *(1, 2)* X *(3, 4)*. The children shown in (12) may be taken as the first child. The second child, being an independent event, will have the same genetic composition with the same probabilities. Hence, the possible sib-sib pairs are as follows:

second child

	(1, 3)	(1, 4)	(2, 3)	(2, 4)
(1, 3)	●●	●	●	○
(1, 4)	●	●●	○	●
(2, 3)	●	○	●●	●
(2, 4)	○	●	●	●●

first child (13)

Of the 16 possible sib-pairs shown above, the first child *(1, 3)* and the second child *(1, 3)* have both genes identical by descent, which is impossible for a parent-child pair in a random-mating population. Consider the next two sib-pairs: first child *(1, 3)*, second child *(1, 4)* or *(2, 3)*. For either pair, they have only one identical gene in common. The genetic relationship for such sib-pairs is exactly the same as that for parent-child pairs. Finally, consider the fourth sib-pair: first child *(1, 3)* and second child *(2, 4)*. They have no identical genes in common, another impossible situation for parent-child pairs. The sib-pairs *(1, 3)* and *(2, 4)* are like unrelated individuals in a random-mating population (as far as this one locus is concerned). The same situation holds for each row of the 4 X 4 array of (13), where (● ●) indicates the sib-pair having two genes in common, (●) the pair having one gene in common, and (○) no gene in common. These three types of sib-pairs occur with probabilities ¼, ½, ¼, respectively. Again, this relationship is purely a familial affair, having nothing to do with gene frequencies in the population. The result of the analysis above is consistent with our daily experience; some sib-pairs are very much alike (to the extent of being identical twins); some sib-pairs are moderately alike (to the extent of parent-child resemblance), and some sib-pairs have no familial resemblance at all. However, the average degree of resemblance between full sibs is about the same as that for parent and child. The resemblance for parent-child pairs is comparatively steady, while that for sib-pairs is more varied. The basic genetic differences between parent-child relationship and sib-sib relationship is exhibited in (12) and (13). The former are called *unilineal* relatives and the latter *bilineal*.

The I T O method

23. In studying the genetic relationships among family members (and other relatives) in a random-mating population, it happens to be more convenient to employ conditional probabilities as a first step and then convert the conditional frequencies into absolute ones. In order to shorten the description, we start with the following three sets of conditional probabilities for random-mating populations:

$$\mathbf{I} = \begin{array}{c} \\ AA \\ Aa \\ aa \end{array} \begin{array}{ccc} AA & Aa & aa \\ \begin{pmatrix} 1 & 0 & 0 \\ 0 & 1 & 0 \\ 0 & 0 & 1 \end{pmatrix} \end{array}, \mathbf{T} = \begin{pmatrix} AA & Aa & aa \\ p & q & 0 \\ \tfrac{1}{2}p & \tfrac{1}{2} & \tfrac{1}{2}q \\ 0 & p & q \end{pmatrix}, \mathbf{O} = \begin{pmatrix} AA & Aa & aa \\ p^2 & 2pq & q^2 \\ p^2 & 2pq & q^2 \\ p^2 & 2pq & q^2 \end{pmatrix} \quad (14)$$

Consider a pair of sibs who have both genes identical by descent; it does not matter whether they are both *(1, 3)*, or both *(1, 4)*, or both *(2, 3)*, or both *(2, 4)*. As long as their two genes are both identical by descent, they must necessarily have the same genotype, whatever the genotype may be. If one sib is AA, the other must also be AA. If one is Aa, the other is certainly Aa. If one is aa, the other is also aa automatically.

ELEMENTS OF POPULATION GENETICS　　　　　　　　　　　　　　207

The first 3×3 array of conditional probabilities in (14), denoted by **I**, describes this situation. These arrays are to be read row by row. The rule will be made clear in the next paragraph.

24. We now consider a pair of individuals who have only one identical gene in common. It has already been noted that a parent-child pair always has one identical gene in common. Further, we noted from (13) that $8/16 = \frac{1}{2}$ of the sib-pairs also have one identical gene in common. The genetic relationship of such sib-pairs is the same as that for parent-child pairs. They are all relatives who have one identical gene in common. To avoid abstract language, however, we refer to the given individual as the "parent" and calculate the conditional probabilities for the various genotypes of the "child." Suppose that the given parent is AA. The child must have one of the A's. That is, all his children must be of the genotype Ax, where the gene x may be A or a. In a random-mating population the total gametic output is $p(A)$ and $q(a)$. That is, the probability that $x = A$ is p and that $x = a$ is q. In the event $x = A$, the resultant child is AA; in the event that $x = a$, the resultant child is Aa. No aa child is possible when one parent is given to be AA. These conditional probabilities are indicated in the first row of the second set of conditional probabilities of (14). The conditional probabilities of each row of the arrays must add up to unity; thus, $p + q + 0 = 1$. Similarly, when the parent is given to be aa (third row), there will be no AA child, and the probabilities that the child should be Aa and aa are p and q, respectively. Finally, if one parent is given to be Aa (second row), his child may inherit the gene A or the gene a with equal probabilities ($\frac{1}{2}, \frac{1}{2}$). In the former event, the conditional probabilities would be exactly like those in the first row. In the latter event, the situation would be exactly like those in the third row. Hence the conditional probabilities in the second row are simply the mean of the first and third rows. These nine probabilities, as arranged in (14), are denoted collectively by **T**. The elements of **T** are the conditional probabilities for the genotype of a child when one parent is given to be AA (1st row), Aa (2nd row), and aa (3rd row). But the initial probabilities for the parents to assume these genotypes are p^2, $2pq$, q^2, respectively. Multiplying the first row of **T** by p^2, the second row by $2pq$, and the third row by q^2, we would obtain the joint distribution of parent-child

pairs in a random-mating population. The student is urged to do this as an exercise and see if his answers check with those given in the lower portion of Table 201. The new point to be emphasized here is that this is true not only for parent-child pairs but also true for any pair of relatives who have one identical gene in common.

25. When two individuals have no identical gene in common, such as the sib-pairs *(1, 3)*, and *(2, 4)* in (13), the conditional probabilities are given in the last set denoted by **O** in (14). When one sib is given to be AA, the two genes of the other sib are simply two random genes in the population, distributed p^2, $2pq$, q^2. This is true no matter what the genotype of the given sib, as long as the two sibs do not share identical genes. The notation **O** for such an array of values may raise some objection by mathematicians, as they have committed the symbol to a matrix with all elements zero. Since we have no occasion to use a matrix with all elements zero, the symbol **O** denoting the three rows of $(p^2, 2pq, q^2)$ in (14) should cause no confusion.

26. Since in a parent-child pair the members always have one identical gene in common, the conditional probabilities of the children are given by the three rows of **T**, from which we obtain the joint distribution of parent-child pairs in a random-mating population. The joint distribution of sib-pairs may be found in a similar way if we can find the conditional probabilities for sibs. Now, we know that the probabilities for sibs to have two, one, or no identical genes in common are ¼, ½, ¼, respectively. Hence the conditional probabilities for sibs are

$$\mathbf{S} = \tfrac{1}{4}\mathbf{I} + \tfrac{1}{2}\mathbf{T} + \tfrac{1}{4}\mathbf{O} \qquad (15)$$

where **I**, **T**, **O** are as given in (14). Having obtained the conditional probabilities we multiply the first row of **S** by p^2, the second row by $2pq$, and the third row by q^2 to obtain the joint distribution of sib-pairs in a random-mating population. Upon simplification, the reader will see that the final results are the same as those in the lower portion of Table 203. To summarize, the parent-child pairs are genetically related by **T**, while the sib-pairs are genetically related by $\mathbf{S} = \tfrac{1}{4}\mathbf{I} + \tfrac{1}{2}\mathbf{T} + \tfrac{1}{4}\mathbf{O}$. It is this difference that

ELEMENTS OF POPULATION GENETICS 209

will help us to understand the algebraic results on covariance and correlation among family members to be studied in the next chapter.

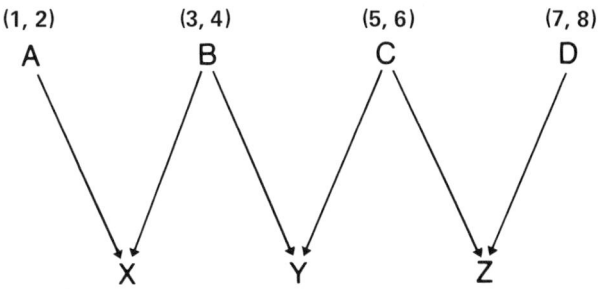

Fig. 209. The parents (A, B, C, D) are uncorrelated in a random-mating population. The half-sibs X and Y are correlated; the half-sibs Y and Z are also correlated. But the individuals X and Z are not correlated; the zig-zag line X-B-Y-C-Z is not a connecting path between X and Z, as it involves a forward-backward (B-Y-C) step.

27. Half-sibs are those who share one common parent. The genetic relationship between half-sibs is illustrated in Fig. 209, in which the children X and Y are half-sibs having the common parent B, and the children Y and Z are also half-sibs because they have the common parent C. In a random-mating population the four parents (A, B, C, D) are uncorrelated with respect to their genotypes. Then their eight genes are eight independent genes of the population, labeled as (1, 2), ..., (7, 8) in Fig. 209. The path coefficient from a parent to a child will be calculated in the next chapter when we study quantitative traits. Here, while we are on the subject of identical genes, we use half-sibs to illustrate the meaning of one of the basic rules in tracing connecting paths. In the previous chapter it was repeatedly said that correlations are not transitive and that in tracing connections there should be no forward-backward step along the connecting pathways. The case of half-sibs provides us with a biological example in which the tracing rules gain an immediate intuitive meaning. When Fig. 209 is viewed as a path diagram, the four parents become the four uncorrelated causes and the three children (X, Y, Z) are the three effects. Then X and Y are correlated because of the common cause B. Similarly, Y and Z are correlated because of the common cause C. But X and Z are not correlated; the zig-zag line X-B-Y-C-Z is *not* a connecting path because it involves the forbidden forward-backward $(B$-Y-$C)$ step. There is no link between B and C in the diagram; they are uncorrelated causes, and they cannot be linked by their effect Y. These rules, as the reader can see, make good biological sense. The children X and Z in Fig. 209 are obviously not correlated; the former is a child of A and B, with possible gene contents *(1, 3), (1, 4), (2, 3), (2, 4)*; the latter is a child of C and D, with possible gene contents *(5, 7), (5, 8), (6, 7), (6, 8)*. They have no common genes at all. The biological conclusion of this analysis is that a blood relative of a blood relative is not necessarily a blood relative.

The genetic relationships among more distant relatives may also be obtained by the **ITO** method. A slight modification of the matrices **(14)** will enable us to obtain the genetic relationships of relatives with respect to sex-linked genes. It is not the purpose of this chapter to delve into such topics. Further details may be found in Li and Sacks (1954) and in textbooks (Li, 1975b, 1961b).

Exercises

Ex. 1. Study Table 200. Then let $(p, q) = (0.6, 0.4)$ and replace all the algebraic expressions in the table by numerical values. Check that the genetic composition of the offspring generation is the same as that of the parents, viz., *(0.36, 0.48, 0.16)*. Repeat the calculations, using $(p, q) = (0.9, 0.1)$. In the latter case, which one of the three types of families *(Aa × Aa), (Aa × aa),* or *(aa × aa)*, contribute most of the recessives to the offspring generation? If gene A is dominant to a, then AA and Aa are indistinguishable by phenotype. Then the family $Aa \times Aa$ is indistinguishable from the families $AA \times AA$ and $AA \times Aa$. Then we would simply observe that most of the recessive individuals in a population are from normal (dominant) parents. Is this consistent with your own observations regarding simple rare recessive hereditary diseases?

Ex. 2. Cruz-Coke, Nagel, and Etcheverry (1964) studied, among other things, various systems of blood types of the native inhabitants on Easter Island. Those who stay on the island permanently are called "Islanders"

ELEMENTS OF POPULATION GENETICS

and those who go back and forth between the island and mainland Chile are called "Continentals." With regard to the MN blood types, they reported the following data:

	Observed number	
Blood type	Islanders	Continentals
M	20	12
MN	50	24
N	43	9
Total	113	45

Are the observed numbers in each group consistent with the Castle-Hardy-Weinberg law? I will carry out the calculations for the Continentals group and the student may do likewise for the Islanders group. Let p, q be the frequencies of the genes for the M, N antigens, respectively. These values have to be estimated from the observed sample. The estimates are

$$p = \frac{12 + \frac{1}{2}(24)}{45} = 0.53\dot{3}; \qquad q = \frac{9 + \frac{1}{2}(24)}{45} = 0.46\dot{6}$$

(We use informally the same symbols p, q for sample estimates. This should not cause any confusion with population parameters which we shall not use at all.) Once the estimates p, q are obtained, the following calculations are self-evident.

Blood type	Observed number	Expected frequency		Expected number
M	12	p^2	= 0.2844	12.8
MN	24	$2pq$	= 0.4978	22.4
N	9	q^2	= 0.2178	9.8
Total	45		1.0000	45.0

The value of $\chi^2 = 0.23$ is negligible. There is only one degree of freedom in the goodness of fit test, as the value of p (or q) has been calculated from the data themselves. It is seen that the observed and expected numbers from the Castle-Hardy-Weinberg law are in good agreement. How about the Islanders group? ($\chi^2 = 0.67$ with one degree of freedom.) The sample size 45 is considered small for population studies. Moreover, some of the 45 individuals are actually related, not independent random individuals. In spite of all this, the observed numbers are in close agreement with the theoretical binomial distribution. This is why we say that the Castle-Hardy-Weinberg law is robust, and may be used, at least for descriptive purposes, for a large number of occasions.

Ex. 3. One way of getting familiar with Tables 201 and 203 is to employ certain numerical values for (p, q) and compare them. For instance, if $p = q = \frac{1}{2}$, the two tables will assume the following form:

	parent-child				sib-sib			
	AA	Aa	aa	Total	AA	Aa	aa	Total
AA	8	8	0	16	9	6	1	16
Aa	8	16	8	32	6	20	6	32
aa	0	8	8	16	1	6	9	16
Total	16	32	16	64	16	32	16	64

where the entries of the two tables have the same common denominator 64. The reader may construct two similar tables using some other values of p; e.g., $p = 3/5 = 0.60$.

Ex. 4. Write out the expression (15) for **S** in full, with the understanding that the order of the rows as well as the columns are always referring to AA, Aa, aa.

$$\mathbf{S} = \tfrac{1}{4}\begin{pmatrix} 1 & 0 & 0 \\ 0 & 1 & 0 \\ 0 & 0 & 1 \end{pmatrix} + \tfrac{1}{2}\begin{pmatrix} p & q & 0 \\ \tfrac{1}{2}p & \tfrac{1}{2} & \tfrac{1}{2}q \\ 0 & p & q \end{pmatrix} + \tfrac{1}{4}\begin{pmatrix} p^2 & 2pq & q^2 \\ p^2 & 2pq & q^2 \\ p^2 & 2pq & q^2 \end{pmatrix}$$

ELEMENTS OF POPULATION GENETICS

where the fraction in front of each array applies to every element of that array. After multiplying the first row by p^2, the second row by $2pq$, the third by q^2, simplify the algebra and see if the frequencies are the same as those given in Table 203. Example: for the Aa-Aa sib-pair,

$$2pq\,(¼ + ½ \cdot ½ + ¼\, 2pq) = pq\,(1 + pq).$$

Again, let $p = q = ½$ as in **Ex. 3**. Show that the sib-sib joint distribution may be expressed as the sum of three separate distributions as follows:

9	6	1
6	20	6
1	6	9

=

4	0	0
0	8	0
0	0	4

+

4	4	0
4	8	4
0	4	4

+

1	2	1
2	4	2
1	2	1

Ex. 5. Consider the sib-pair AA-aa derived from $Aa \times Aa$ families. Is it possible that these two individuals share one or two identical genes in common? If not, then the four genes involved must be four random genes in the population. With these facts in mind, can you give a direct interpretation of the probability $¼p^2q^2$ for such sib-pairs? [The probability that a pair of sibs does not share any identical genes is ¼; and the probability that the first two random genes be AA and the second two random genes be aa is p^2q^2.]

Chapter 8

QUANTITATIVE TRAITS; HERITABILITY

WE CONTINUE to study random-mating populations with respect to a pair of autosomal genes, except that now a measurement value of a quantitative trait is associated with each genotype. In the beginning we treat the subject as if each genotype can be recognized. This provides the foundation for the more realistic situation where a quantitative trait is determined by more than one locus and by environmental conditions. In the latter case, the exact genotype of any individual is unknown. What we actually study is the phenotype (measurement values), the various components of variance and covariance, and the various correlations among family members. We shall see how the application of path coefficients helps us to clarify the concept of heritability in studying quantitative inheritance. Needless to say, we have to limit ourselves to the very fundamentals of the vast and difficult field of quantitative genetics.

1. Additive gene effects. To avoid abstractness we may think of the quantitative trait as height or any other readily observable and measurable character of individuals. Suppose that each allele asserts a certain effect on height; say, the effect of allele A is $\alpha_1 = 6$ cm and the effect of allele a is $\alpha_2 = -9$ cm when measured from the mean value $\mu = 32$ cm of the population. If these gene effects are additive, then the measurements of the genotypes would be as follows:

genotype:	AA	Aa	aa	
value (L):	$\mu + 2\alpha_1$,	$\mu + \alpha_1 + \alpha_2$,	$\mu + 2\alpha_2$	(1)
example (cm):	44	29	14	

In such a case, the three L values form an arithmetic progression and hence the difference between two successive genotype values is a constant:

$$2\alpha_1 - (\alpha_1 + \alpha_2) = (\alpha_1 + \alpha_2) - 2\alpha_2 = \alpha_1 - \alpha_2 = \beta$$

$$44 - 29 \quad = \quad 29 - 14 \quad = \quad 6 - (-9) \quad = 15 \text{ cm} \qquad (2)$$

The deduction of a constant from each measurement makes no difference to variance or linear regression coefficients; thus we may replace the actual measurements (44, 29, 14) by (30, 15, 0). Further, the division of the measurements by a constant makes no difference to correlation and we may further replace (30, 15, 0) by (2, 1, 0). In other words, when the gene effects are additive, the genotype measurements may be taken as

$$X: \quad \begin{array}{ccc} AA & Aa & aa \\ 2, & 1, & 0 \end{array} \quad (3)$$

without affecting its correlations with other variables.

The coded value $X = 2, 1, 0$ is also the number of A genes in a genotype. This leads to still another interpretation of X. If we take the effect of gene A as unity and that of gene a as zero, then the values of AA, Aa, aa would be 2, 1, 0, respectively. The variances of L (44, 29, 14) and X (2, 1, 0) are related in the following manner:

$$\sigma_L^2 = \sigma_X^2 (\alpha_1 - \alpha_2)^2 = \sigma_X^2 \beta^2 \quad (4)$$

The correlation between L and X is of course unity, as X differs from L only with respect to origin and unit of measurement. Furthermore, if Y is any other variable, then the correlations $r(Y, L) = r(Y, X)$, so that X and L may be regarded as one and the same variable as far as the correlation with other variables is concerned. Finally, it should be realized that the so-called "gene effect" is a concept derived from statistical analysis of genetic data, not a directly observed quantity. The term facilitates our description of the data and the results of our analysis. The method of path coefficients permits us to introduce a variable even though it is not directly observable.

In the following few paragraphs we assume additive gene effects and take the effect of gene A as 1 and that of gene a as 0. In a random-mating population the three genotypes (AA, Aa, aa) and therefore the three measurement values (2, 1, 0) are distributed according to the binomial law $(pA + qa)^2$. For the time being, the environmental effects are ignored and we shall study the pure genetic relationships first. Later, however, the environmental effects will be introduced into the analysis.

Genetic correlations

2. Correlation between genotype and its gametes. The gametic output of the three genotypes and their joint distribution in a random-mating population are as follows:

	gamete produced			joint distribution		
	(A)	(a)		(A) $\quad$ 1	(a) $\quad$ 0	
given parents AA	1	0	(AA) 2	p^2	0	p^2
Aa	½	½	(Aa) 1	pq	pq	$2pq$
aa	0	1	(aa) 0	0	q^2	q^2
				p	q	1.00

(5)

The table on the left gives the pattern of gamete formation by each genotype. Multiplying the genotypes by their frequencies in the population, we obtain the joint distribution (correlation table) of the genotypes and the gametes produced by them. Using the additive values (2, 1, 0) indicated in (5), the correlation may be found easily. The gamete values have mean p and variance pq; the genotype values have mean $2p$ and variance $2pq$. The covariance is $2p^2 + pq - p(2p) = pq$. Hence the correlation between the genotypes and the gametes produced by them is

$$r\binom{\text{genotype}}{\text{gamete}} = b = \frac{pq}{\sqrt{pq}\sqrt{2pq}} = \frac{1}{\sqrt{2}} = 0.7071 \quad (6)$$

The letter b is used to denote this special correlation coefficient, not to be confused with the ordinary regression coefficient. The reason for this is to conform with the established notation of Wright in his path analysis.

3. Correlation between gametes and genotypes formed by them. We shall now study the correlation between the parental gametes and the resulting offspring formed by them. Note carefully the difference between this and the previous situation. The difference will be illustrated by diagrams when we come to the application of path coefficients. For the time being we shall find the correlation by longhand algebra. In constructing the following correlation table, it should be remembered that the total gametic output of the given parent (say, mother) is $p(A)$ and $q(a)$, and for any given mother's gamete, the contribution of the other parent (father) is also $p(A)$ and $q(a)$. Hence:

		conditional offspring				joint distribution			
		AA	Aa	aa		AA 2	Aa 1	aa 1	
given parental gamete	(A)	p	q	0	1	p^2	pq	0	p
	(a)	0	p	q	0	0	pq	q^2	q
						p^2	$2pq$	q^2	1.00

(7)

Comparing (7) with (5), we see that the two tables on the left (conditional probabilities) are different, but that the ones on the right (joint distribution) are the same except that they are turned 90 degrees. Hence the correlation is the same as before:

$$r\left(\genfrac{}{}{0pt}{}{\text{gamete}}{\text{genotype}}\right) = a = \frac{pq}{\sqrt{pq}\,\sqrt{2pq}} = \frac{1}{\sqrt{2}} = 0.7071 \quad (8)$$

Although the numerical values of (8) and (6) are the same, they have very different biological meaning and are denoted by different letters: a for (8) and b for (6). That $a = b = \sqrt{1/2}$ is true only in random-mating populations.

4. Parent-child correlation. In the previous two sections we have seen the conditional probabilities for parent-gamete and then gamete-offspring relations. Combining these two steps we will obtain the conditional probabilities of offspring for given parents:

$$\text{given parent}\begin{array}{c} \\ AA \\ Aa \\ aa \end{array}\begin{pmatrix} (A) & (a) \\ 1 & 0 \\ \tfrac{1}{2} & \tfrac{1}{2} \\ 0 & 1 \end{pmatrix}\begin{array}{c} \\ (A) \\ (a) \end{array}\begin{pmatrix} AA & Aa & aa \\ p & q & 0 \\ 0 & p & q \end{pmatrix} = \begin{pmatrix} AA & Aa & aa \\ p & q & 0 \\ \tfrac{1}{2}p & \tfrac{1}{2} & \tfrac{1}{2}p \\ 0 & p & q \end{pmatrix} = \mathbf{T}$$

which, the reader will recall, is exactly the **T** matrix of Chapter 7 **(14)**. Multiplying the first row of **T** by p^2, the second row by $2pq$, and the third by q^2, we obtain the joint distribution of parent-child pairs (Table 201). Assigning the values $X = 2, 1, 0$ to the genotypes AA, Aa, aa, respectively, we may calculate the correlation coefficient between parent and child in a random-mating population with respect to quantitative traits controlled by additive gene effects. Thus:

[from Table 201]

		\multicolumn{3}{c}{Offspring, X}		
		2	1	0	
	2	p^3	p^2q	0	p^2
Parent, X'	1	p^2q	pq	pq^2	$2pq$
	0	0	pq^2	q^3	q^2
		p^2	$2pq$	q^2	1.00

For both marginal distributions, the mean is $\bar{X} = 2p$ and the variance is $\sigma_X^2 = 2pq$. The parent-offspring covariance is calculated from the cell frequencies of the table above. Using a prime to indicate the parent values,

$$Cov(X', X) = 4p^3 + 4p^2q + pq - (2p)^2 = pq = \tfrac{1}{2}\sigma_X^2 \qquad (9)$$

The parent-child correlation is

$$r_{PO} = r(X', X) = \frac{pq}{\sqrt{2pq}\sqrt{2pq}} = \left(\frac{1}{\sqrt{2}}\right)\left(\frac{1}{\sqrt{2}}\right) = \frac{1}{2} \qquad (10)$$

or

$$r(X', X) = \frac{\tfrac{1}{2}\sigma_X^2}{\sigma_X^2} = \frac{1}{2} \qquad (10')$$

This is one of the most basic results in quantitative population genetics. The two expressions of **(10)** permit two verbal descriptions of the result. The first says that the *PO* correlation consists of two steps:

QUANTITATIVE TRAITS; HERITABILITY   219

$$r(parent\text{-}offspring) = r(parent\text{-}gamete) \cdot r(gamete\text{-}offspring) = b\,a \quad (11)$$

where b and a are given by (6) and (8), respectively. The second expression emphasizes the fact that $Cov(X', X) = \frac{1}{2}\sigma_X^2$ which is true for all linear scales in a random-mating population. We shall make use of this fact again when we deal with dominance.

5. Sib-sib correlation. Similarly, if we assign the values 2, 1, 0 to AA, Aa, aa, respectively in Table 203, we will be able to calculate the sib-sib correlation. The marginal variables remain the same for all random-mating populations and need no new calculation. The covariance is (Table 203)

$$p^2(1+p)^2 + 2p^2\,q(1+p) + pq(1+pq) - (2p)^2 = pq$$

so that

$$r(sib\text{-}sib) = r_{00} = \frac{pq}{2pq} = \frac{1}{2} \quad (12)$$

which is the same as r_{P0} for additive gene effects. This correlation may be decomposed into separate components. The decomposition in terms of b and a will be given when we apply path coefficients. For the time being, we shall use a subdivision according to the number of identical genes the sibs share in common, viz., by the **ITO** method (Chapter 7). It will be recalled that the entries of Table 203 may be split into three components. For clarity, those three components can be rewritten in the following form:

$$\frac{1}{4}\begin{array}{|c|c|c|}\hline p^2 & 0 & 0 \\\hline 0 & 2pq & 0 \\\hline 0 & 0 & q^2 \\\hline\end{array} + \frac{1}{2}\begin{array}{|c|c|c|}\hline p^3 & p^2q & 0 \\\hline p^2q & pq & pq^2 \\\hline 0 & pq^2 & q^3 \\\hline\end{array} + \frac{1}{4}\begin{array}{|c|c|c|}\hline p^4 & 2p^3q & p^2q^2 \\\hline 2p^3q & 4p^2q^2 & 2pq^3 \\\hline p^2q^2 & 2pq^3 & q^4 \\\hline\end{array} \quad (13)$$

These are obtained by first writing out the elements of $\mathbf{S} = \frac{1}{4}\mathbf{I} + \frac{1}{2}\mathbf{T} + \frac{1}{4}\mathbf{O}$, as was done in Ex. 3 of Chapter 7, and then multiplying the 1st, 2nd, 3rd rows by p^2, $2pq$, q^2, respectively. Then it is clear that the sib-sib correlation table is a mixture of three separate correlation tables with the same marginal distributions. If we let r_1, r_2, r_3 be the correlation coefficients of the 1st, 2nd, 3rd table, respectively, in (13) above, then the overall sib-sib correlation would be the weighted average:

$$r_{00} = \tfrac{1}{4} r_1 + \tfrac{1}{2} r_2 + \tfrac{1}{4} r_3 \tag{14}$$

Now, for the first table, the correlation is obviously unity $(r_1 = 1)$, no matter what the genotype measurements may be (additive or otherwise). The second table is identical with the parent-child correlation table, so that $r_2 = \tfrac{1}{2}$ for additive gene effects. The third table shows independent distribution of the two sibs, so that $r_3 = 0$ no matter what the genotype measurements may be (additive or otherwise). Combining the three results, we obtain

$$r_{00} = \tfrac{1}{4}(1) + \tfrac{1}{2}(\tfrac{1}{2}) + \tfrac{1}{4}(0) = \tfrac{1}{2} \tag{15}$$

The three tables of (13) will be used again when we study dominance.

The reader must have noticed the remarkable fact that all four correlations, $r(parent\text{-}gamete)$, $r(gamete\text{-}offspring)$, $r(parent\text{-}offspring)$, and $r(sib\text{-}sib)$, are independent of gene frequencies (p, q) when the gene effects are additive. When the numbers *2, 1, 0* are associated with the genotypes *AA, Aa, aa,* respectively, these correlation coefficients give the basic genetic relationships among the family members. They are applicable to all random-mating populations and to all traits controlled by additive gene effects. Although the correlation tables for parent-child and sib-sib (Tables 201 and 203) are different, they have the same correlation coefficient with respect to additive gene effects. If the quantitative trait is influenced by a certain number of loci and there are no interactions between the genes of different loci which are independently distributed in the population, then the correlation coefficients $r_{PO} = \tfrac{1}{2}$ and $r_{OO} = \tfrac{1}{2}$ will remain the same, because both the variance and the covariance are then simply the sum of the contributions from the individual loci.

6. Now we shall examine some empirical studies and see how they fare with theory. The dermal patterns and the number of ridges on palms, fingers, soles, and toes are known to remain constant throughout life with no environmental intervention. The total number of ridges on the ten fingers of an

individual is known as the "total (finger) ridge-count," a quantitative trait of the individual. The number of loci influencing the formation of dermal ridges is unknown, but the correlations between relatives may be studied empirically. Some of the results of Holt (1968) are given in Table 221. First, we note that the correlation for the 200 pairs of parents is smaller than its standard error, implying random mating with respect to the total finger ridge-count. Next, the correlations between parent and child are all very close to the theoretical value of 0.50. This suggests that the gene effects are essentially additive with respect to the trait under study (total ridge-count). The correlation for the 642 pairs of full sibs is 0.50, as expected from additive gene effects. Dizygotic twins are genetically the same as full sibs, and their correlation also turns out to be very close to 0.50. In spite of their common milieu during pregnancy and greater similar environment after birth, the correlation for dizygotic twins is not a bit higher than that for ordinary full sibs. All of these are in perfect agreement with the genetic theory developed in the preceding paragraphs. The correlation for monozygotic twins is 0.95, totally unlike that for dizygotic twins, although they are also twins. The bottom line of Table 221 shows that the correlation between the midparent (mean of two parents) and child is 0.66. Its theoretical value for a random-mating population is $\sqrt{0.50} = 0.707$, as will be explained in a later paragraph.

Table 221. The correlation among family members with respect to the total finger ridge-count (Holt, 1968).

number of pairs	relationship	observed correlation
200	parent-parent	$r = 0.05 \pm 0.07$
405	mother-child	$r = 0.48 \pm 0.04$
405	father-child	$r = 0.49 \pm 0.04$
810	parent-child	$r = 0.48 \pm 0.03$
642	sib-sib	$r = 0.50 \pm 0.04$
92	dizygotic twins	$r = 0.49 \pm 0.08$
80	monozygotic twins	$r = 0.95 \pm 0.01$
405	midparent-child	$r = 0.66 \pm 0.03$

Analysis of genetic variance by families

7. Population variance. When gene effects are additive, we shall continue to use the scales $2, 1, 0$, as they have no effect on ratios or proportions. Now consider the random-mating population:

$$\begin{array}{lccc} \text{genotype:} & AA & Aa & aa \\ \text{frequency:} & p^2 & 2pq & q^2 \\ \text{additive scale, } X: & 2 & 1 & 0 \end{array} \qquad (16)$$

It is shown in textbooks that for a binomial distribution of the nth degree, $(p + q)^n$, the mean is $\bar{X} = np$ and the variance is $\sigma_X^2 = npq$. In distribution (16) above, $n = 2$; hence the mean and variance are, respectively,

$$\bar{X} = 2p, \qquad \sigma_X^2 = 2pq \qquad (17)$$

The only warning to the reader is that he must not confuse the variance $2pq$ in (17) with the heterozygote frequency $2pq$ in (16). Although they happen to have the same apparent value in this particular case, they are entirely different things. (This coincidence happens only when $n = 2$ for a binomial distribution.)

In an equilibrium population under random mating, the frequencies of genotypes remain constant from generation to generation and so will the mean and variance of the quantitative trait. If an environmental improvement increases the measurement of all genotypes by the same magnitude (e.g., replacing the old values $44, 29, 14$ by the new values $45, 30, 15$) the variance of the trait remains the same.

8. One-way classification. The individuals of the population may be subdivided into six groups according to the six types of parental combinations (Table 200). When a population is subdivided into several groups according to whatever criterion, it is called a "one-way" classification. Then the population variance $\sigma_X^2 = 2pq$ may be subdivided into two components; one component is the variance between the group means and one is the average variance within the groups. Table 223 classifies the individuals of the population into six family groups. The total frequency of individuals in a group is called the group "size." The six groups are thus of unequal size, being

QUANTITATIVE TRAITS; HERITABILITY

Table 223. **One-way classification.** Analysis of genetic variance due to additive gene effects by family groups in a random-mating population (based on Table 200).

familial source	additive scale X			group "size"	group mean	within-group variance
	2	1	0			
AA × AA	p^4	0	0	p^4	2.00	0
AA × Aa	$2p^3q$	$2p^3q$	0	$4p^3q$	1.50	$4p^3q\,(¼)$
Aa × Aa	p^2q^2	$2p^2q^2$	p^2q^2	$4p^2q^2$	1.00	$4p^2q^2\,(½)$
AA × aa	0	$2p^2q^2$	0	$2p^2q^2$	1.00	0
Aa × aa	0	$2pq^3$	$2pq^3$	$4pq^3$	0.50	$4pq^3\,(¼)$
aa × aa	0	0	q^4	q^4	0	0
Total	p^2	$2pq$	q^2	1.00	$2p$	pq

proportional to the frequency of the various types of mating in the population. Each group has a mean and a variance of its own, as indicated in Table 223. Before proceeding further, we may observe from the table that the mean of each group of individuals is equal to the mean value of the two parents producing such a group of individuals. For instance, the mean of parents AA and Aa is $½(2 + 1) = 1.50$ and so is the average value of their children.

9. Between and within. The variance of the group means may be readily calculated from Table 223. The following argument, however, gives us a shortcut. For a binomial variable $M = 4, 3, 2, 1, 0$ with the terms of $(p + q)^4$ as frequencies, its mean is $\overline{M} = np = 4p$ and its variance is $\sigma_M^2 = npq = 4pq$. The group means shown in Table 223 are

$$m = ½M = 2.0,\ 1.5,\ 1.0,\ 0.5,\ 0$$

with binomial frequencies. Hence, its mean value is $\overline{m} = ½(4p) = 2p$ and its variance is

$$\sigma_m^2 = (½)^2(4pq) = pq = ½\sigma_X^2 \qquad (18)$$

which is the *between-family* variance. Next, we calculate the average within-family variance. In the first family group of Table 223, all individuals are AA

with measurement 2; the within-group variance is zero. In the second family group half of the individuals have the measurement 2 and half have the value 1; the within group variance is $\frac{1}{2} \cdot \frac{1}{2} = \frac{1}{4}$. It is then weighted by $4p^3q$, the size of the group, as indicated in the last column of Table 223. Proceeding this way throughout the six family groups and summing, we obtain the weighted average *within-family* variance. It turns out to be pq also. To summarize:

total variance		between families		within families	
$2pq$	$=$	pq	$+$	pq	(19)
σ_X^2	$=$	$\frac{1}{2}\sigma_X^2$	$+$	$\frac{1}{2}\sigma_X^2$	

In words, in a random-mating population, if a quantitative trait is influenced by additive gene effects, then half of the population variance is due to genetic differences between the families and half due to gene segregation within the families. The 50-50 subdivision of the genetic variance is independent of gene frequencies in the population. It is a property of Mendelian segregation and random mating.

Genetic path analysis

10. The various genetic correlations have been obtained in the previous sections by actually constructing the various correlation tables for one autosomal locus. Now we shall obtain the results by the method of path coefficients for its succinctness as well as for its generality. The path method does not depend on the number of loci involved; the only requirements are that the gene effects are additive and the loci are independently distributed in the population. We begin with the simplest case, of course.

Consider the two gametes (g_1 and g_2) that unite to form a new genotype (Fig. 225). If the gene effects are additive, then the measurement value (Z) of the new genotype is always the sum of those of the two uniting gametes:

$$Z = g_1 + g_2 \qquad (20)$$

and there is no residual involved. That is to say, Z is completely determined by g_1 and g_2. In a random-mating population the uniting gametes are uncorrelated, so that $r(g_1, g_2) = 0$. For autosomal genes, the male and female

QUANTITATIVE TRAITS; HERITABILITY 225

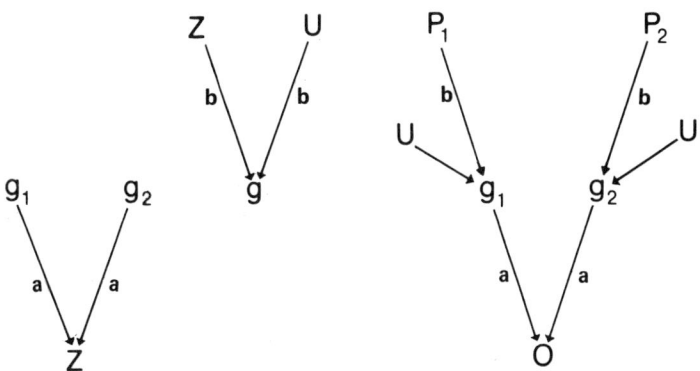

Fig. 225. Path coefficients from uniting gametes to zygote and from a parental zygote to the gametes produced by them. *Left:* two uncorrelated gametes unite at fertilization to form a new genotype of the zygote. *Middle:* a gamete is determined by the parental genotype Z and chance at segregation U. *Right:* a complete diagram showing the path relations between parents (P_1 and P_2) and offspring (O) in a random-mating population.

gametes are equally important (i.e., g_1 and g_2 have the same mean and variance). Let a_1 be the path coefficient from the gamete variable g_1 to a resulting genotype Z and a_2 the path coefficient from g_2 to Z. Since these two gametes are equally important in determining the value of Z, we put $a_1 = a_2 = a$, as indicated in Fig. 225. (The symbol a here is a path coefficient, not to be confused with allele a.) Then from the theorem of complete determination (Chapter 6), we have immediately

$$a^2 + a^2 = 2a^2 = 1$$
$$a^2 = \tfrac{1}{2}; \quad a = \sqrt{\tfrac{1}{2}} = 0.7071 \tag{21}$$

Since the "causes" g_1 and g_2 are uncorrelated, the path coefficient $a = \sqrt{\tfrac{1}{2}}$ is also the correlation coefficient between a gamete and a resulting genotype, in agreement with our previous result (8) derived from (7). This is the very first result in applying the path method to genetics problems. The reader may begin to appreciate the power of the path method; it arrives at the correct correlation in one stroke. In more complicated problems, the construction of a complete correlation table is extremely tedious, if not impossible. But the path method will yield the correct correlation in short order.

11. The path coefficient *(b)* from a parental genotype to a gamete produced by the parent is not so obvious, as a genotype may produce many different types of gametes by the random process of Mendelian segregation. We may argue along several lines. First, we have already obtained from direct calculation [(5), (6), (7), (8)] that in a random-mating population

$$b = a = \sqrt{\tfrac{1}{2}} \tag{22}$$

Or, we may use the result of the analysis of variance which shows that half of the variance is due to chance of segregation and half due to parental genetic differences (19). If we visualize (Fig. 225) that a gamete variable g is completely and equally determined by a segregation variable U (analogous to residual) and the parental genotype Z, then we may write

$$g = Z + U, \quad \text{where} \quad r(Z, U) = 0$$

Since Z and U are equally important in determining g, the path coefficients from them to g must be the same. So, as in (21), we have

$$b^2 + b^2 = 1, \quad b = \sqrt{\tfrac{1}{2}} \tag{23}$$

A more general argument that applies to non-random-mating populations as well is the following. Imagine a genotype as an urn, into which we put marbles and from which we take marbles out. The correlations (not necessarily path coefficients) between the urn content and the in-marbles and the out-marbles must be the same. In the special case of random mating, these correlations are also path coefficients. Hence $b = a$. (For non-random-mating populations, $b \neq a$, as will be shown in the next chapter.)

12. Parent-offspring. When the two basic path coefficients *(b* and *a)* are known, the correlations and path coefficients between relatives may be obtained immediately. The path from one parent $(P_1$ or $P_2)$ to one offspring *(O)* is, from Fig. 225,

$$r_{PO} = ab = \sqrt{½}\sqrt{½} = ½ \qquad (24)$$

which we have verified by longhand algebra (10). That the compound path ab is also the correlation r_{PO} is due to the fact that the "causes" (P's and U's) are uncorrelated. Note that the chances of segregation play a role in the stage of producing the gametes (from P to g); but once the gametes have been produced, the two uniting gametes will completely determine the resulting genotype (O). Consequently, the two parents do not completely determine the genotype of the offspring on account of the segregation variables U involved. The determination of offspring O by P_1 and P_2 is $(ab)^2 + (ab)^2 = ¼ + ¼ = ½$. The rest is determined by the two U's. To emphasize this point, the two U's are explicitly shown in Fig. 225. However, in the interest of simplicity, the U's will be omitted from all subsequent diagrams, as is usually done in genetic literature, but their presence should always be kept in mind.

13. Sib-sib. The correlation between full sibs will be obtained as soon as the appropriate path diagram is constructed with practically no calculation. Of course, this is the advantage of using the method of path coefficients. From Fig. 228 (U's being omitted), we see that the two children are connected by two paths, one through the male parent and one through the female parent. The two parents are the two common causes in the language of path method. The correlation between the two children is the sum of the two connecting paths:

$$r_{OO} = abba + abba = 2a^2b^2 = 2(½)(½) = ½ \qquad (25)$$

The correlation is the same as that for parent-offspring, but the path analysis shows that each parent contributes ¼ toward the total correlation of ½. Again, in order to simplify the diagram, the intermediate gamete variables may be omitted as shown in the right side of Fig. 228. In such a case, the path from a parent to a child is $ba = ½$. Note that Fig. 225 is only a part of Fig. 228, except for the simplified notations. If we omit one of the two parents from Fig. 228, then the resulting diagram gives the correlation between half-sibs who have only one connecting path:

$$r(\text{half-sibs}) = r_{O/O} = abba = (ab)^2 = ¼ \qquad (26)$$

We shall not carry the analysis to more remote relatives in this book.

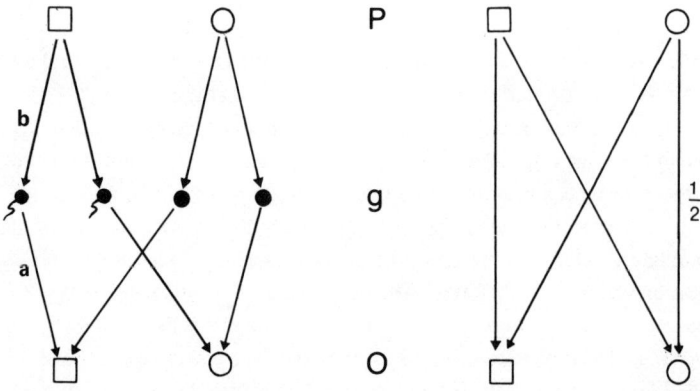

Fig. 228. The path relations between parents P and offspring O in a random-mating population. *Left:* complete diagram showing the gametes g. *Right:* simplification of diagram by omission of gametes. The chance factor at segregation has also been omitted for simplicity.

14. Midparent and average offspring. Let the measurements of the two parents be P_1 and P_2, and those of their offspring be $O_1, \ldots, O_s$, where s is the size of the sibship (number of children in the family). We often wish to know the correlation between the average of the two parents and the average of their children. The correlation between the two averages is the same as the correlation between the sum of the two parents and the sum of their children. Thus, there is no need to actually calculate the "midparent" $\bar{P} = \frac{1}{2}(P_1 + P_2)$ and average child. The following calculation is based on the two sums:

$$\Sigma P = P_1 + P_2; \qquad \Sigma O = O_1 + \ldots + O_s \qquad (27)$$

In an equilibrium population the variance of these individuals is all the same:

$$\sigma^2_{P_1} = \sigma^2_{P_2} = \sigma^2_{O_i} = \sigma^2 \qquad (28)$$

which may be taken as unity, as in path analysis. Then the covariances are

QUANTITATIVE TRAITS; HERITABILITY

$$Cov(P_i, O_j) = r_{PO} = \tfrac{1}{2}$$

$$Cov(O_i, O_j) = r_{OO} = \tfrac{1}{2}$$

and $Cov(P_1, P_2) = 0$ under random mating. Based on these results the variances and covariance between the two sums may be found easily:

$$\sigma^2_{\Sigma P} = \sigma^2_{P_1} + \sigma^2_{P_2} = 2\sigma^2 = 2$$

$$\sigma^2_{\Sigma O} = s\sigma^2_0 + s(s-1)r_{OO} = s + \tfrac{1}{2}s(s-1) = \tfrac{1}{2}s(s+1) \quad (29)$$

$$Cov(\Sigma P, \Sigma O) = E(P_1 + P_2)(O_1 + \ldots + O_s) = 2s\,r_{PO} = s$$

so that the correlation between the average parent $\bar{P}$ and the average offspring $\bar{O}$ is

$$r(\Sigma P, \Sigma O) = \frac{s}{\sqrt{2}\sqrt{\tfrac{1}{2}s(s+1)}} = \sqrt{\frac{s}{s+1}} \quad (30)$$

in agreement with Li (1954) through a different method. Note again that these results are independent of gene frequencies. The correlation (30) depends only on the size of the sibship. As a particular case, the correlation between the midparent and one child $(s = 1)$ is $\sqrt{1/(1+1)} = \sqrt{\tfrac{1}{2}} = 0.7071$. The observed correlation for 405 families with respect to the total finger ridge-count (bottom line of Table 225) is 0.66 ± 0.03, slightly lower than expected but not significantly so. The correlation between one parent and the average of s children may be obtained simply by omitting one parent from the expressions above. We leave this as an exercise for the reader (**Ex. 1**).

Alleles with dominance

15. All of the foregoing results are independent of gene frequencies and applicable to all quantitative traits controlled by genes without dominance (i.e., additive gene effects). As exemplified by the three L values in (**1**), the measurements of the three genotypes are 44, 29, 14, where the measurement of the heterozygote is the midpoint between the two homozygotes. Now, suppose that the actual measurements of the genotypes are as follows:

genotype:	AA	Aa	aa	
measurement value:	40	35	5	
general notation, Y:	Y_2	Y_1	Y_0	(31)
frequency, f:	p^2	$2pq$	q^2	

What would be the correlations between the family members? As the example above shows, the heterozygote measurement is closer to one homozygote than to the other. In the example above, the allele A is partially dominant to allele a. (Rather, the allele with dominance is called A). This section investigates the effects of dominance on the genetic correlations. For the time being, we continue to assume that each genotype has only one value (Y_2 for AA, etc.). Environmental fluctuations will be introduced afterwards.

16. Population variance. When the three genotypes (AA, Aa, aa) take on the measurements (Y_2, Y_1, Y_0) as shown in (31), it is not clear what would be the correlation between, say, parent and offspring. At first sight it seems doubtful that a general and meaningful expression can be found. We are indebted to the profound insights of the pioneers W. Weinberg, R. A. Fisher, and S. Wright for a neat solution of the problem. The writer regards the solution as a major breakthrough in quantitative genetics. It also paves the way to the concept of heritability. Once the answer becomes known, however, it may be explained in various other ways, for hindsight is easier than foresight. For the purpose of exposition by elementary means, we shall not follow exactly the original arguments of the pioneers but resort to simpler ones. The methods we shall employ require only a knowledge of linear regression (Chapter 3) and some basic knowledge of population genetics (Chapter 7). According to the notation of (31), the population mean and variance are

$$\mu \;=\; \overline{Y} \;=\; p^2 Y_2 \;+\; 2pq\, Y_1 \;+\; q^2 Y_0 \quad (32)$$

$$\sigma_Y^2 \;=\; \Sigma f Y^2 - \overline{Y}^2 \;=\; p^2 Y_2^2 \;+\; 2pq\, Y_1^2 \;+\; q^2 Y_0^2 - \overline{Y}^2 \quad (33)$$

A key step in finding the correlation between relatives is an appropriate subdivision of the variance of Y. This we shall do in the next few paragraphs.

QUANTITATIVE TRAITS; HERITABILITY

17. The linear component. Probably the most obvious and easiest way of subdividing the variance of Y is by linear regression of Y on X, where $X = 2, 1, 0$, representing the number of A-genes in a genotype. In reading the following algebra, the reader may from time to time consult Table 231 for numerical illustrations. The numerical results of Table 231 are no different from those of Table 61, except that now genetic contents and terminology (and thus notation) have been injected. The regression coefficient

Table 231. The genetic composition of a random-mating population with quantitative trait and analysis by linear regression (also see Table 61).

genotype	frequency f	number genes X	measurement Y	linear value L	dominance deviation D
AA	$p^2 = 0.36$	2	$Y_2 = 40$	$L_2 = 44$	$D_2 = -4$
Aa	$2pq = 0.48$	1	$Y_1 = 35$	$L_1 = 29$	$D_1 = +6$
aa	$q^2 = 0.16$	0	$Y_0 = 5$	$L_0 = 14$	$D_0 = -9$
Total or mean	1.00	$\bar{X} = 1.2$	$\bar{Y} = 32$	$\bar{L} = 32$	$\bar{D} = 0$
Variance		$\sigma_X^2 = 0.48$	$\sigma_Y^2 = 144$	$\sigma_L^2 = 108$	$\sigma_D^2 = 36$

of Y on X will be called β, where $\beta = Cov(X, Y)/Var(X)$. But $V(X) = 2pq$ as given by (17). So we need only to find the covariance. From Table 231,

$$\begin{aligned} Cov(X, Y) &= \Sigma f XY - \bar{X}\bar{Y} \\ &= 2p^2 Y_2 + 2pq\, Y_1 - 2p(p^2 Y_2 + 2pq\, Y_1 + q^2 Y_0) \\ &= 2pq\, [p(Y_2 - Y_1) + q(Y_1 - Y_0)] \end{aligned} \quad (34)$$

noting that $p^2 - p^3 = p^2 q$ and $pq = p^2 q + pq^2$. Dividing the covariance (34) by $Var(X) = 2pq$, we obtain the linear regression coefficient of Y on X:

$$\beta = p(Y_2 - Y_1) + q(Y_1 - Y_0) \quad (35)$$

which is a very important quantity in population genetics. It is the weighted mean of the differences $(Y_2 - Y_1)$ and $(Y_1 - Y_0)$, the weights being the gene frequencies. The linear regression equation is

$$\hat{Y} = L = \overline{Y} - \beta \overline{X} + \beta X = \alpha + \beta X \qquad (36)$$

where $\alpha = \overline{Y} - \beta \overline{X}$ is a constant and $X = 2, 1, 0$. In our numerical example (31) with $(p, q) = (0.6, 0.4)$,

$$\beta = 0.6(40 - 35) + 0.4(35 - 5) = 3 + 12 = 15$$

in agreement with (50) in Chapter 3. The reader should now proceed to calculate the three L values for the genotypes AA, Aa, aa from equation (36) and see that they are 44, 29, 14, respectively. Note that there is no dominance in these L values and their mean $\overline{L}$ is equal to the population mean $\overline{Y}$. Since X varies by unit step, the difference between two adjacent L values is also the regression coefficient. Thus, $\beta = 44 - 29 = 29 - 14 = 15$. The variance of L is called the "linear component" of the variance of Y, and it is equal to

$$\sigma_L^2 = \sigma_X^2 \beta^2 = 2pq\beta^2 \qquad (37)$$

where β is given by (35). This component is due to the influence of X on Y, as reflected by the regression coefficient β. In genetic literature this component is also known as the "additive component" denoted by σ_A^2, because the L values (44, 29, 14) exhibit no dominance and they are as if determined by additive gene effects.

18. The dominance component. The deviation of the actual Y from the calculated linear value L is

$$D = Y - L \qquad (38)$$

This deviation has been called "residual" in Chapter 3 as in all statistics books; but, in view of its genetic meaning here, we will call them "dominance deviations," as indicated in the last column of Table 231. If the Y values were linear to begin with, all these D values would be zero. That they are not is due to the presence of nonlinearity (dominance) in the original Y measurements. In Chapter 3, it was shown that L and D are uncorrelated; hence

QUANTITATIVE TRAITS; HERITABILITY

$$Y = L + D, \qquad \sigma_Y^2 = \sigma_L^2 + \sigma_D^2 \qquad (39)$$

where σ_D^2 is called the dominance component of the variance of Y. In practice there is no need to find an expression for σ_D^2, because it may be calculated by subtraction, $\sigma_Y^2 - \sigma_L^2$. In this particular genetic context, however, a meaningful expression for the dominance component may be obtained. Writing out the expressions for variance of Y (33) and the linear component (37) and subtracting, we find

$$\sigma_D^2 = p^2 q^2 (Y_2 - 2Y_1 + Y_0)^2 = p^2 q^2 \delta^2 \qquad (40)$$

where $\delta = Y_2 - 2Y_1 + Y_0$. The longhand algebraic subtraction procedure adopted here is tedious but requires no additional knowledge beyond high school algebra. Note that if the original measurements were linear, $\delta = Y_2 - 2Y_1 + Y_0 = 0$ and $\sigma_D^2 = 0$. Then $\sigma_Y^2 = \sigma_L^2$. The quantity $\delta = Y_2 - 2Y_1 + Y_0$ is sometimes also called "dominance deviation," because it measures the departure from linearity. Whatever we choose to call such quantities, strict distinction must be made between the individual deviations $D = Y - L$ for each genotype, and the single index $\delta = Y_2 - 2Y_1 + Y_0$ based on all three genotypes. The reader should calculate the variance of D from the last column of Table 231 and then check his answer with that given by (40).

19. Degree of dominance. For descriptive purposes we need an index to measure the degree of dominance. Various indexes may be constructed. In view of the reduced scale $X = 2, 1, 0$ for the case of no dominance, two convenient indexes (θ and ω) are as follows:

	genotype	AA	Aa	aa	
(present)	scale I	2	$1 + \theta$	0	(41)
(Wright)	scale II	$2 - \omega$	1	0	(42)

For both scales, $\theta = \omega = 0$ for no dominance and $\theta = \omega = 1$ for complete dominance. The relationships between the two measures are

$$\frac{2}{1 + \theta} = 2 - \omega, \qquad \omega = \frac{2\theta}{1 + \theta}, \qquad \theta = \frac{\omega}{2 - \omega} \qquad (43)$$

When the measurements of genotypes are given Y_2, Y_1, Y_0, they are first reduced to $(Y_2 - Y_0)$, $(Y_1 - Y_0)$, 0. If we use scale I, let $(Y_2 - Y_0) : (Y_1 - Y_0) = 2 : 1 + \theta$. If we use scale II, let $(Y_2 - Y_0)/(Y_1 - Y_0) = 2 - \omega$. Solving,

$$\theta = \frac{2Y_1 - Y_2 - Y_0}{Y_2 - Y_0}, \qquad \omega = \frac{2Y_1 - Y_2 - Y_0}{Y_1 - Y_0} \qquad (44)$$

In our numerical example the three Y-values are 40, 35, 5, so that $2Y_1 - Y_2 - Y_0 = 70 - 45 = 25$, and the degrees of dominance are

$$\theta = \frac{25}{35} = 0.7143, \qquad \omega = \frac{25}{30} = 0.8333$$

Wright (1952) used ω to measure the degree of dominance. We shall use the θ index in the next paragraph.

20. Effects of gene frequency and dominance on variance components. Through linear regression the population variance of the measurements has been subdivided into linear and dominance components. There is no fixed relationship between the two components except that they add up to the total variance. Each of the two components may vary from zero to the full magnitude of the variance of Y. The relative magnitudes of the two components depends on gene frequencies of the population as well as on the degree of dominance. The two extreme situations are easy to see. One extreme is when $\delta = Y_2 - 2Y_1 + Y_0 = 0$, and $\sigma_D^2 = 0$; hence, $\sigma_Y^2 = \sigma_L^2$. The other extreme is when the regression coefficient is zero, viz.,

$$\beta = p(Y_2 - Y_1) + q(Y_1 - Y_0) = 0 \qquad (45)$$

Solving, $\quad p = \dfrac{Y_1 - Y_0}{(Y_1 - Y_2) + (Y_1 - Y_0)}, \quad q = \dfrac{Y_1 - Y_2}{(Y_1 - Y_2) + (Y_1 - Y_0)} \quad (46)$

In such a case, $\sigma_L^2 = 2pq\beta^2 = 0$, and $\sigma_D^2 = \sigma_Y^2$. Since p and q are positive fractions, an examination of (46) shows that the two differences $(Y_1 - Y_2)$ and $(Y_1 - Y_0)$ must be both positive or both negative. That is, the solution (46) of the equation (45) is possible only when the heterozygote measurement Y_1 is larger than Y_2 *and* than Y_0, or smaller than Y_2 *and* than Y_0. Or, briefly, Y_1 is outside the range of Y_2 and Y_0.

If the Y's are the "reproductive value" of the genotypes, the result (46) is a very important one in the theory of natural selection and evolution, a subject that is out of the scope of this volume. Readers may consult Li (1967) for a review of "selection genetics." For a quantitative trait influenced by several or many pairs of genes, the situation (46) is unlikely to happen. Even if it happens for a certain single locus, it is still a remote possibility that many loci have similar situations. Hence we shall not consider the situation (45) any further except to point out that the linear component σ_L^2 will be relatively small if Y_1 is outside the range of Y_2 and Y_0. So we shall consider only the usual case in which Y_1 is inside the range of Y_2 and Y_0.

21. It will be clear later that the proportions of the two components of variance are important in calculating the correlation between relatives. The two proportions will be denoted by

$$g^2 = \frac{\sigma_L^2}{\sigma_Y^2}, \qquad d^2 = \frac{\sigma_D^2}{\sigma_Y^2}, \qquad g^2 + d^2 = 1 \qquad (47)$$

where d^2 is known as the "dominance ratio" (Fisher, 1918). In our numerical example (Table 231), $g^2 = 108/144 = ¾$ and $d^2 = 36/144 = ¼$. As long as Y_1 is between Y_2 and Y_0, there will always be a linear component. The proportions g^2 and d^2 depend on the degree of dominance and gene frequencies, as evident from (35) and (40). In order to tabulate the values of g^2, we use the scale (41): $Y = 2, 1+\theta, 0$. Then the regression (35) becomes

$$\beta = p(1-\theta) + q(1+\theta) = 1 + (q-p)\theta \qquad (48)$$

and
$$\delta = Y_2 - 2Y_1 + Y_0 = -2\theta \qquad (49)$$

Substituting in (37), (40), and (47), we have

$$g^2 = \frac{[1 + (q-p)\theta]^2}{1 + 2(q-p)\theta + (p^2+q^2)\theta^2} \qquad (50)$$

which is the value listed in Table 236. The first column $(\theta = 0)$ of Table 236 shows $g^2 = 1$ for all gene frequencies. This is merely a repetition of

Table 236. Values of the proportion $g^2 = \sigma_L^2 / \sigma_Y^2$ for various gene frequencies and degrees of dominance in a random-mating population.

gene frequencies (p, q)		degree of dominance, θ				
		0	0.25	0.50	0.75	1.00
0.90,	0.10	1.00	0.983	0.889	0.612	0.182
0.70,	0.30	1.00	0.969	0.859	0.675	0.462
0.50,	0.50	1.00	0.970	0.889	0.780	0.667
0.30,	0.70	1.00	0.979	0.932	0.877	0.824
0.10,	0.90	1.00	0.992	0.978	0.962	0.947

what we have noted previously, viz., $\sigma_Y^2 = \sigma_L^2$ when there is no dominance. For moderate dominance, the proportion g^2 is still high, being greater than 95% for $\theta = 0.25$ and greater than 85% for $\theta = 0.50$. For complete dominance, we put $\theta = 1$ in the previous expressions:

$$\beta = 1 + q - p = 2q, \qquad \delta = -2$$

and
$$\sigma_L^2 = 2pq\beta^2 = 2pq(2q)^2 = 8pq^3$$

$$\sigma_D^2 = p^2 q^2 \delta^2 = p^2 q^2 (-2)^2 = 4p^2 q^2$$

$$\sigma_Y^2 = \sigma_L^2 + \sigma_D^2 = 4pq^2(1+q)$$

so that
$$g^2 = \frac{2q}{1+q}, \qquad d^2 = \frac{1-q}{1+q}$$

When $q = 0.10$, $g^2 = 0.20/1.10 = 0.182$. When $q = 0.90$, $g^2 = 1.80/1.90 = 0.947$. These are the values listed in the last column of Table 236. The conclusion is that in general the value of g^2 is fairly high, except for rare recessive genes. (When the heterozygote measurement is outside the range of the two homozygote measurements, g^2 is invariably low.) It should be emphasized that since g^2 depends on both gene frequency and degree of dominance, its value will vary from population to population and from trait to trait.

Genetic correlation with dominance

22. Now we are ready to tackle the main problem of finding the correlation between parent and offspring when the measurements of the genotypes AA, Aa, aa are Y_2, Y_1, Y_0, respectively. There are about half a dozen different ways of arriving at the desired result. In this chapter we have developed the subject by linear regression to prepare us for future representation of path diagrams. Hence we shall derive the desired correlation using only what we have already obtained in the preceding paragraphs. By linear regression the Y values have been split into two uncorrelated parts, L and D. Our method will be based entirely on the properties of L and D, some of which are reiterated here or put into a slightly different form. First, we note that in calculating the covariance of parent and offspring it is unnecessary to use the full value (36), $L = \alpha + \beta X$. The constant α may be deleted without affecting either the variance or the covariance. Hence we shall simply regard L as βX in calculating any correlation coefficient. As $X = 2, 1, 0$, the L values will be regarded as $(L_2, L_1, L_0) = (2\beta, \beta, 0)$, which is essentially the same as $(2, 1, 0)$ except for the unit β.

The fact that $Cov(L, D) = 0$ may also be much simplified:

$$\Sigma fLD = \Sigma f(\alpha + \beta X)D = \alpha \Sigma fD + \beta \Sigma fXD = 0$$

Since the first sum $\Sigma fD = 0$, the second sum ΣfXD must also be zero; hence, substituting $X = 2, 1, 0$, and $f = p^2, 2pq, q^2$, we have

$$\Sigma fXD = p^2 \cdot 2D_2 + 2pq\, D_1 = 0$$

that is,
$$pD_2 + qD_1 = 0 \qquad (51)$$

Also, write
$$\bar{D} = p^2 D_2 + pq D_1 + pq D_1 + q^2 D_0 = 0$$
$$= p(pD_2 + qD_1) + q(pD_1 + qD_0) = 0$$

Since the first term is zero by (51), the second term must also be zero:

$$pD_1 + qD_0 = 0 \qquad (52)$$

Before proceeding further, the reader may wish to verify these relations using the numerical values of Table 231.

$$(D_2, D_1, D_0) = (-4, +6, -9), \text{ and } (p, q) = (0.6, 0.4)$$

By (51), $\qquad 0.6(-4) + 0.4(6) = 0$

By (52), $\qquad 0.6(6) + 0.4(-9) = 0$

We shall rely on the relationships (51) and (52) heavily in deriving the parent-offspring correlation.

23. Parent-offspring correlation. Table 201, giving the parent-offspring combinations, is reproduced as Table 239 with one important addition: each genotype is now associated with a measurement value $Y = L + D$. The covariance of parent and offspring is calculated from the entries of the entire table. For clarification, we shall write the covariance as $Cov(Y', Y)$ where Y' is the measurement of a parent and Y is that of a child. The marginal variance of Table 239 is σ_Y^2 for both parent and offspring in an equilibrium population. Hence we need only to find the covariance of parent and offspring with respect to the Y's. The following formula is true in general, whether L and D are correlated or not.

$$\begin{aligned} Cov(Y', Y) &= Cov(L' + D', L + D) \\ &= Cov(L', L) + Cov(L', D) + Cov(D', L) + Cov(D', D) \end{aligned} \quad (53)$$

These four covariances may be found separately from Table 239. For instance, if we omit the L's (and Y's, of course) from the measurements of parents and offspring, then Table 239 becomes the correlation table between the parents' D's and the children's D's.

Proceeding with Table 239 row by row and attending the D's only, we have

$$\begin{aligned} Cov(D', D) \quad = \quad & p^2 D_2' \, (pD_2 + qD_1) \\ + \quad & pq \, D_1' \, [(pD_2 + qD_1) + (pD_1 + qD_0)] \\ + \quad & q^2 D_0' \, (pD_1 + qD_0) = 0 \end{aligned} \quad (54)$$

QUANTITATIVE TRAITS; HERITABILITY

Table 239. Correlation between parent and offspring in a random-mating population when the measurements of the genotypes are Y_2, Y_1, Y_0.

Parent	Offspring AA $Y_2 = L_2 + D_2$	Aa $Y_1 = L_1 + D_1$	aa $Y_0 = L_0 + D_0$	Total frequency
AA: $Y_2' = L_2' + D_2'$	p^3	$p^2 q$	0	p^2
Aa: $Y_1' = L_1' + D_1'$	$p^2 q$	$pq^2 + p^2 q$	pq^2	$2pq$
aa: $Y_0' = L_0' + D_0'$	0	pq^2	q^3	q^2
Total frequency	p^2	$2pq$	q^2	1.00

The sum of each row of D's is zero by the relations (51) and (52). The obvious conclusion from (54) is that the parent and offspring dominance deviations are uncorrelated. Furthermore, if the parents' D_2', D_1', D_0' in (54) are replaced by parents' L_2', L_1', L_0', the result is still zero, as $(pD_2 + qD_1) = 0$, etc. Conversely, if we retain the parents' D' but use offspring L, the result is again zero because of the symmetry of the table. The important conclusion from these observations is that three of the four covariances of (53) are zero:

$$Cov(L', D) = Cov(D', L) = Cov(D', D) = 0$$

hence, $\quad Cov(Y', Y) = Cov(L', L) \quad\quad\quad (55)$

The parent-offspring covariance with respect to measurements Y does not involve the dominance deviations at all. We have noted previously that in calculating a covariance the L values may be taken as βX, and $Cov(X', X) = pq$, as given by (9) for additive gene effects. Thus, (55) becomes

$$Cov(Y', Y) = Cov(L', L)$$
$$= Cov(\beta X', \beta X) = \beta^2 Cov(X', X) = pq\beta^2 \quad (56)$$

Recalling from (37) that $\sigma_L^2 = \beta^2 \sigma_X^2 = 2pq\beta^2$, we reach the final form for covariance:

$$Cov(Y', Y) = Cov(L', L) = \frac{1}{2} \sigma_L^2 \qquad (57)$$

The covariance is half of the linear component of the variance of Y and does not contain any part of the dominance component. The parent-offspring correlation with respect to the Y's is

$$r_{PO} = r(Y', Y) = \frac{Cov(Y', Y)}{Var(Y)} = \frac{1}{2} \left(\cdot \frac{\sigma_L^2}{\sigma_Y^2} \right) = \frac{1}{2} g^2 \qquad (58)$$

which is the desired formula (without environmental influence). Although this is not the shortest nor the most elegant way of arriving at the result, it makes use exclusively of the properties of L and D obtained by simple linear regression and it shows in detail how the D values drop out of covariance completely, so that $Cov(Y', Y) = Cov(L', L)$. The correlation formula (58) shows that when there is no dominance, $g^2 = 1$ and $r_{PO} = \frac{1}{2}$, which is the basic quantity. When there is dominance, only the fraction $g^2 = \sigma_L^2/\sigma_Y^2$ goes to the parent-offspring correlation. This situation is usually described by saying that only the linear component σ_L^2 has been transmitted from parent to offspring. For this reason, the linear component σ_L^2 is said to be "heritable" (while σ_D^2 is not) and thus the L values are sometimes called the "breeding values" of the individuals by animal and plant breeders.

24. Sib-sib correlation. The sib-sib correlation may be found by a similar procedure, substituting the sib-sib combination table (Table 203). The mathematics involved, however, become much longer and heavier. Fortunately, our **ITO** method (Chapter 7) renders a separate calculation of sib-sib correlation unnecessary. Once the parent-offspring correlation is known, the sib-sib correlation may be written down immediately without further calculation. The **ITO** method enables us to split the sib-sib combination frequencies into three parts (13) according to the number of identical genes a sib-pair shares in common. Writing Y_2, Y_1, Y_0 on the top and on the side of each of the three component tables of (13), we may calculate a

QUANTITATIVE TRAITS; HERITABILITY

correlation coefficient for each of them, and then the sib-sib correlation will be given by the weighted average (14). Now, the correlation for the first table of (13) is always unity ($r_1 = 1$), no matter what values Y_2, Y_1, Y_0 may take. That means, $(L' + D')$ of parents and $(L + D)$ of offspring have perfect correlation for the sib-pairs who share two identical genes by descent. On the other extreme, the correlation for the third table of (13) is always zero ($r_3 = 0$), no matter what values Y_2, Y_1, Y_0 may take, because the two marginal variables are independently distributed ($f_{ij} = f_i f_j$). The correlation for the middle table of (13) is exactly the same as that for parent-offspring which we have already obtained (58). Thus, the sib-sib correlation is, by (14) and (58),

$$r(sib\text{-}sib) = r_{00} = \tfrac{1}{4}(1) + \tfrac{1}{2}(r_{PO}) + \tfrac{1}{4}(0)$$

$$= \tfrac{1}{4} + \tfrac{1}{2}(\tfrac{1}{2}g^2) = \tfrac{1}{4} + \tfrac{1}{4}g^2 \tag{59}$$

This expression shows that r_{00} has a minimum value of ¼, no matter what values Y_2, Y_1, Y_0 may take, on account of the I-component, the first table of (13). If the heterozygote measurement Y_1 is outside the range of Y_2 and Y_0, the proportion of the linear component $g^2 = \sigma_L^2/\sigma_Y^2$ will be low and could even be zero. In the latter event, $r_{PO} = 0$, but $r_{00} = \tfrac{1}{4}$.

The form (59) for r_{00} follows directly from the ITO method, but it is not the usual form the reader encounters in other books. A more conventional form of (59) is, due to $g^2 + d^2 = 1$,

$$r_{00} = \tfrac{1}{4}(g^2 + d^2) + \tfrac{1}{4}g^2$$

$$= \tfrac{1}{2}g^2 + \tfrac{1}{4}d^2 \tag{60}$$

This implies that the sib-sib covariance is

$$Sib\text{-}sib \quad Cov(Y', Y) = \tfrac{1}{2}\sigma_L^2 + \tfrac{1}{4}\sigma_D^2 \tag{61}$$

It is seen that with dominance, r_{00} is always greater than r_{PO} by the amount $\tfrac{1}{4}d^2$. The ITO method shows that the extra term $\overline{\tfrac{1}{4}d^2}$ comes from the I-component when the two siblings share two identical genes by descent, causing the dominance deviations D to be correlated between the two siblings.

25. Half-sibs. Either by the path method or by the **ITO** method, we obtain $r(half\text{-}sibs) = r_{0/0} = \frac{1}{4}$ as given in **(26)**, when there is no dominance. Since a pair of half-sibs involves two parent-offspring steps, their dominance deviations are not correlated. Hence, with dominance,

$$r(half\text{-}sibs) = r_{0/0} = \frac{1}{4}\left(\frac{\sigma_L^2}{\sigma_Y^2}\right) = \frac{1}{4} g^2 \qquad (62)$$

The correlation values for parent-child, sib-sib, and half-sibs have been plotted against the values of g^2 in Fig. 242 as a summary. We shall not develop the subject beyond this point. The only relatives we mentioned outside the immediate family circle are the half-sibs, because they may be a potential additional source of research material for social scientists. Not only will they increase in number, but they will be reared either in common or different environments, so that the social scientists will be able to obtain additional equations to evaluate the relative importance of heredity and environment with respect to specified traits.

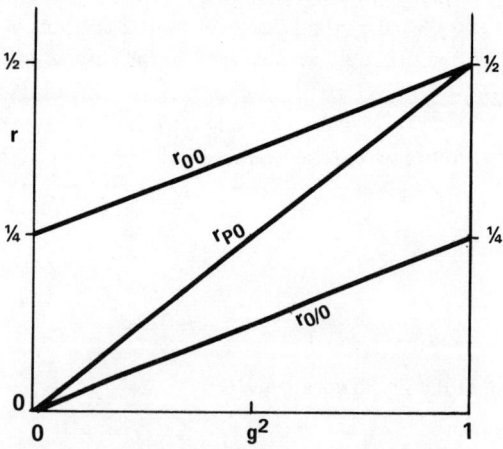

Fig. 242. Graphs for correlations $r_{P_0} = \frac{1}{2} g^2$, $r_{00} = \frac{1}{4} + \frac{1}{4} g^2$, and $r_{0/0} = \frac{1}{4} g^2$, where $g^2 = \sigma_L^2/\sigma_Y^2$ ranges from 0 to 1.

26. Multiple loci. A quantitative trait is usually influenced by a number of loci. The preceding analysis is based on *one* pair of genes. How could the results for one locus also be used for *multiple* loci? In general, they cannot, of course. For practical reasons, however, two basic assumptions are usually made: one is that the effects of the various pairs of genes of a quantitative trait are additive. For instance, the measurement of the genotype *AaBB* is assumed to be the sum of those of *Aa* and *BB*. The second assumption is that the various pairs of genes are independently distributed in the population. Thus, in a random-mating population, if the frequency of *Aa* is *2pq*, and that of *BB* is u^2, then the frequency of *AaBB* is $2pq\, u^2$. The independent distribution of the genes of the various loci make the contributions to variance by the separate loci additive. That is, if σ_1^2 is the variance caused by the first pair of genes (say, *A, a*) and σ_2^2 is that caused by the second pair (say, *B, b*), then the variance caused by the two pairs is $\sigma_1^2 + \sigma_2^2$. For each pair the variance may be subdivided into linear and dominance components. The total linear components of all the loci is called the "linear component" and the total dominance components is called the "dominance component." The independent distribution of the loci also makes the parent-offspring *(PO)* covariances additive. The situation is as follows:

	variance	PO covariance
1st pair,	$\sigma_{Y(1)}^2 = \sigma_{L(1)}^2 + \sigma_{D(1)}^2$,	$Cov(Y'(1),\ Y(1)) = \tfrac{1}{2}\,\sigma_{L(1)}^2$
	$\ldots$	$\ldots$
i th pair,	$\sigma_{Y(i)}^2 = \sigma_{L(i)}^2 + \sigma_{D(i)}^2$,	$Cov(Y'(i),\ Y(i)) = \tfrac{1}{2}\,\sigma_{L(i)}^2$
	$\ldots$	$\ldots$ (63)
Total	$\sigma_Y^2 = \sigma_L^2 + \sigma_D^2$,	$Cov(Y',\ Y) = \tfrac{1}{2}\,\sigma_L^2$
Proportion	$1 = g^2 + d^2$,	$r_{PO} = r(Y',\ Y) = \tfrac{1}{2}\,g^2$

In expression (63) above, we simply use σ_Y^2, etc. to denote the sum of the contributions to variance or covariance from the individual loci. Likewise, the proportion of the linear component is calculated from the respective totals, viz., $g^2 = \sigma_L^2/\sigma_Y^2$. Hence we see that the g^2 in the formula $r_{P_0} = \frac{1}{2} g^2$ really refers to the overall proportion of the linear components for the loci involved, and is not necessarily equal to $g_i^2 = \sigma_{L(i)}^2/\sigma_{Y(i)}^2$ for any particular locus. In conclusion, the correlation formulas (58) and (60) are applicable to multiple loci as long as the loci contributions are additive and the genes are independently distributed in the population.

27. Epistasis. If the two basic assumptions (additive loci effects and independent distribution of genes) are both abolished, the situation becomes almost unmanageable. One compromise approach is to relax one assumption at a time. If, instead of random mating, there is inbreeding, but the loci effects are still additive, the situation is not too bad. However, if the loci effects are not additive, the partitioning of genetic variance becomes very complicated even in a random-mating population. When the loci effects are not additive, it is said to be "epistatic;" or there is epistasis (epistacy) or interaction among the loci effects. If there is epistasis between two pairs of genes (but no epistasis of higher order), the variance may be subdivided by the same general procedure of analysis-of-variance for a two-way classification with unequal cell frequencies. Cockerham (1954) and Kempthorne (1954, 1955, 1955a, 1957) give the following partitioning for two loci:

$$\sigma_Y^2 = \sigma_L^2 + \sigma_D^2 + \sigma_{LL}^2 + \sigma_{LD}^2 + \sigma_{DD}^2 \qquad (64)$$

where the double subscripts LL, LD, DD denote the variance components due to linear X linear, linear X dominance, and dominance X dominance interactions between the two loci. The analytical development of the subject is beyond the scope of the present volume. We mention it here to show the type of complications we will run into with epistasis. The expression (64) is only the beginning, with merely two-factor interactions. If there are three-locus interactions as well, four more terms, viz., $\sigma_{LLL}^2 + \sigma_{LLD}^2 + \sigma_{LDD}^2 + \sigma_{DDD}^2$, should be added to (64), and so on. Sometimes all the interaction terms are pooled together and written as σ_{epis}^2. This practice is deceptive; it shortens our writing but does not solve any practical problems, because the covariances between relatives consist of different fractions of the various interaction components, not a single fraction of σ_{epis}^2. For instance, if there are only two-factor interactions,

$Cov(P, O) = \tfrac{1}{2}\sigma_L^2 \qquad\qquad\qquad + (\tfrac{1}{2})^2 \sigma_{LL}^2$

$Cov(O, O) = \tfrac{1}{2}\sigma_L^2 + \tfrac{1}{4}\sigma_D^2 + (\tfrac{1}{2})^2 \sigma_{LL}^2 + (\tfrac{1}{2})(\tfrac{1}{4})\sigma_{LD}^2 + (\tfrac{1}{4})^2 \sigma_{DD}^2$ **(65)**

$Cov(O/O) = \tfrac{1}{4}\sigma_L^2 \qquad\qquad\qquad + (\tfrac{1}{4})^2 \sigma_{LL}^2$

Even if we limit ourselves to two-locus interactions only, there will be five variance components involved. To solve for the four unknowns, we need four independent equations derived from observed correlations or covariances between relatives. At the present time no suitable data exist to permit the estimation of all these components. There is a huge gap between theoretical analysis and practical applications. As a compromise, however, we may perhaps retain only the (usually) largest term of the interaction components, viz., σ_{LL}^2 and assume all the rest to be negligible. In most applications we assume additive loci effects and use the simple formulas.

Environmental effects; heritability

28. All of the analyses up to this point are based on genetic variations. Since each genotype has a measurement, the entire variance of Y is due to the differences between the genotypes. Hence, σ_Y^2 is called the "genotype variance" or "genetic variance." The verbal terminology is far from being standardized, and therefore it is better to rely on symbols rather than on words. The linear component σ_L^2, based on the calculated linear values L_2, L_1, L_0, is sometimes also called the "genetic variance," although usually known as the "additive component." We specify them by symbols such as σ_Y^2 and σ_L^2, or, simply the Y-variance and the L-variance. We have seen that the hereditary behavior of these two variances are different. Although the entire Y-variance is determined by genotype differences, only the smaller L-variance is transmitted from parent to offspring. In view of this fact, the meaning of the conventional phrase "due to heredity" becomes ambiguous. Both σ_Y^2 and σ_L^2 are of genetic origin. Although only the L values are transmitted from parent to offspring (which is equivalent to saying that the genes are transmitted), the variance in the offspring generation will be σ_Y^2 again, due to the formation of *new* genotypes with stationary frequencies.

The phrase "due to heredity" becomes more ambiguous when we introduce *environment* into consideration. The situation may be depicted as follows, where "normal" means absence of disease:

Environment

	Env. I	Env. II	Env. III	
genotypes G_1	normal	diseased	diseased	(66)
G_2	normal	diseased	normal	

When the population is located in Environment I, all individuals are normal and genetic differences are not observable. Environment I does not make the genetic differences disappear (they still exist) but renders those differences irrelevant. On the other hand, if the population is located in Env. II, everybody acquires the disease, again obscuring the genetic differences. If the population is exposed to a mixture of Env. I and II, epidemiologists would describe the disease as an *environmental* one: under Env. I, people do not have the disease, but under Env. II, they do. Under Env. III, however, individuals of genotype G_1 develop the disease but those of genotype G_2 remain normal. If the population is actually located in Env. III, a medical geneticist would have described the condition as a "genetic disease," correctly but directly opposite to the conclusion of the epidemiologist who studied Env. I and II. Now, is the disease an environmental or a genetic one? Both are correct or wrong, depending on the portion of the "elephant" (66) we are looking at. Taking all the situations of (66) as a whole, we say that there is *interaction between environment and heredity*. In such a case, the concepts of "due to heredity" and "due to environment" break down because they are inadequate to describe an interacting system.

When we come to consider *quantitative traits*, an equally interacting situation may exist. The following example (Li, 1970) is hypothetical and has grossly exaggerated the point to make the meaning of interaction clear.

		Env. I	Env. II	Env. III	Env. IV	Total	
	G_1	8	3	5	4	20	
genotypes	G_2	4	5	3	8	20	(67)
	G_3	3	7	7	3	20	
Total		15	15	15	15	60	

The numbers in the table above are hypothetical measurements of a quantitative trait. It is seen that as far as the marginal totals (or means) are concerned, there is no difference between the four environmental conditions, nor is there any difference between the various genotypes. Yet there is considerable variation in the body of the table. The performances of the genotypes are different under different conditions. The entire variation of this set of data is due to *interaction* between environment and heredity. Again, the concepts of "due to heredity" and "due to environment" break down in the face of a strongly interacting system.

The previous examples serve to remind us of the possible difficulties in evaluating the relative importance of heredity and environment with respect to specified traits. In practice we shall only study the situation in which the environmental and hereditary effects are additive, or at least, without strong interaction, so that the ordinary linear statistical model still gives the first approximation to the true situation. Note carefully that we did not assume that environmental and hereditary effects are *always* additive in nature; we merely said that we shall study the cases in which they are.

29. Random and systematic environmental effects. Environmental effects are of various types and influence the measurement of a quantitative trait in various ways. Roughly speaking, they may be classified into two major types: random and systematic. By random environmental effects we mean that they produce random deviations about a certain true value of the measurement. In our previous example in which the measurement of the genotype AA is $Y_2 = 40$, without mentioning environmental effects. Now, environmental effects will make the measurements of AA not always at 40 but possibly vary over a range between, say, 30 and 50. Then the variation about the true value is said to be due to random environmental effects. On the other hand, a set of common environmental factors for siblings living in the same household (i.e., systematic environmental effects) may raise the sib-sib correlation to a level beyond the maximum value of $r_{oo} = \frac{1}{2}$ on a genetic basis. These common household factors produce common effects on the siblings, so that the actual correlation may be as high as, say, 0.60, with respect to the traits so influenced. Briefly, *random environmental effects tend to reduce the correlations between relatives while systematic environmental effects tend to increase them.* In the following we shall study the random environmental effects first. Common home environmental effects will be treated in the last paragraph of the chapter.

30. Phenotype measurements, Z. With random environmental effects, the measurements of the genotypes AA, Aa, aa will not be steadily Y_2, Y_1, Y_0; each of these genotypes may take a series of values (Z) about their respective means Y_2, Y_1, Y_0. A numerical example will clarify the situation. The following is an extension of Table 231.

genotype	frequency	X	Y	phenotype values					
AA	$p^2 = 0.36$	2	40	$f:$ $Z:$	0.04 50	0.12 46	0.12 34	0.08 35	
Aa	$2pq = 0.48$	1	35	$f:$ $Z:$	0.04 43	0.12 39	0.20 34	0.08 31	0.04 28
aa	$q^2 = 0.16$	0	5	$f:$ $Z:$		0.08 8		0.08 2	

(68)

For each genotype there is not one but a series of values, whose mean is the genotype value. Thus, for the genotype AA, the phenotype values Z are 50, 46, 34, 35, whose weighted mean is 40. The phenotype values Z are the measurements actually observed in practice and therefore contain environmental effects. They are usually denoted by P, rather than by Z. In order to avoid the possible confusion with a parent P, we use the letter Z that follows Y. For each genotype the deviation of Z from Y is then called the "(random) environmental effect," $\in$. That is,

$$Z = Y + \in \qquad (69)$$
$$\text{(phenotype)} \quad \text{(genotype)} \quad \text{(environment)}$$

From the numerical example (68), it is also obvious that Z and Y are positively correlated; but Y and $\in$ are uncorrelated because for each fixed genotype group,

$$\Sigma f Y \in \; = \; Y \Sigma f \in \; = \; 0 \qquad (70)$$

It follows that X is also uncorrelated with $\in$. The example (68) specifies the statistical model by which we study the environmental and genetical effects on a metrical trait. The two main assumptions are that the environmental and genetical effects are additive (69) and that they are uncorrelated (70). When either of these two basic assumptions are not fulfilled, there is really no satisfactory statistical analysis available at the present time.

31. Scattering of points. The reader may recall that there is a section by the same title in the last section of Chapter 3. In fact, our present example (68) is the same one as in Table 66 except for terminology and symbols. In Chapter 3, we studied the statistical relationships among the three variables, $x, y, z,$ without any subject matter content. Here, we use the same numerical example but injected with genetic meaning. The scattered points are now the phenotypes. The statistical relationships among the variables do not change no matter what specific meaning has been attached to the numbers. Since all the detailed calculations have been given in Chapter 3, the reader should review that section before proceeding. Briefly, the results established in that section are as follows, in terms of our present symbols. First,

$$\sigma_Z^2 = \sigma_Y^2 + \sigma_\in^2 \tag{71}$$

In our numerical example, $168 = 144 + 24$. The phenotype variance is the sum of the genotype variance and the environmental variance. This arises from the fact that Y and $\in$ are uncorrelated. It follows that

$$Cov(Y, Z) = Cov(Y, Y + \in) = \sigma_Y^2 \tag{72}$$

$$Cov(X, Z) = Cov(X, Y + \in) = Cov(X, Y) \tag{73}$$

The last relationship implies that the regression line of Z on X is the same as that of Y on X, for $\overline{Z} = \overline{Y}$, and

$$b_{YX} = \frac{Cov(Y, X)}{Var(X)} = \frac{Cov(Z, X)}{Var(X)} = b_{ZX} \tag{74}$$

The situation is illustrated in Fig. 67. The results on correlations are (the former g is our present L):

$$r(X, Y) = r(L, Y) = \frac{\sigma_L^2}{\sigma_L \sigma_Y} = \frac{\sigma_L}{\sigma_Y} \tag{75}$$

$$r(Y, Z) = \frac{Cov(Y, Z)}{\sigma_Y \sigma_Z} = \frac{\sigma_Y^2}{\sigma_Y \sigma_Z} = \frac{\sigma_Y}{\sigma_Z} \tag{76}$$

$$r(X, Z) = r(L, Z) = \frac{\sigma_L}{\sigma_Z} = r(L, Y) \cdot r(Y, Z) \tag{77}$$

Translated into genetic language, the last result states that

$$r(linear, phenotype) = r(linear, genotype) \cdot r(genotype, phenotype) \tag{77'}$$

This relationship clearly leads to applications of path coefficients to the genetic problem under consideration. It is advisable that the reader insert the corresponding numerical values of these expressions (obtained in Chapter 3) for further reference.

32. Random deviations do not enter into covariance. We digress from the genetic problem to state a general theorem about random deviations and covariance. Let a and b be two variables with covariance $Cov(a, b)$. Then let

$$x = a + \in, \quad \text{and} \quad y = b + \eta$$

where the random deviations $\in$ and η have mean zero and are uncorrelated with variable a or b or among themselves. The general theorem is

$$Cov(x, y) = Cov(a, b) \tag{78}$$

for $Cov(x, y) = Cov(a + \in, b + \eta)$

$$= Cov(a, b) + Cov(a, \eta) + Cov(b, \in) + Cov(\in, \eta)$$

wherein the last three covariances are all zero by assumption. The theorem will be employed in the next paragraph in dealing with random environmental effects.

A clear distinction must be made between the situation described here and the situation in which the dominance deviations drop out of the covariance between parent and offspring. The two situations are as follows:

$$
\begin{array}{c}
L + D \\
L' + D' \quad \boxed{\begin{array}{c} PO \text{ joint} \\ \text{distribution} \end{array}} \quad a + \in \quad \boxed{\begin{array}{c} \text{any joint} \\ \text{distribution} \end{array}}
\end{array} \quad (79)
$$

In the parent-offspring *(PO)* case, each L is associated with a particular value of D. The fact that the D's do not enter into the parent-offspring covariance is due to the parent-offspring joint distribution. If we substitute the sib-sib joint distribution, the D's will not drop out of the sib-sib covariance. In the case of random deviations $\in$ and η, they will not enter into the covariance (x, y), whatever the joint distribution. This is due to the random nature of $\in$ and η.

33. Parent-offspring correlation with random environmental effects.
Now we return to our genetic topic. When a genotype value Y becomes a series of phenotype values $Z = Y + \in$ due to environmental effects $\in$, what would be the correlation between parent and offspring in a random-mating population? Intuitively, the correlation will be lower than that without environmental effects, but we wish to know how much lower. As before, let a prime indicate the values of a parent. The parent-offspring covariance is

$$Cov(Z', Z) = Cov(Y' + \in', Y + \in) = Cov(Y', Y) \quad (80)$$

by the general theorem (78), assuming that the environmental effects $\in'$ of parents are uncorrelated with those $\in$ of the offspring. The variance of Z are the same for parents and for offspring in an equilibrium population; $\sigma_{Z'}^2 = \sigma_Z^2 = \sigma_Y^2 + \sigma_\in^2$ as stated in (71). Hence the parent-offspring *(PO)* correlation with respect to phenotype values (including environmental effects) is

$$r_{PO} = r(Z', Z) = \frac{Cov(Z', Z)}{\sigma_Z^2} = \frac{Cov(Y', Y)}{\sigma_Y^2} \cdot \frac{\sigma_Y^2}{\sigma_Z^2}$$

that is,

$$r_{PO} = r(Y', Y) \cdot \frac{\sigma_Y^2}{\sigma_Z^2} \qquad (81)$$

where $r(Y', Y)$ is the parent-offspring correlation with respect to genotype values (without environmental effects). It is seen that the correlation has been decreased by the fraction σ_Y^2 / σ_Z^2.

34. Heritability (environmental). Since the value Y is determined by the genotype and Z value determined by genotype *and* environment, the ratio σ_Y^2/σ_Z^2 gives the proportion of the total (phenotype) variance that is due to genotype differences among the individuals of the population. This ratio is defined as the heritability (h^2) of the trait under consideration, viz.,

$$h^2 = \frac{\sigma_Y^2}{\sigma_Z^2}; \qquad e^2 = \frac{\sigma_\in^2}{\sigma_Z^2}, \qquad h^2 + e^2 = 1 \qquad (82)$$

In our numerical example, **(68)** and Table 66, $h^2 = 144/168 = 6/7 = 0.857$ and $e^2 = 24/168 = 1/7 = 0.143$. Then the expression **(81)** may be written

$$r(Z', Z) = r(Y', Y) \cdot h^2 \qquad (83)$$

The heritability (h^2) so defined proves very useful in both analytical and descriptive work. It measures two things at the same time. First, it gives the proportion σ_Y^2/σ_Z^2, showing the fraction of the observed variance that is due to heredity. Second, the same proportion also serves to show to what extent a genetic correlation has been diluted or attenuated by random environmental deviations.

35. Further analysis of heritability. The meaning of the expression **(83)** $r(Z', Z) = r(Y', Y) \cdot h^2$, is conceptually simple and clear but not immediately useful in practice, because $r(Y', Y)$ is not directly observable in general.

QUANTITATIVE TRAITS; HERITABILITY

We will carry the analysis one step further so that the presumably observed parent-child correlation is related to a known quantity. Substituting (57) and (58) in (81) or (83), we have

$$Cov(Z', Z) = Cov(Y', Y) = Cov(L', L) = \tfrac{1}{2}\sigma_L^2 \qquad (84)$$

and

$$r_{PO} = r(Z', Z) = r(Y', Y) \cdot b^2 = \tfrac{1}{2}g^2 \cdot b^2 \qquad (85)$$

where $g^2 = \sigma_L^2/\sigma_Y^2$ as defined by (49) and $b^2 = \sigma_Y^2/\sigma_Z^2$ as defined by (82), so that

$$g^2 b^2 = \frac{\sigma_L^2}{\sigma_Z^2} = \frac{\sigma_L^2}{\sigma_Y^2 + \sigma_\in^2} = \frac{\sigma_L^2}{\sigma_L^2 + \sigma_D^2 + \sigma_\in^2} \qquad (86)$$

For instance, if an observed parent-child correlation turns out to be r_{PO} = 9/28 = 0.3214, then we know from (85), $g^2 b^2 = 2 \times 9/28 = 0.643$, which is the attenuation factor for the ideal case, $r_{PO} = \tfrac{1}{2}$, without dominance and without environmental deviations. Therefore, $g^2 b^2$ is also a type of heritability, if we use $r_{PO} = \tfrac{1}{2}$ as the starting point, just as b^2 is the heritability if we use $r_{PO} = r(Y', Y) = \tfrac{1}{2}g^2$ as the starting point. In genetic literature and recently in social science literature, the fraction $b^2 = \sigma_Y^2/\sigma_Z^2$ is called the "heritability in the broad sense" and the smaller fraction $g^2 b^2 = \sigma_L^2/\sigma_Z^2$ is called the "heritability in the narrow sense." In many instances, however, the writer fails to mention explicitly which heritability he really meant. Also, the words "broad" and "narrow" do not reflect the true meaning of the two types of heritabilities. In our treatment, the fractions g^2 and b^2 are introduced separately and kept distinct, because they have very different meanings. The fraction $g^2 = \sigma_L^2/\sigma_Y^2$ is entirely due to dominance, or deviation from linearity, and has nothing to do with the environmental effects $\in$. Conversely, the fraction $b^2 = \sigma_Y^2/\sigma_Z^2$ is entirely due to the environmental effects and has nothing to do with dominance, as its numerator $\sigma_Y^2 = \sigma_L^2 + \sigma_D^2$ is the total genotype variance, regardless of the relative magnitudes of σ_L^2 and σ_D^2. The two fractions g^2 and b^2 are due to two unrelated reasons and have no necessary relationships with each other. In view of the discussion above, probably the proposed terms in the following are more descriptive and may cause less confusion.

fraction		conventional	proposed name
h^2	$= \sigma_Y^2 / \sigma_Z^2$	heritability (broad sense)	*environmental heritability*
g^2	$= \sigma_L^2 / \sigma_Y^2$	$d^2 = 1 - g^2$ (dominance ratio)	*dominance heritability* (87)
$g^2 h^2$	$= \sigma_L^2 / \sigma_Z^2$	heritability (narrow sense)	*combined heritability*, or *net heritability*

The writer is not in favor of denoting the product $g^2 h^2$ by a single letter which may make us lose sight of the two distinct processes involved. As will be shown in a later section, g and h are path coefficients pertaining to different steps in the path diagram.

36. Sib-sib correlation with random environmental effects. Let Z°, Z and Y°, Y be the phenotype and genotype measurements, respectively, of two full sibs, where $Z = Y + \in$ and $\in$ is a random environmental effect. By the general theorem (79) that random deviations do not enter into covariance, we have from (60) and (61)

$$Cov(Z^\circ, Z) = Cov(Y^\circ, Y) = \tfrac{1}{2}\sigma_L^2 + \tfrac{1}{4}\sigma_D^2 \qquad (88)$$

and
$$r(Z^\circ, Z) = \frac{Cov(Z^\circ, Z)}{\sigma_Z^2} = \frac{Cov(Y^\circ, Y)}{\sigma_Y^2} \cdot \frac{\sigma_Y^2}{\sigma_Z^2}$$

i.e.,
$$r_{00} = r(Y^\circ, Y) \cdot h^2 = \left\{ \tfrac{1}{2} g^2 + \tfrac{1}{4} d^2 \right\} h^2 \qquad (89)$$

Again, the expression (89) emphasizes the fact that the factor $\tfrac{1}{2}g^2 + \tfrac{1}{4}d^2$ is due to heredity and the factor h^2 is due to environment. With random environmental deviations, r_{00} still remains larger than r_{PO}, but not to the same extent as in the case without environmental effects. In the latter, the difference is $\tfrac{1}{4} d^2$, while in the former, the difference is $\tfrac{1}{4} d^2 h^2$. Substituting $d^2 = 1 - g^2$, the sib-sib correlation (89) becomes

$$r_{00} = \left\{ \tfrac{1}{4} + \tfrac{1}{4} g^2 \right\} h^2 \qquad (89')$$

which is a more convenient form to use in the next paragraph.

37. Separate estimates of g^2 and h^2. When both r_{PO} and r_{OO} are observed, two equations are available to solve for the two unknowns g^2 and h^2. Writing the correlations (85) and (89') as

$$2\, r_{PO} = g^2 h^2, \qquad\qquad 4\, r_{OO} = h^2 + g^2 h^2 \qquad (90)$$

we see that their difference is the environmental heritability

$$h^2 = 4\, r_{OO} - 2\, r_{PO} \qquad (91)$$

Substituting in the first equation of (90), we obtain the dominance heritability

$$g^2 = \frac{2\, r_{PO}}{h^2} = \frac{r_{PO}}{2\, r_{OO} - r_{PO}} \qquad (92)$$

The expression given by Penrose (1971) for h^2 is the same as our (91) but his other expression is equivalent to $d^2 = 1 - g^2 = (2r_{OO} - 2r_{PO}) / (2r_{OO} - r_{PO})$. We repeat once more that the expression (91) is purely environmental and independent of dominance, while (92) depends solely on dominance and is independent of environment. Both g^2 and h^2 vary from population to population and from trait to trait.

Summary by path analysis

If the materials covered in the preceding sections are not well organized and need further clarification, the writer knows no better way than introducing a path diagram (Fig. 256) which will put everything in its right place. In this respect a path diagram is very much like an organization chart of a corporation, or a flow chart of a program. Even if one does not use path coefficients as a research tool, he will see that a path diagram such as that of Fig. 256 is a highly succinct, systematic, and meaningful way of arranging the various pieces of a puzzle into a picture that is amenable to interpretation. Once we learn the rules of reading a path diagram (Chapter 6), almost every formula developed in this chapter follows from Fig. 256 automatically. As a review of this chapter as well as the path method, we now proceed to study that diagram in some detail.

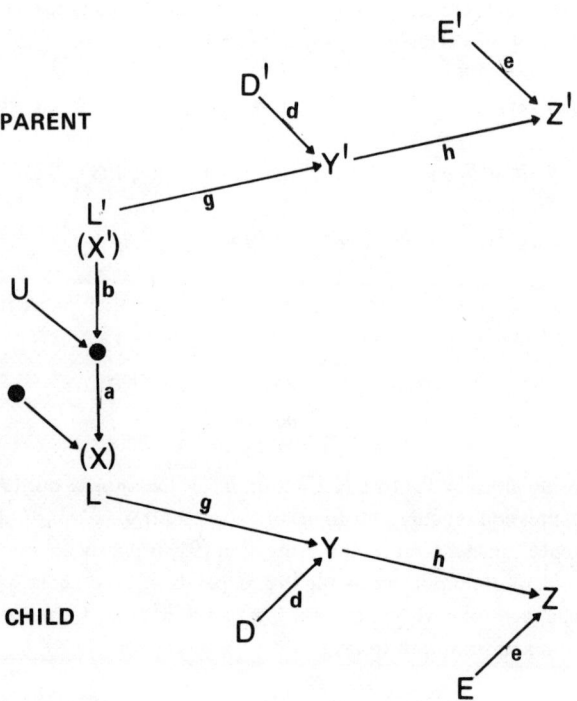

Fig. 256. Path diagram showing the correlation between parent (upper part) and offspring (lower part) at various stages of development. *First stage:* L = linear value of a genotype based on additive gene effects which may be reduced to X scale. $r_{PO} = r(L', L) = ba = \frac{1}{2}$. *Second stage:* genotype value with dominance $Y = L + D$, and g is the path coefficient from L to Y. $r_{PO} = r(Y', Y) = g(ba)g = \frac{1}{2}g^2$. *Third stage:* phenotype values with environmental influence, $Z = Y + E$, and h is the path coefficient from Y to Z. $r_{PO} = r(Z', Z) = hg(ba)gh = (\frac{1}{2})g^2h^2$.

The upper part of Fig. 256 refers to the parent (with the prime). The lower part refers to the offspring. We study the diagram from left to right in three stages.

QUANTITATIVE TRAITS; HERITABILITY

First stage: If the gene effects are additive, the genotype values L may always be reduced to the scale $X = 2, 1, 0$, without affecting its correlation with any other variable; $r(X, L) = 1$. Hence in the diagram X is shown in parentheses underneath L. The path coefficient from a parent genotype to his (or her) gamete is $b = \sqrt{½}$ in a random-mating population. The path coefficient from a uniting gamete to the resultant offspring is $a = \sqrt{½} = b$. This part of Fig. 256 has already been represented by Fig. 225, where U is the chance factor of Mendelian segregation. At this stage, the parent-offspring correlation is

$$r_{PO} = r(L', L) = r(X', X) = ba = ½ \quad (10)$$

Second stage: When there is dominance the genotype values are Y which are no longer linear, $Y_2 - 2Y_1 + Y_0 = \delta$. In such a case, the Y value may be split into two uncorrelated parts, the linear value and the dominance deviation, $Y = L + D$. Since L and D are uncorrelated and completely determine Y, we have the path coefficients (Chapter 6),

$$p_{YL} = g = \frac{\sigma_L}{\sigma_Y}, \qquad p_{YD} = d = \frac{\sigma_D}{\sigma_Y}, \qquad g^2 + d^2 = 1 \quad (49)$$

At the Y stage, the parent-offspring correlation is the connecting path

$$r_{PO} = r(Y', Y) = g(ba)g = ½g^2 \quad (58)$$

Third stage: When the measurements are influenced by random environmental effects which are uncorrelated with the genotypes and act additively with those of genotypes, the observed (phenotype) value is $Z = Y + \mathcal{E}$, where Y and $\mathcal{E}$ are uncorrelated. In Fig. 256, the environmental variable is denoted by E, not to be confused with the same letter for the "expected value of." Again, since Y and E are uncorrelated and completely determine Z, the path coefficients from the respective causes to Z are

$$p_{ZY} = h = \frac{\sigma_Y}{\sigma_Z}, \qquad p_{ZE} = e = \frac{\sigma_\mathcal{E}}{\sigma_Z}, \qquad h^2 + e^2 = 1 \quad (82)$$

At the phenotype level, the parent-offspring correlation is the connecting path

$$r_{PO} = r(Z', Z) = hg(ba)gh = ½g^2h^2 \quad (85)$$

Thus we have reached all the important results in a few steps by using basic path theorems established in Chapter 6. Another important advantage of arranging the variables in systematic order is that the concept of heritability is made directly visible by the path diagram. The influence from L to Y and then from Y to Z are measured by the path coefficients g and h, respectively. The L-to-Y part is due to dominance, and the Y-to-Z part is due to environment.

The situation for sib-sib correlation may be summarized in a similar way by a path diagram (Fig. 259). Again, we shall study it from left to right in three stages, ignoring for the time being the possible common environmental effects C (indicated by dotted lines).

First stage: At the linear value level, $r_{00} = r(L, L) = \tfrac{1}{2}$. Although numerically $r_{00} = r_{PO} = \tfrac{1}{2}$ at the linear level, they are due to different biological reasons as revealed by path analysis. In the parent-offspring case, it is simply $ba = \sqrt{\tfrac{1}{2}}\sqrt{\tfrac{1}{2}} = \tfrac{1}{2}$. In the sib-sib case, it is $abba + abba = \tfrac{1}{4} + \tfrac{1}{4} = \tfrac{1}{2}$, as shown in Fig. 228. For compactness, the parents (Fig. 228) are not shown in Fig. 259. In path analysis the remote "causes" may be cut off by inserting a correlation linking the variables to be treated as the initial ones.

Second stage: It is in this stage Fig. 259 differs from Fig. 256. In Fig. 256 the parent's dominance deviations D are not correlated with the D's of offspring. In the sib-sib diagram Fig. 259 the dominance deviations D of the two siblings are correlated with a coefficient of $\tfrac{1}{4}$. Hence the genotype values Y of the two siblings are connected by two paths:

$$r_{00} = g(\tfrac{1}{2})g + d(\tfrac{1}{4})d = \tfrac{1}{2}g^2 + \tfrac{1}{4}d^2 \qquad (60)$$

where g and d are the path coefficients, having the same meaning as before. It should be noted that there is no connecting path between L and D, whether they belong to the same sib or to different sibs, so that the Y's of the two siblings have only two connecting paths.

Third stage: The rest of the diagram, showing random environmental effects, is the same as the one for parent-offspring. Thus, at the phenotype level,

$$r_{00} = hg(\tfrac{1}{2})gh + hd(\tfrac{1}{4})dh = \left[\tfrac{1}{2}g^2 + \tfrac{1}{4}d^2\right]h^2 \qquad (89)$$

QUANTITATIVE TRAITS; HERITABILITY

Again, all the important results about sib-sib correlation may be read off the diagram Fig. 259 with ease and clarity.

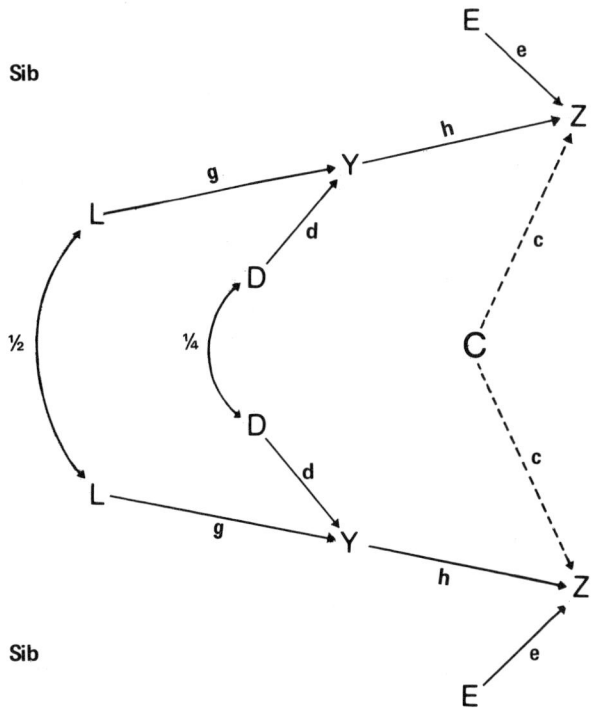

Fig. 259. Path diagram showing the correlation between full sibs at various stages of development. *First stage:* linear value, $r_{00} = r(L, L) = \frac{1}{2}$. *Second stage:* genotype value, $r_{00} = r(Y, Y) = (\frac{1}{2}) g^2 + (\frac{1}{4}) d^2$, as the dominance deviations D of the sibs are correlated with a coefficient of $\frac{1}{4}$. *Third stage:* phenotype value, $r_{00} = r(Z, Z) = r(Y, Y) h^2$, ignoring the possible common factor C. *Fourth stage:* if the common factor C exists, add c^2 to $r(Z, Z)$.

39. Common household environmental factors. The random environmental effects as we have seen, tend to reduce or attenuate the true genetic correlation between relatives. On the other hand, the common and systematic environmental factors that exist in a household may increase the correlation among the children raised in the same household. Let C be such common environmental factors existing in a household and c be the path coefficient from C to phenotype value Z, as shown in Fig. 259. We postulate that the existence of common factors in a household does not nullify the simultaneous existence of the omnipresent random environmental effects. The contribution of such common factors to the correlation between the sibling phenotypes is c^2. Now we need three equations to solve for the three unknowns, g^2, h^2, and c^2. The correlation between half-sibs reared in the same house will provide the third equation. Thus,

$$\begin{aligned} \text{parent-offspring,} \quad & r_{PO} = \tfrac{1}{2} g^2 h^2 \\ \text{full-sibs,} \quad & r_{OO} = \tfrac{1}{4} g^2 h^2 + \tfrac{1}{4} h^2 + c^2 \\ \text{half-sibs,} \quad & r_{O/O} = \tfrac{1}{4} g^2 h^2 + c^2 \end{aligned} \quad (93)$$

Solving,

$$h^2 = 4(r_{OO} - r_{O/O}), \qquad g^2 = \frac{r_{PO}}{2(r_{OO} - r_{O/O})}, \qquad c^2 = r_{O/O} - \tfrac{1}{2} r_{PO} \quad (94)$$

These give us separate estimates of the importance of general environment, gene dominance, and common home factors for children reared together. In the presence of the common factor C, the complete determination of the phenotype Z is $h^2 + e^2 + c^2 = 1$.

Exercises

Ex. 1. The correlation between one parent and several children. Let the parent be P and children total $\Sigma O = O_1 + \ldots + O_s$. Then take $\sigma_P^2 = \sigma_O^2 = 1$. The variance of ΣO remains the same as that given in the text (29). The covariance is

$$Cov(P, \Sigma O) = E P(O_1 + \ldots + O_s) = s \, r_{PO} = (\tfrac{1}{2}) s$$

$$r(P, \Sigma O) = r(P, \overline{O}) = \sqrt{\frac{s}{2(s+1)}}$$

QUANTITATIVE TRAITS; HERITABILITY

Compare this with (30). What is the correlation between one parent and two children? Putting $s = 2$, we have $r(P, O_1 + O_2) = \sqrt{2/6} = 0.577$.

Ex. 2. Show that the correlations (58) and (59) may also be written as

$$r_{PO} = \tfrac{1}{2}(1 - d^2)$$

$$r_{OO} = \tfrac{1}{2}(1 - \tfrac{1}{2}d^2)$$

Ex. 3. Fill out the cell frequencies of the following table for parent-offspring combinations in a random-mating population. Then show that $Cov(Y', D) = 0$, where the prime indicates the measurements of the parents.

		Offspring		
		D_2	D_1	D_0
Parent	Y'_2			
	Y'_1			
	Y'_0			

$$Cov(Y', D) = p^2 Y'_2 (pD_2 + qD_1)$$
$$+ pq Y'_1 (pD_2 + qD_1 + pD_1 + qD_0)$$
$$+ q^2 Y'_0 (pD_1 + qD_0)$$

Why is this quantity zero?

Ex. 4. Consider Table 239 once more. If we do not introduce L and D but use the Y values *directly* for parent and offspring, could we also arrive at the key result that

$$Cov(Y', Y) = pq\beta^2 = \tfrac{1}{2}\sigma_L^2 \ ?$$

To shorten the algebraic expressions we let

$$pY_2 + qY_1 = u_1, \qquad pY_1 + qY_0 = u_2$$

so that the population mean takes the form:

$$\overline{Y} = p(pY_2 + qY_1) + q(pY_1 + qY_0) = pu_1 + qu_2$$

The covariance between parent and offspring is then

$$\begin{aligned}
Cov(Y', Y) &= p^2 Y_2 u_1 + pq Y_1 u_1 + pq Y_1 u_2 + q^2 Y_0 u_2 - \overline{Y}^2 \\
&= pu_1^2 + qu_2^2 - (pu_1 + qu_2)^2 \\
&= pq(u_1 - u_2)^2 = pq\beta^2 = \tfrac{1}{2}\sigma_L^2
\end{aligned}$$

because

$$u_1 - u_2 = (pY_2 + qY_1) - (pY_1 + qY_0) = p(Y_2 - Y_1) + q(Y_1 - Y_0) = \beta \quad (35)$$

In a sense this is the easiest and most direct way of arriving at the desired result, but it is a piece of pure algebra and does not show explicitly how the dominance deviations drop out completely from the parent-offspring covariance, so that $Cov(Y', Y) = Cov(L', L)$. The latter shows that only the linear values L are transmitted from parent to offspring. It is this fact that enables us to say that L and its variance are "heritable" while D and its variance are not.

Ex. 5. Taking $p = 3/5$ and $q = 2/5$, as we did in Table 231, construct the correlation tables for parent-offspring and sib-sib. (Maybe these two tables are already available to you if you have done the exercises of Chapter 7.) For ease of visual comparison, the two tables may be adjusted to a common total, such as those shown in the following:

	Parent-offspring					Sib-sib			
	(40)	(35)	(5)			(40)	(35)	(5)	
Y	8	7	1		Y	8	7	1	
8	108	72	0	180	8	115.2	57.6	7.2	180
7	72	120	48	240	7	57.6	148.8	33.6	240
1	0	48	32	80	1	7.2	33.6	39.2	80
	180	240	80	500		180	240	80	500

The genotype Y values in Table 231 are 40, 35, 5, which, for the purpose of calculating correlation, may be taken as 8, 7, 1, respectively. Also, this reduced scale does not affect the proportions $g^2 = 3/4$ and $d^2 = 1/4$. Calculate r_{PO} and r_{OO} directly from the numerical tables above and then check your answers with those given by the formulas:

$$r_{PO} = \tfrac{1}{2} g^2 = \tfrac{1}{2}(3/4) = 3/8 = 0.3750$$

$$r_{OO} = \tfrac{1}{2} g^2 + \tfrac{1}{4} d^2 = \tfrac{1}{2}(3/4) + \tfrac{1}{4}(1/4) = 7/16 = 0.4375$$

Ex. 6. Cruz-Coke, Nagel, and Etcheverry (1964) measured the diastolic blood pressure (mm Hg.) of the three genotypes of the MN blood system in two groups of Easter Islanders and reported the following findings:

genotype	Islanders		Continentals	
	number	diastolic Y	number	diastolic Y
NN	43	85.3	9	93.7
MN	50	84.6	24	89.7
MM	20	81.7	12	77.0
Total or mean	113	84.35	45	87.11

264   QUANTITATIVE TRAITS; HERITABILITY

The observed numbers have been given in **Ex. 2** of Chapter 7, but the order of the genotypes here has been reversed, so that we shall use p to denote the frequency of gene N in this exercise. The formulas given in the text apply only when the frequencies are p^2, $2pq$, q^2. Of course, empirical regression analysis can always be done for any observed distribution. However, in order to give students an opportunity to review and check the formulas in the text, we shall use the theoretical binomial distribution based on the estimated gene frequency from the sample. In other words, we shall treat the sample as a population purely as an exercise. I shall do the genetic analysis for the group of Continentals and the student is urged to do likewise for the group of Islanders.

The gene frequencies have already been estimated in **Ex. 2** of the previous chapter. Here we shall use the exact value to avoid accumulated rounding off of errors. The estimates are:

$$p = \frac{9 + \frac{1}{2}(24)}{45} = \frac{7}{15}, \qquad q = \frac{12 + \frac{1}{2}(24)}{45} = \frac{8}{15}$$

so that the theoretical frequencies of the three genotypes are $(49, 112, 64)/(15)^2$. In calculating the linear and dominance components of the genetic variance, a constant 70 mm Hg. has been deducted from the real values, as was done by Cruz-Coke, et al.

genotype	f	X	Y	fXY	L	D
NN	49	2	23.7	2322.6	26.1747	−2.4747
MN	112	1	19.7	2206.4	17.5347	+2.1653
MM	64	0	7.0	0	8.8947	−1.8947
Total or mean	225	$\bar{X} = \frac{14}{15}$	$\bar{Y} = 16.9587$	4529.0	16.9587	0
Variance or covariance		$\sigma_X^2 = \frac{112}{225}$	$\sigma_Y^2 =$ 41.848	$\sigma_{XY} =$ 4.3008	$\sigma_L^2 =$ 37.159	$\sigma_D^2 =$ 4.689

The calculation of the variances need no comment. The covariance of X and Y is

$$\sigma_{XY} = \frac{4529}{225} - \bar{X}\,\bar{Y} = 4.3008$$

and the linear regression coefficient is $\beta = 4.3008/\sigma_X^2 = 8.6400$. This checks with the formula

$$\beta = p(Y_2 - Y_1) + q(Y_1 - Y_0) = \frac{7(4.0) + 8(12.7)}{15} = 8.6400$$

The regression equation is $L = (\bar{Y} - \beta\bar{X}) + \beta X = 8.8947 + 8.64X$. Putting $X = 0$, we obtain $L_0 = 8.8947$. The other L values are obtained by adding 8.64 successively. The D values are obtained by subtraction. The variance components may then be calculated from such L and D values listed above. These again agree with the formulas ($\delta = Y_2 - 2Y_1 + Y_0 = -8.7$)

$$\sigma_L^2 = \sigma_X^2 \beta^2 = \frac{112}{225}(8.64)^2 = 37.159; \qquad g^2 = 0.888$$

$$\sigma_D^2 = p^2 q^2 \delta^2 = \frac{49}{225} \cdot \frac{64}{225}(-8.7)^2 = 4.689; \qquad d^2 = 0.112$$

Total, $\sigma_Y^2 = \sigma_L^2 + \sigma_D^2 \qquad = 41.848; \qquad g^2 + d^2 = 1.000$

Environmental variance may be estimated by the standard errors given by Cruz-Coke for each genotype. The environmental heritability h^2 turned out to be small.

Chapter **9**

INBREEDING AND ASSORTATIVE MATING

DEPARTURES FROM random mating introduce a number of complications, and we have to restrict ourselves to simple cases without striving for generalizations. First, we assume throughout the chapter that the population under study is already in an equilibrium state. Next, only positive assortative mating will be considered, as negative assortative mating sometimes implies selection and hence changes the gene frequency. The inbreeding and assortative mating systems to be considered in this chapter are "pure" mating systems which, by themselves, do not change gene frequency. It is hoped that, in spite of these restrictions, this chapter will be of some relevance to social scientists studying human assortative matings with respect to specified quantitative traits.

Populations with Inbreeding

Our main purpose is to study positive assortative mating in human populations. Inbreeding per se is of little interest in the social sciences. The only reason for mentioning inbreeding in this chapter is to establish a few preliminary relationships among the family members which we shall use in studying assortative mating in human populations. Inbreeding means matings between genetically-related individuals. The results of inbreeding may be obtained by direct algebraic methods. For the purpose of this chapter, however, we shall summarize the relationships among family members by the method of path coefficients. The notation for the path coefficient remains the same as that of the last chapter.

Since inbreeding involves mates who are genetically related, their quantitative trait will be correlated to the extent that the trait is genetically determined. Additive gene effects on the quantitative trait will be assumed. Let m be the correlation between the mates (also known as the marital correlation). In an equilibrium population with inbreeding, the marital correlation m remains constant through the generations. In a random-mating population, $m = 0$.

In the first diagram (left) of Fig. 267, the correlation m is shown at the top linking the two parents. Otherwise, it is the same as Figs. 225 and 228.

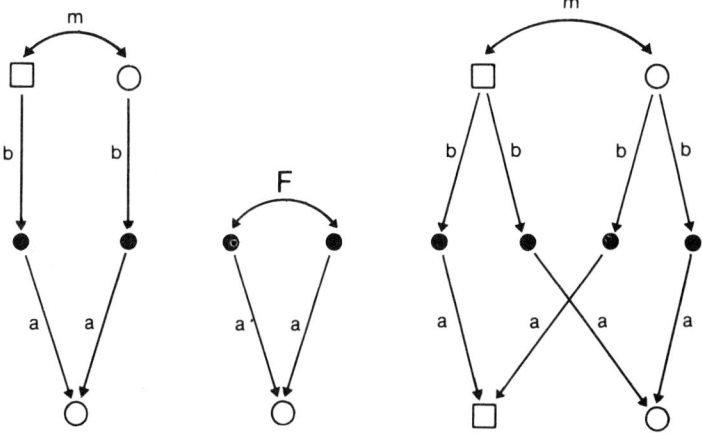

Fig. 267. Correlation between family members with inbreeding. *Left:* parent-offspring correlation; m is the genetic correlation between parents. *Middle:* correlation between uniting gametes is F; this is an abbreviation of the left diagram. *Right:* correlation between full-sibs is the sum of four connecting paths.

Let us concentrate on one parent (say, the father symbolized by a square). In a random-mating population, a parent and a child are connected by only one path, viz., ab. Now, with m linking the two parents, a parent and a child are connected by two paths, so that the parent-child correlation is

$$r_{PO} = ab + abm = ab(1 + m) \tag{1}$$

The method of path coefficients permits us to delete remote "causes" and summarize their influences by a correlation (Chapter 6). Since the parents are correlated, their gametes will also be correlated. Let F be the correlation between the uniting gametes. The parents shown in the left diagram of Fig. 267 may thus be omitted and replaced by the correlation F as shown in the middle diagram. The numerical value of F is smaller than m,

$$F = bmb = b^2 m \tag{2}$$

The two uniting gametes completely determine the genotype (and thus the quantitative measurement) of the child. By the theorem of complete determination, we have

$$a^2 + a^2 + 2aFa = 2a^2 + 2a^2 F = 1$$

$$a^2 = \frac{1}{2(1+F)}, \qquad a = \sqrt{\frac{1}{2(1+F)}} \qquad (3)$$

In the last chapter it was also stated that the correlation between a genotype and the gamete produced by it is the same as the correlation between the genotype and the gamete that united to produce it. Hence, from the two diagrams of Fig. 267 and substituting (3), we have

$$b = a + aF = a(1+F) = \sqrt{\frac{1+F}{2}} \qquad (4)$$

Combining (3) and (4), we obtain the path coefficient from parent to child

$$ab = \sqrt{\frac{1}{2(1+F)}} \sqrt{\frac{1+F}{2}} = \frac{1}{2} \qquad (5)$$

It is a remarkable fact that the path coefficient from parent to offspring is $ab = \frac{1}{2}$ in an equilibrium population, whether it is random mating or inbreeding. It will be shown later that it is also $ab = \frac{1}{2}$ in an equilibrium population with assortative mating.

Substituting (4) in (2) we obtain for an equilibrium population

$$F = b^2 m = \left(\frac{1+F}{2}\right) m$$

yielding $\qquad m = \dfrac{2F}{1+F}, \qquad F = \dfrac{m}{2-m} \qquad (6)$

Again, the relationship between the marital correlation and the uniting gametes correlation is not limited to inbreeding populations. In an equilibrium population with assortative mating, the relationship (6) holds also.

Substituting (5), viz., $ab = \frac{1}{2}$ in (1), we have

$$r_{PO} = ab(1 + m) = \tfrac{1}{2}(1 + m) \tag{7}$$

Finally, from Fig. 267, we see that there are four connecting chains between two full sibs: two direct routes like those in a random-mating population and two additional routes via the marital correlation m.

$$r_{OO} = 2\,abba + 2\,ab(m)ba$$

$$= 2a^2b^2(1 + m) = \tfrac{1}{2}(1 + m) \tag{8}$$

When there are random environmental effects, let $h^2 = \sigma_Y^2/\sigma_Z^2$ be the environmental heritability, where $Z = Y + \epsilon$ is the phenotype measurement. Then each of the correlations (7) and (8) should be multiplied by h^2. Further, if there are common household factors which tend to increase the similarity between sibs, a term γ^2 should be added so that $r_{OO} = \tfrac{1}{2}(1 + m)h^2 + \gamma^2$.

All of the previous results involving inbreeding are valid for multiple loci as long as the gene effects on the quantitative trait are additive. Space does not permit us to consider the situation with dominance.

In studying assortative mating in the rest of the chapter, we shall introduce another type of correlation, viz., the correlation between alleles, to be denoted by f and to be distinguished from F which applies to whole gametes. The genotype distribution for a single locus is given in terms of f. Since the uniting gametes are correlated, the situation for a single locus may be depicted as follows:

uniting alleles		A (1)	a (0)	allele frequency	
A	(1)	$p^2 + c$	$pq - c$	p	
a	(0)	$pq - c$	$q^2 + c$	q	(9)
allele frequency		p	q	1	

The marginal mean is p and the marginal variance is pq. The covariance is $cov = p^2 + c - p^2 = c,$ so that the correlation between this pair of uniting alleles is

$$f = \frac{c}{pq}, \qquad\qquad c = fpq \qquad (10)$$

The genotype distribution for this pair of genes is, upon substituting (10) in (9),

genotype	general case	when $p = q = \frac{1}{2}$	when $f = 0.20$	
AA	$p^2 + fpq$	$\frac{1}{4}(1 + f)$	0.30	
Aa	$2pq - 2fpq$	$\frac{1}{2}(1 - f)$	0.40	(11)
aa	$q^2 + fpq$	$\frac{1}{4}(1 + f)$	0.30	
total	1.00	1.00	1.00	

This must suffice as an introduction to the study of assortative mating in the following sections.

Assortative mating

Assortative mating based on phenotypic resemblance is a realistic type of mating in human populations with respect to a number of quantitative traits, but the analysis of the consequences of assortative mating is much more difficult than that for the consequences of inbreeding. Fisher (1918) gives the first comprehensive treatment of assortative mating, which, unfortunately remains obscure to most biologists (including the writer). Kempthorne's (1957) chapter on assortative mating seems to be entirely based on Fisher and is an attempt to make Fisher's results on assortative mating more understandable by statistically-minded geneticists. The commentary by Moran and Smith (1966) is another attempt to make Fisher's original contribution more accessible to the general readership. It is out of the scope of this book to go into the subject at the level of the references mentioned above. We shall only point out the general characteristics of

INBREEDING AND ASSORTATIVE MATING

assortative mating under simple genetic models. Two expository articles, Crow and Felsenstein (1968) and Li (1968), will be helpful in this respect. The general theme, however, follows the work of Wright (1921, 1969 II), omitting certain details of derivation.

The difference between inbreeding and assortative mating. The crucial difference lies in the cause for the marital correlation. Inbreeding is based on the genetic relationship of the two mates (e.g., first cousins), regardless of their phenotypic appearance. Assortative mating is based on the phenotypic resemblance of the two mates (e.g., both tall), in which case the phenotypic measurements usually include environmental effects. Fig. 272 is intended to make the difference clear. In the case of inbreeding (left diagram), we say that the *primary* marital correlation is $m = r(L, L)$, and the other correlations $r(Y, Y)$ and $r(Z, Z)$ are secondary, being the consequences of m. The left diagram should be read from the top down. As in Chapter 8, let $g = \sigma_L/\sigma_Y$ and $h = \sigma_Y/\sigma_Z$ be the path coefficients from L to Y and from Y to Z, respectively. Then the correlation between mates at different stages are

$$\begin{aligned} r(L, L) &= m \\ r(Y, Y) &= m\,g^2 \\ r(Z, Z) &= m\,g^2\,h^2 \end{aligned} \tag{12}$$

The right diagram of Fig. 272 depicts the situation for assortative mating with respect to the phenotypic values Z of the mates. Here the primary correlation is $r_{PP} = r(Z, Z)$ and the other correlations $r(Y, Y)$ and $r(L, L)$ are the consequences of r_{PP}. The right diagram of Fig. 272 should be read from the bottom up. The correlations between mates at different stages are

$$\begin{aligned} r(Z, Z) &= r_{PP} \\ r(Y, Y) &= r_{PP}\,h^2 \\ r(L, L) &= r_{PP}\,h^2\,g^2 \end{aligned} \tag{13}$$

and are the reverse of (**12**). To summarize, the primary correlation is always greater than the secondary correlations, the extent depending on the degree of dominance and the magnitude of the environmental effect. In practical

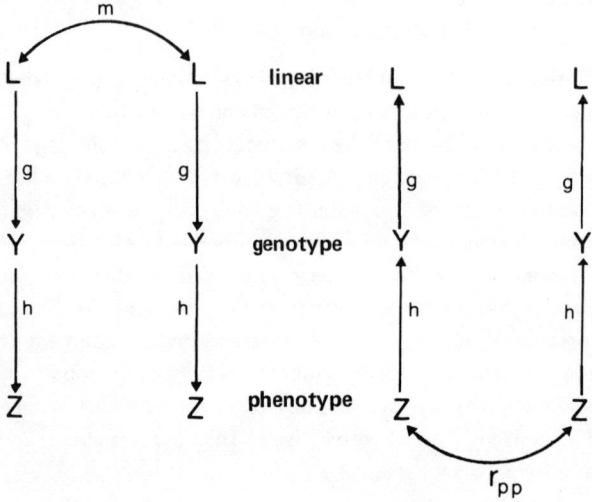

Fig. 272. Correlation between mates for inbreeding and for assortative mating. *Left diagram:* the primary correlation is between the genetic constitution of the mates as in populations with inbreeding; the diagram should be read from the top down. *Right diagram:* the primary correlation is between the phenotype of the mates as in populations with assortative mating; the diagram should be read from the bottom up.

studies of assortative mating, the observed correlation between parents is their phenotypic correlation $r_{PP} = r(Z, Z)$. Their genetic relationship depends on the magnitude of the combined heritability $g^2 h^2$. Hence it is necessary to convert the observed phenotypic correlation r_{PP} into genetic correlation m by **(13)** before we obtain the correlation between relatives. In this chapter, no dominance is assumed $(Y = L)$.

Two-locus model. In order to make the genetic model for assortative mating more realistic, we dispense with the usual one-locus model and go directly to the two-locus situation, as a phenotype (a given measurement), in any real case, usually consists of many genotypes. To reduce the problem to its simplest form, we shall also assume additive effects of genes both within and between loci. There is no dominance and no epistasis. Further, if we take the gene effects to be unity for both loci (symbolically, $A = B = 1$, $a = b = 0$), then the nine genotypes fall into five phenotype classes, ignoring environmental deviations for the time being:

genotype:	aabb	Aabb aaBb	AAbb aaBB AaBb	AABb AaBB	AABB
measurement:	0	1	2	3	4

(14)

where the "measurement" is simply the number of capital letters in a genotype. The assortative mating is assumed to be with respect to these five phenotype groups. To further simplify subsequent arguments, we assume that A and B have equal frequencies; $p = freq\ (A) = freq\ (B)$, and $q = freq\ (a) = freq\ (b)$, where $p + q = 1$.

Correlation between homologous and nonhomologous genes. Consider two individuals P_1 and P_2, each possessing two pairs of genes. If they are to be mates with a certain degree of preference for a measurement class, the "assortativeness" or selectiveness will be based solely on the measurements shown in (14) and not on loci per se. For example, when one prefers a mate of measurement 2, it does not matter whether this mate is of genotype $AAbb$ or of genotype $aaBB$, which are indistinguishable in their measurements. The simple model (14) has obliterated the distinction between loci (i.e., between letters) and only takes into account the phase of the genes (capital or lower case). Hence the correlation between the *genes* of mates (not gametes) induced by assortative mating is the same whether the two genes belong to the same locus (homologous) or to different loci (nonhomologous). To be more specific, let the two mates be symbolized by $P_1 = A_1 A_2 B_1 B_2$ and $P_2 = A_3 A_4 B_3 B_4$, where $A_1 A_2$ denotes any two alleles of the same locus, etc. Let f be the correlation between homologous genes of the two mates, e.g., $r(A_1, A_3)$; and j be the correlation between their nonhomologous genes, e.g., $r(A_1, B_4)$. Since these genes have the same effect (1 or 0) on the measurements which are the basis for assortative mating, the result in an equilibrium population is

$$f = r(A, A) = r(B, B) = r(A, B) = j \quad (15)$$

where $r(A, A)$ is an informal notation for the correlation between homologous genes of the two mates, and $r(A, B)$ for that between their nonhomologous genes. Although our model contains two loci only, the argument and the result (15) are true for any number of loci, provided the gene effects are equal and additive both within and between loci. In inbreeding, there is correlation between the genes of the same locus $(f = r(A, A))$ but no correlation between the genes of different loci $(j = r(A, B) = 0)$. This is a crucial difference between assortative mating based on measurements, and inbreeding based on blood relationship. This also accounts for the phenomenon that assortative mating may cause high marital correlation with little increase in homozygosis of loci—a phenomenon to be illustrated later in the section on population composition.

The population variance. The fact that all genes are equally correlated in an equilibrium population greatly simplifies the mathematics involved in finding the consequences of assortative mating. Assuming assortative mating with respect to a certain trait in human populations is not a new phenomenon; we are essentially interested in the equilibrium state and thus able to bypass many of the details of path analysis. As a preliminary step, let us find the variance of the individuals of the population. To simplify the notation, we write the measurement of a genotype $A_1 A_2 B_1 B_2$ as

$$Y = A_1 + A_2 + B_1 + B_2 \tag{16}$$

where the symbol A, for instance, has a dual meaning; it represents a gene at the (A, a) locus and at the same time it is also a binomial variable taking the value 1 with probability p, and the value 0 with probability q. Then the measurement Y of an individual is the sum of four correlated random variables. The correlation between any two of the four variables (genes) is f in an equilibrium population. The variance of each of the four binomial variables is pq and the covariance between any two of them is pqf. This being the case, the variance of Y is

$$\begin{aligned}
\sigma_Y^2 &= V(A_1) + \ldots + V(B_2) + 2\,Cov(A_1, A_2) + 2\,Cov(A_1, B_1) + \ldots \\
&= pq + \ldots + pq + 2pqf + 2pqf + \ldots \\
&= 4pq + 12pqf = 4pq(1 + 3f)
\end{aligned} \tag{17}$$

INBREEDING AND ASSORTATIVE MATING

Every expression in this and subsequent sections may be generalized to n loci with $2n$ genes. The expressions (16) and (17) in the general case would become, respectively,

$$Y = A_1 + A_2 + B_1 + B_2 + C_1 + C_2 + \ldots \tag{16'}$$

$$\sigma_Y^2 = 2n\,pq + 2n(2n-1)\,pq\,f \tag{17'}$$

We shall work with two loci $(n = 2)$ specifically. Once the principle is understood, generalization to n loci follows.

The genetic correlation between mates. Let the genotype of the two parents be $P_1 = (A_1 A_2 B_1 B_2)$ and $P_2 = (A_3 A_4 B_3 B_4)$. The covariance between these two individuals is

$$Cov(P_1, P_2) = E(A_1 + A_2 + B_1 + B_2)(A_3 + A_4 + B_3 + B_4) = 16\,pq\,f \tag{18}$$

as each of the $4 \times 4 = 16$ covariances is $pq\,f$. If there are n loci, the covariance will be

$$Cov(P_1, P_2) = (2n)^2\,pqf. \tag{18'}$$

The genetic correlation between mates is, from (17) and (18),

$$m = \frac{Cov(P_1, P_2)}{Var\,P} = \frac{16\,pq\,f}{4pq(1+3f)} = \frac{4f}{1+3f} \tag{19}$$

yielding

$$f = \frac{m}{4 - 3m} \tag{20}$$

For n loci, the last two expressions are

$$m = \frac{2n\,f}{1 + (2n-1)f} \tag{19'}$$

$$f = \frac{m}{2n - (2n-1)\,m} \tag{20'}$$

as originally given by Wright (1921) through path analysis.

The correlation between uniting gametes. Once the genetic correlation between mates is found, the subsequent treatment of assortative mating becomes very similar to that of inbreeding. So far, the lower case f has been used to denote the correlation between the genes. Now, as in the case of inbreeding, let the capital F denote the correlation between uniting gametes, each gamete consisting of 2 genes, one from each locus, and n genes in the general case. The procedure of finding F is exactly the same as that of finding m. Let a gamete be represented by G. We obtain:

	2 loci	n loci	
Value:	$G = A_1 + B_1$	$G = A_1 + B_1 + C_1 + \ldots$	(21)
Variance:	$\sigma_G^2 = 2pq + 2pqf$	$\sigma_G^2 = npq + n(n-1)pqf$	(22)
Covariance:	$Cov(G_1, G_2) =$	$Cov(G_1, G_2) =$	
	$E(A_1 + B_1)(A_2 + B_2) = 4pqf$	$E(A_1 + \ldots)(A_2 + \ldots) = n^2 pqf$	(23)
Correlation:	$r(G_1, G_2) =$	$r(G_1, G_2) =$	
	$F = \dfrac{2f}{1+f}$	$F = \dfrac{nf}{1 + (n-1)f}$	(24)

From the expressions for m and F, whether for two loci or n loci, it may be immediately established that

$$m = \frac{2F}{1+F}, \qquad F = \frac{m}{2-m} \qquad \text{i.e., (6)}$$

The conclusion is that as far as the genetic relations are concerned, F and m are related in the same way whether by assortative mating or by inbreeding.

Path coefficient from parent to offspring. When the genetic correlation m between mates is specified, the remaining properties are the same as in inbreeding. Again, let b be the path coefficient from a parent to the gamete produced by that parent, and let a be that from a uniting gamete to the resulting offspring. From Fig. 267, $F = b\, m\, b = b^2 m$. Substituting (19) and (24),

$$b^2 = \frac{F}{m} = \frac{1 + 3f}{2(1+f)} \qquad (25)$$

Since two uniting gametes completely and equally determine the value of the resulting genotype, $2a^2 + 2a^2 F = 1$. Hence, using **(24)**,

$$a^2 = \frac{1}{2(1 + F)} = \frac{1+f}{2(1 + 3f)}, \qquad (26)$$

so that $\qquad a^2 b^2 = \frac{1}{4}, \qquad ab = \frac{1}{2} \qquad (27)$

in an equilibrium population, as in the case of inbreeding and random-mating populations. It may be easily verified that **(27)** is true for any number of loci. Jencks, et al. (1972) use g (our ab) for the path coefficient from a parent genotype to a resulting offspring genotype in their Fig. A-1 (p. 268). Apparently they have assumed an equilibrium state (although there is no such explicit statement), as the path coefficients in their diagram are the same for the two generations. But they have also entertained the possibilities that $g = ab = 0.45, 0.40, 0.35, 0.30, 0.25$ in their Table A-5 (p. 281) without explanations. At equilibrium, $g = ab = \frac{1}{2}$, regardless of the magnitude of the correlations m, F, or f.

Population composition

Partial or imperfect assortative mating would lead to an equilibrium condition, as partial inbreeding does, but the genetic composition is different from that of inbreeding. The main difference is in the homozygosis of loci. Inbreeding increases the homozygosis of loci with little effect on genes belonging to different loci. Assortative mating induces correlation between all genes (homologous or not) but has much less effect on homozygosis. A small f may yield a high value of m if the number of loci n is large. For example, if $n = 12$ and $f = 0.04$, we have from **(19′)**

$$m = \frac{24(0.04)}{1 + 23(0.04)} = \frac{0.96}{1.92} = 0.50$$

A small correlation between genes $(f = 0.04)$ causes very little increase in homozygosis of loci, but twelve such loci will give a very high correlation between mates.

The composition of a population with assortative mating depends on the precise mating patterns. Since various mating patterns may yield the same correlation between mates, no general solution may be given. However, for purposes of illustration, we shall deal with the unlinked two-locus case in some detail under simplifying conditions. Of course, different assumptions lead to different equilibrium compositions of the population. Let the gene frequencies of both loci be $p = q = \frac{1}{2}$, so that the distribution of the genotypes and phenotypes of (14) are symmetrical, continuing to assume additive gene effects. Purely for arithmetic convenience, we assume that the genetic (not phenotypic) correlation between the mates is $m = \frac{1}{2}$. This is too high a value for human populations with respect to any trait but we merely use it for numerical illustration. From (20), $f = m/(4 - 3m) = 0.20$, which measures the increase in "homozygosity" of gene pairs, homologous *(AA, aa)* or nonhomologous *(AB, ab)*. Let the equilibrium symmetrical genotype frequencies in the population be as follows (in Wright's notation):

	BB	Bb	bb	Total [see (11)]	
AA	x	z	y	¼ $(1 + f)$ =	0.30
Aa	z	w	z	½ $(1 - f)$ =	0.40	(28)
aa	y	z	x	¼ $(1 + f)$ =	0.30

The gametic output of this population is calculated and given in Table 279. In our assortative-mating model, there is no distinction between the combinations *AB* and *AA*, nor any distinction between *Ab* and *Aa*. The total frequency of *(AB)* gametes is therefore also ¼ $(1 + f)$, the same as that of the *AA*-genotype. From (28) and the sum of *(Ab) + (aB)* gametes in Table 279, we obtain three independent equations

$$
\begin{aligned}
&(i) & x + y + z \qquad\qquad &= \tfrac{1}{4}(1+f) = 0.30 \\
&(ii) & 2z + w \qquad &= \tfrac{1}{2}(1-f) = 0.40 \\
&(iii) & 2y + 2z + \tfrac{1}{2}w &= \tfrac{1}{2}(1-f) = 0.40
\end{aligned}
\qquad (29)
$$

INBREEDING AND ASSORTATIVE MATING

We observe that $2(i) + (ii)$ yields the equation for total frequency; and $(ii) = (iii)$ yields a relationship between y and w, viz.,

$$\begin{aligned}(i') \quad & 2x + 2y + 4z + w = 1 \\ (ii') \quad & 4y = w\end{aligned} \tag{29'}$$

From now on we are free to use these relations in subsequent manipulations, but we need a fourth equation for the complete solution of x, y, z, w.

Table 279. Gametes produced by the nine genotypes of (28).

class	genotype	frequency	(AB)	(Ab)	(aB)	(ab)
4	AABB	x	x	0	0	0
3	AABb	z	½z	½z	0	0
	AaBB	z	½z	0	½z	0
2	AAbb	y	0	y	0	0
	AaBb	w	¼w	¼w	¼w	¼w
	aaBB	y	0	0	y	0
1	Aabb	z	0	½z	0	½z
	aaBb	z	0	0	½z	½z
0	aabb	x	0	0	0	x
	Total	1.00	¼$(1+f)$	¼$(1-f)$	¼$(1-f)$	¼$(1+f)$
			0.30	0.20	0.20	0.30

The $(1-m)$ and (m) mating system. The fourth equation may be provided by imposing certain mating properties on the population. The simplest system is that a fraction $(1-m)$ of the matings are at random and m of the matings are between mates within the same phenotype class. This will yield a marital correlation m. The frequency of offspring from these matings may then be calculated. It is simplest to confine our attention to genotype $AABB$ (or $aabb$). The total frequency of the $AABB$ offspring must be equal to x, the frequency of $AABB$ parents, thus providing the needed equation.

Continuing with our numerical example, we assume that $1 - m = m = \frac{1}{2}$. Random mating is equivalent to random union of gametes (Chapter 7). We first calculate the $AABB$ offspring from the random matings. The bottom line of Table 279 gives the total gametic output from the population. Thus, $(0.30\,AB)^2$ gives $0.09\,AABB$ from the random-mating portion of the population. But there are $1 - m$ random matings. Hence, the contribution to the offspring is $0.09(1 - m) = 0.09(\frac{1}{2}) = 0.045$, as listed in Table 280.

Table 280. Contributions to $AABB$ offspring from various types of matings.

type of mating	$AABB$ produced			contribution to next generation		
random mating	$(0.30\,AB)^2$	=	0.09	$0.09(1-m)$	=	0.045
within class 4	$(1\,AB)^2$	=	1.0	$1\,mx$	=	$\frac{1}{2}x$
within class 3	$(\frac{1}{2}AB)^2$	=	$\frac{1}{4}$	$\frac{1}{4}m(2z)$	=	$\frac{1}{4}z$
within class 2	$(\frac{1}{6}\,AB)^2$	=	$\frac{1}{36}$	$\frac{1}{36}m(6y)$	=	$\frac{1}{12}\,y$
				Total		x

The contributions to $AABB$ from the within-class matings are calculated the same way. We use the matings within class 2 as an example. Using the known relation $w = 4y$, we see that the gametes produced by individuals of class 2 are in the proportion $(AB) : (Ab) : (aB) : (ab) = 1 : 2 : 2 : 1$. That is, $1/6$ of the gametes are (AB), producing $1/36\,AABB$ offspring. But there are $m(2y + w) = m(6y) = 3y$ matings within class 2. Hence, the contribution to $AABB$ offspring is $3y(1/36) = y/12$, as listed in Table 280. Classes 1 and 0 produce no (AB) gametes and contribute no $AABB$ offspring. The last column of Table 280 should add up to x, yielding the equation $x = 0.045 + (1/2)x + (1/4)z + (1/12)y$. Eliminating w by putting $w = 4y$, we obtain from this and (29) the three independent equations for x, y, z:

$$\begin{aligned} x + y + z &= 0.30 \\ 2y + z &= 0.20 \\ 6x - y - 3z &= 0.54 \end{aligned} \quad (30)$$

INBREEDING AND ASSORTATIVE MATING

Solving, we obtain

$$(x, y, z, w) = (82, 27, 56, 108)/550 \qquad (31)$$

or, putting the solutions in the form of **(28)**,

	BB	Bb	bb	Total	frequency
AA	82	56	27	165	0.30
Aa	56	108	56	220	0.40
aa	27	56	82	165	0.30
				550	1.00

(32)

It will be instructive for the reader to repeat the calculations with these numbers and see that the population is actually in equilibrium under the $(1 - m)$ and (m) assortative mating system (where $m = \frac{1}{2}$).

Numerical examples may be used to verify a number of relations established previously by analytical considerations. I shall mention only two of them. According to **(24)**, the correlation between uniting gametes is $r(A_1B_1, A_2B_2) = F = 2f/(1 + f)$ for two pairs of genes. When $m = 1/2$, $f = 1/5$, and $F = 1/3$. Since there is no distinction between A and B, or between a and b, the correlation between pairs of genes of the two loci, $r(A_1A_2, B_1B_2)$, should be the same as F. That this is true may be verified readily:

locus,	gamete		2	1	0
AA	AB	2	82	56	27
Aa	Ab + aB	1	56	108	56
aa	ab	0	27	56	82

(33)

correlation = $1/3$ = F

Next, collecting the genotypes of **(33)** into phenotypic classes according to **(14)**, we obtain the following distribution:

class Y :	0	1	2	3	4	Total	
Population (32)	82	112	162	112	82	550	
	0.149	0.204	0.295	0.204	0.149	1.00	(34)
If random mating	1	4	6	4	1	16	
	0.0625	0.2500	0.3750	0.2500	0.0625	1.00	

Direct numerical calculation yields $\sigma_Y^2 = 5.60 - (2)^2 = 1.60$ for populations (32). According to (17), the population variances should be

$$\sigma_Y^2 = 4pq + 12pqf = 4(1/4) + 12(1/4)(1/5) = 1.60 \qquad (35)$$

correctly. The reader may verify other analytical relationships established earlier in this chapter. In the case of random mating, $f = 0$ and the variance is simply $4pq = 1$.

Parent-offspring combinations

Now we wish to calculate the frequency of parent-offspring pairs under assortative mating, using the same numerical example with marital correlation $m = 1/2$ and gene frequencies $p = q = 1/2$. For arithmetic convenience, the common denominator 550 in (31) may be ignored and we will simply take $x = 82$, $y = 27$, $z = 56$, $w = 4y = 108$, so that the frequency in class 2, for instance, is $2y + w = 162$. We shall use class 2 to illustrate the procedure of calculation. As shown in Table 279, the gametic output of class 2 is $1 : 2 : 2 : 1$ in the order $(AB), (Ab), (aB), (ab)$; and the total gametic output of the population is 0.30, 0.20, 0.20, 0.30, in the same order. The gamete (AB) contributes a value of 2 to the offspring measurement; the gamete (Ab) or (aB) contributes a value of 1; and the gamete (ab) contributes 0 to the offspring measurement. As far as measurement is concerned, the gametes (Ab) and (aB) may be combined, yielding 1/6, 4/6, 1/6 from class 2, and 0.30, 0.40, 0.30 from the entire population.

Since half of the matings are at random and half within the same class, we may imagine that 81 individuals in class 2 practice random mating and the other 81 practice within-class mating. The offspring produced by these two types of mating are indicated in the upper half of Table 283. Collecting

INBREEDING AND ASSORTATIVE MATING

Table 283. Calculation of the frequency of parent-child pairs in an assortative mating population with marital correlation $m = \frac{1}{2}$, gene frequency $p = q = \frac{1}{2}$ for two loci, and correlation between genes $f = 0.20$. (2), (1), (0) are "measurements" of gametes (AB), $(Ab + aB)$, (ab), respectively.

No. parents	gametes produced		offspring produced by random mating			offspring produced by within-class mating		
			(2) 0.30	(1) 0.40	(0) 0.30	(2) 1/6	(1) 4/6	(0) 1/6
class 2								
81 ×	1/6 = 13.5	(2)	4.05	5.40	4.05	2.25	9.00	2.25
	4/6 = 54.0	(1)	16.20	21.60	16.20	9.00	36.00	9.00
	1/6 = 13.5	(0)	4.05	5.40	4.05	2.25	9.00	2.25

Collecting according to measurement

offspring measurement	4	3	2	1	0	total
from random mating	4.05	21.6	29.7	21.6	4.05	81
from within-class mating	2.25	18.0	40.5	18.0	2.25	81
total offspring of class 2	6.3	39.6	70.2	39.6	6.3	162

			(2) 0.30	(1) 0.40	(0) 0.30	(2) 0	(1) 0.50	(0) 0.50
class 1								
56 ×	1/2 = 28	(1)	8.4	11.2	8.4	0	14.0	14.0
	1/2 = 28	(0)	8.4	11.2	8.4	0	14.0	14.0

Collecting according to measurement

offspring measurement	4	3	2	1	0	total
from random mating	0	8.4	19.6	19.6	8.4	56
from within-class mating	0	0	14.0	28.0	14.0	56
total offspring of class 1	0	8.4	33.6	47.6	22.4	112

according to the measurement of the offspring (i.e., taking the diagonal sums of the 3 × 3 arrays) we obtain the array of offspring produced by the 162 parents of class 2, which is diagrammatically represented in Fig. 284.

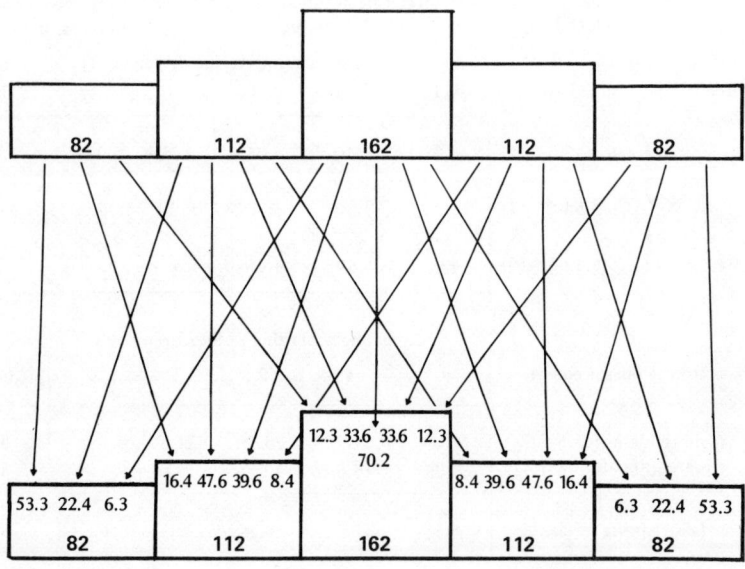

Fig. 284. Equilibrium populations with gene frequencies $p = q = 1/2$ for both loci and additive unit gene effects. The genetic correlation between mates is assumed to be $m = 1/2$. The five measurement classes are, from left to right, 0, 1, 2, 3, 4. The arrows show the segregation of the parental classes. For instance, the 112 parents in class 1 will produce 112 offspring distributed in four classes: 22.4, 47.6, 33.6, 8.4., in classes 0, 1, 2, 3, respectively.

The offspring array produced by parents of the other classes may be obtained in a similar way. Due to symmetry, we need only to find the descendants of parents of classes 1 and 0. Of the 112 individuals in class 1, we may imagine 56 practice random mating and 56 practice within-class mating. The genotype $Aabb \rightarrow \frac{1}{2}(Ab) + \frac{1}{2}(ab)$, and $aaBb \rightarrow \frac{1}{2}(aB) + \frac{1}{2}(ab)$. As far as the measurement values are concerned, these two genotypes and their gametic output are identical: being $\frac{1}{2}(1)$ and $\frac{1}{2}(0)$. The offspring produced by the two types of mating are shown in the lower half of Table 283, also illustrated in Fig. 284.

INBREEDING AND ASSORTATIVE MATING

The descendants of parents of class 0 (not shown in Table 284) are easy to obtain, as the gametic output of *aabb* is all *(ab)*, contributing (0) to the measurement of offspring. The ½(82) = 41 individuals practicing random mating and 41 practicing within-class mating will produce the following offspring:

		random mating					within class 0 mating		
		0.30	0.40	0.30			0	0	1.0
		(2)	(1)	(0)			(2)	(1)	(0)
41	(0)	12.3	16.4	12.3	41	(0)	0	0	41.0

totaling 53.3 in class 0, 16.4 in class 1, and 12.3 in class 2. The full picture is shown in Fig. 284. The resulting offspring generation has, of course, the same distribution as the parents. The purpose of presenting a diagram such as Fig. 284 is to provide a visual aid demonstrating the shifting of classes between the two generations and thus to dispel the common mistaken notion of "like begets like" in heredity. The pattern of shifting (lines with arrows) from a parent class to an offspring class is due to gene segregation. This shifting happens even when the environment is uniform and there is strong assortative mating. Further comments will follow when we compare this situation with that of random mating.

A diagram such as Fig. 284 is nothing more than a correlation table. Collecting from Fig. 284 the frequency of parent-offspring pairs under strong assortative mating, we obtain the conventional correlation table (Table 286). The marginal variance of Table 286 has already been calculated (35): $\sigma_Y^2 = 1.60$. The covariance will be found to be $Cov = 2860/550 - 2^2 = 5.20 - 4 = 1.20$, so that the parent-offspring correlation is $r_{PO} = 1.2/1.6 = 0.75$, in agreement with the formula

$$r_{PO} = \tfrac{1}{2}(1 + m) = \tfrac{1}{2}(1 + \tfrac{1}{2}) = \tfrac{3}{4} \qquad \text{i.e., (7)}$$

It is tedious to calculate numerically the frequency of the various sib-pairs for two loci. Suffice it to say the sib-sib correlation is the same; $r_{OO} = \tfrac{1}{2}(1 + m) = 0.75$.

Table 286. Joint distribution of parent-child pairs in an assortative-mating population with very high marital correlation $(m = \frac{1}{2})$, equal gene frequencies, and additive gene effects.

		offspring measurement class					
		0	1	2	3	4	total
	0	53.3	16.4	12.3	0	0	82
	1	22.4	47.6	33.6	8.4	0	112
parent	2	6.3	39.6	70.2	39.6	6.3	162
	3	0	8.4	33.6	47.6	22.4	112
	4	0	0	12.3	16.4	53.3	82
	total	82	112	162	112	82	550

Joint distribution of parent-offspring pairs in a random-mating population under the same conditions, except that $m = 0$.

		0	1	2	3	4	total
	0	1	2	1	0	0	4
	1	2	6	6	2	0	16
parent	2	1	6	10	6	1	24
	3	0	2	6	6	2	16
	4	0	0	1	2	1	4
	total	4	16	24	16	4	64

Comparison with a random-mating population. Li (1971) constructed a diagram similar to Fig. 284 for a random-mating population with the same gene frequencies $(p = q = \frac{1}{2})$ and additive gene effects. (For an analogous diagram for a continuous distribution, see Li, 1970.) Instead of reproducing that diagram, it is presented in the lower portion of Table 286 for easy comparison. It will be found that the parent-offspring correlation is $r_{PO} = \frac{1}{2}$ in the random-mating population.

Now we are interested in the relative magnitude of the shift from a parental to an offspring class, for each parental class. Dividing the entries of each row by its own row total, we obtain from Table 286, the following conditional probabilities:

INBREEDING AND ASSORTATIVE MATING

	assortative mating offspring class						random mating offspring class					
parent	0	1	2	3	4	total	0	1	2	3	4	total
0	.650	.200	.150	0	0	1	.250	.500	.250	0	0	1
1	.200	.425	.300	.075	0	1	.125	.375	.375	.125	0	1
2	.039	.244	.433	.244	.039	1	.042	.250	.417	.250	.042	1 (36)
3	0	.075	.300	.425	.200	1	0	.125	.375	.375	.125	1
4	0	0	.150	.200	.650	1	0	0	.250	.500	.250	1

Due to symmetry, we need only to examine the first three rows of the tabulations above. In the assortative-mating population, if the parent is in class 2, then 43.3% of the offspring will remain in class 2, only slightly higher than the corresponding 41.7% for a random-mating population. As a whole, the percentages in the third row (parent class 2) of the two populations do not differ much. But the second row (parent class 1) shows that in the assortative-mating population, 42.5% of the offspring will remain in class 1, while in the random-mating population only 37.5% will do so. But the greatest difference is in the first (and last) row. In the assortative-mating population, 65% of the offspring of class 0 parents will remain in class 0, while in the random-mating population, only 25% will do so. The percentages in each row are called the "transition" probabilities from one state to another in one unit-time (one generation in our case). The transition probabilities are symmetrical ($p = \frac{1}{2}$ for both loci) in each row for the random-mating population, but in the assortative-mating population each row differs in pattern from the others. While the middle row remains symmetrical, row 2 is skewed to the left as row 3 is skewed to the right. In the extreme classes, 0 and 4, the skewness is very marked toward the parental class.

In drawing these general conclusions, two important points should be kept in mind. One is that in our example the genetic correlation between mates, $m = \frac{1}{2}$, is entirely too high for human populations with respect to a quantitative trait of any interest. So the situation exemplified probably represents a limiting rather than a typical situation. Second, there are only two pairs of genes in our example, so that the population has an appreciable proportion of the extreme homozygotes (*AABB, aabb*). When the number of loci increases, the frequency of such extreme genotypes (*AABBCCDD* ...)

will decrease. Our conclusions above should be taken as qualitative rather than quantitative. It is worth repeating to nongenetics students that the transitions from a parental class to an offspring class, whether under assortative or random mating, are entirely due to gene segregation (different gametes produced by the same parental genotype). There are $5 \times 5 = 25$ transition probabilities (zero is a transition probability) in our example, all being the result of one and the same mechanism during gametogenesis, viz., gene segregation. The transition of classes may perhaps be explained by environmental factors (plural) to a certain extent. There is no single environmental factor that can account for the 25 transitions, some of which are in opposite directions.

Ancestors and descendants. While still on the subject of transition of classes from the parental generation to the offspring, we may extend Fig. 284 to several generations as shown in Fig. 289 in order to obtain a more complete view of the endless transitions. Between any two given generations, the transition probabilities are given by (36). These transition probabilities are repeated again and again over the generations. Imagine that we concentrate our attention on one parent in class 1 of a certain generation and follow the descendants over the generations. A child may be found in class 3; a grandchild in class 2, a great-grandchild in class 0, etc. It is intuitively clear that the correlation between an ancestor and his descendant several generations away will be small and will approach zero in due time, the approach being faster for random-mating than for assortative-mating populations. If the relationship is to be described in terms of path coefficients, we may summarize all the connections between parent and offspring into one path, viz., r_{PO} for any one generation. Then, by the theorem on chain of causes, the correlation between an ancestor and his descendant n generations away will be r_{PO}^n. Thus, in populations with $m = 0, 1/4, 1/2$, we have

when $m = 0$, $r_{PO} = 0.500$, $r_{PO}^3 = 0.1250$

when $m = 1/4$, $r_{PO} = 0.625$, $r_{PO}^5 = 0.0954$ (37)

when $m = 1/2$, $r_{PO} = 0.750$, $r_{PO}^8 = 0.1001$

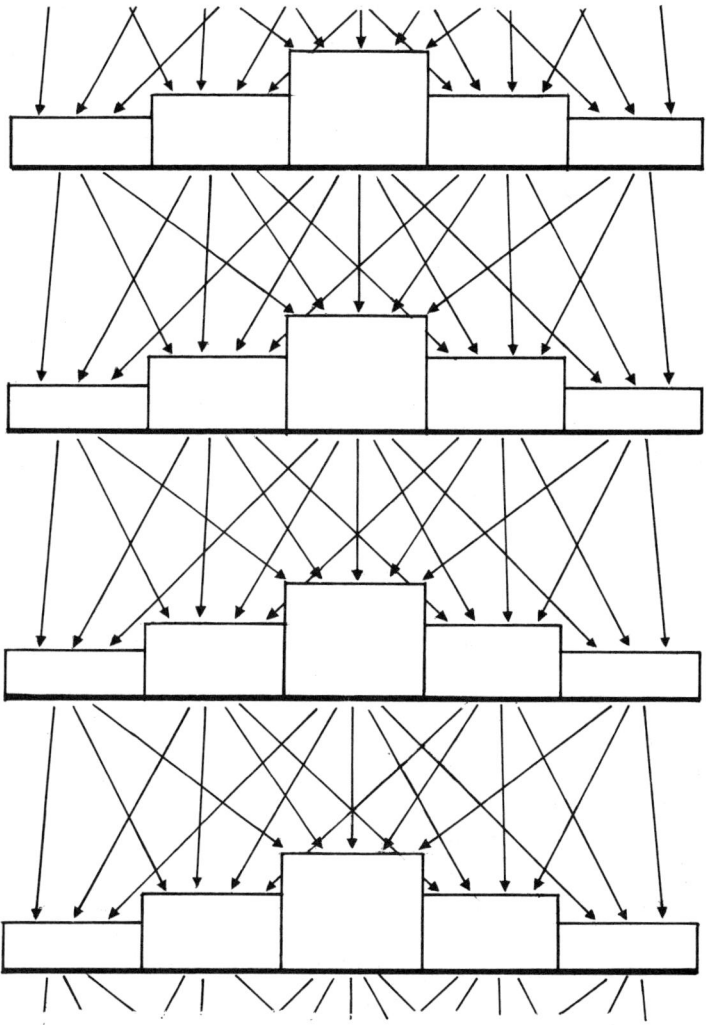

Fig. 289. An extension of Fig. 284 to include several generations to show the endless transitions of classes (genotype measurements) over the generations. As a result of these transitions, there is no correlation between an ancestor and his distant descendants. In other words, the line of family name has no long-reaching effect.

We see that in a random-mating population the ancestor-descendant correlation reduces to a low of 0.125 in only three generations. Even with moderately strong assortative mating $(m = ¼)$, the ancestor-descendant correlation reduces to a level below 0.100 in five generations. With very strong genetic correlation between mates $(m = ½)$, the ancestor-descendant correlation becomes 0.10 in eight generations. In all cases the limiting correlation value is zero. It is clear that the Mendelian mechanism of heredity (gene segregation and recombination) has no long-term or long-reaching effects on distant relatives and descendants. Genes know no family names and they do not stay with any family for long. While all of these properties of heredity are well known to population geneticists, they may surprise those who believe in family lines and are proud of their family names. With respect to certain socio-economic variables not determined by genotypes in this way, man-made rules (laws) and artificial controls (business management) may be able to keep the wealth in a family longer than they can keep the genes.

It is beyond the scope of this book to study the transition probabilities (36) analytically. A brief verbal description of their properties must suffice here. In spite of the complicated appearance of the transition lines in Fig. 289, the transition probabilities between any number of generations may be readily calculated in terms of the elementary transition probabilities of (36). In fact, if we let T be the matrix of transition probabilities (36), then the transition probabilities for ancestor-descendant n generations apart are simply given by the elements of T^n. When the number of generations (n) becomes large, the five rows of T^n will be nearly alike; and in the limit $(n \to \infty)$, the five rows of T^∞ will be identical. If we use the transitional probabilities (36) for random-mating populations, the limiting five rows of T^∞ will be each $(1, 4, 6, 4, 1)/16$. If the transition probabilities for the assortative-mating population are used, each of the limiting five rows of T^∞ will be $(82, 112, 162, 112, 82)/550$. This means: no matter the class to which the ancestor may belong, the probability distribution of his distant descendants over the five classes is the same. After a sufficiently large number of generations, the descendants of an original class 0 ancestor will have the same probability distribution over the five classes as the descendants of an original class 4 ancestor. In other words, there will be no correlation between an ancestor and his distant descendants.

INBREEDING AND ASSORTATIVE MATING

This agrees with our previous conclusion: $r_{PO}^n \to 0$, as n increases. Further discussion may be found in Li (1970, 1971).

Since we are primarily interested in obtaining the limiting form of T^n, we use the 5 x 5 transition matrix for simplicity, as if each individual in any class has the same transition probabilities from that class. Although this is the usual model in probability textbooks, it is not strictly true for our genetic problem. Class 2 consists of three genotypes: *AAbb, AaBb, aaBB*. The segregation pattern of *AAbb* and *aaBB* is the same, but that of *AaBb* is different. Strictly pursued, we should use a 9 x 9 transition matrix, and then pool the genotypes into five classes according to their phenotypic measurement. Calculations show that the results of using $T_{(5 \times 5)}$ and $T_{(9 \times 9)}$ are approximately but not exactly the same for the first few generations, but they become increasingly alike as time goes on. At the limit, they give the same result.

Environmental effects on correlations

All the previous calculations are based on the genotype measurements with additive gene effects $(Y = L)$. The random environmental effects upon the genotype measurements may be introduced in the same manner as in Chapter 8. Let $\in$ be the random effects on the measurements, so that the observed phenotype value is $Z = Y + \in$ and $\sigma_Z^2 = \sigma_Y^2 + \sigma_E^2$, assuming environmental effects and genotype are uncorrelated. The path coefficients from genotype and from environment to the phenotype are, respectively, $h = \sigma_Y/\sigma_Z$ and $e = \sigma_E/\sigma_Z$, so that $h^2 + e^2 = 1$. Not all problems concerning the correlation between relatives are solved under assortative mating based on phenotype values Z. Wright, in his forthcoming Volume III of the series entitled *Evolution and the Genetics of Populations*, will present a comprehensive treatment which requires the parental phenotypes to appear twice in the path diagram. For the time being we must be content with the simplest model, which assumes no dominance, no epistasis, and independence of heredity and environment.

The phenotypic correlation between mates. Before calculating the parent-offspring and sib-sib correlation, we have to investigate the nature of the phenotypic correlation between mates. Let r_{PP} be the correlation between mates with respect to their phenotypic (Z) values (including environmental effects), which are to be distinguished from m, the correlation between mates with respect to their genotypic (Y) values (not including environmental effects). Under the system of phenotypic assortative mating, the two mates tend to have similar phenotypes, whatever the reason for the similarity of their phenotypes. This similarity may arise from the similarity of their genotypes, from the similarity of their environments, or from the similarity caused by the environment of one mate and the genotype of the other. The phenotypic correlation between mates, which is the primary correlation, leads to the four correlations between the genotypes (geno) and environments (env) of the two mates, as shown in Fig. 292. The right-hand diagram shows that the primary correlation r_{pp} between the phenotype (pheno) of the two mates leads to the following four correlations:

geno-geno,	geno-env,	env-geno,	env-env	
$m = r_{pp} h^2,$	$r_{pp} he,$	$r_{pp} eh,$	$r_{pp} e^2$	(38)

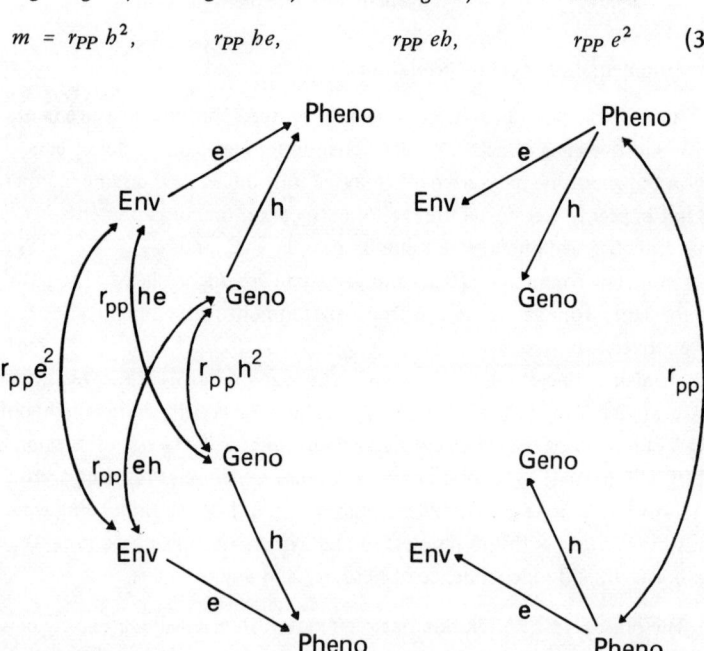

Fig. 292. The phenotypic correlation between parents with assortative mating: Pheno = phenotype, Geno = genotype, Env = environment. *Right diagram:* the primary phenotypic correlation r_{pp} leads to other correlations; e.g., the correlation between the genotype of one parent and the environment of the other parent is $r(geno', env) = h(r_{pp})e$, etc. *Left diagram:* reverse of the right diagram. The path coefficients e and h remain the same. The correlations $r_{pp}e^2$, $r_{pp}he$, $r_{pp}eh$, $r_{pp}h^2 = m$, are derived from the right diagram. (Based on Wright, 1969.)

To prepare us to calculate the parent-offspring and sib-sib correlation subsequently, the right diagram is reversed, becoming the left diagram of Fig. 292 and showing the four correlations of (**38**). In the reversing process, the path coefficients e and h are unaffected, because the correlations $r(pheno, env)$ and $r(pheno, geno)$ for each parent remain the same. (In Chapter 3 it was shown that whether regression of Y on X or regression of X on Y, the degree of determination of Y by X and the degree of determination of X by Y are the same, both being r_{XY}^2.) The consistency of the two diagrams of Fig. 292 may be seen by working from the left diagram:

$$r_{PP} = h(r_{PP}h^2)h + e(r_{PP}he)h + h(r_{PP}eh)e + e(r_{PP}e^2)e$$
$$= r_{PP}(h^4 + 2h^2e^2 + e^4) = r_{PP}(h^2 + e^2)^2 = r_{PP} \tag{39}$$

where $h^2 + e^2 = 1$. Thus the total r_{PP} is the sum of four connecting paths from the viewpoint of the left diagram of Fig. 292. The four connecting paths may of course be summarized into one path to simplify an otherwise complicated diagram. This will be done in the next paragraph.

Phenotypic correlation between relatives. Fig. 294 shows the relationship between parents *(left)* and offspring *(right)* with respect to both genotypes and phenotypes. The left portion (parents) is the same as the left diagram of Fig. 292, except that the four connections between the parental phenotypes have been summarized into one correlation r_{PP}. In Fig. 294 we note that the genotypic correlation between mates, $m = r_{PP}h^2$, is indicated by dotted lines, as it is a part of r_{PP} and therefore partially redundant. In calculating the phenotypic parent-offspring correlation we use r_{PP} as a connecting link between one parental phenotype and another. In calculating the phenotypic sib-sib correlation, however, we can only use $m = r_{PP}h^2$ as a connecting link between the genotypes of the two parents, because only this link contributes to the correlation between full sibs. The important point about redundant lines is that we may not use both for calculating any one correlation. But, in separate calculations, we may use one for one purpose and the other for another purpose.

From Fig. 294 we see that the phenotypes of a parent and an offspring are connected by two routes: one is the direct route $h(½)h = ½h^2$; the other is the route through the phenotype of the other parent, $h(½)h \cdot r_{PP} =$

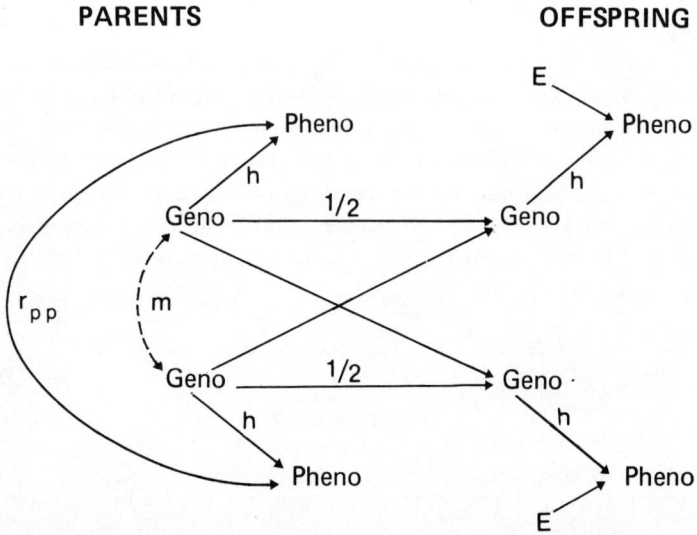

Fig. 294. Phenotypic parent-offspring and sib-sib correlations in a population with assortative mating. Pheno = phenotype, Geno = genotype, Env = environment. The *left* portion represents parents and the *right* portion represents offspring. The path from parent to offspring is $ba = \frac{1}{2}$ in an equilibrium population (gametes not shown). The overall phenotypic parental correlation r_{PP} is the sum of the four correlations shown in the left diagram of Fig. 292. The genetic correlation between mates, $m = r_{PP}h^2$, is indicated in dotted lines; it is a part of r_{PP}, being partially redundant.

$\frac{1}{2}h^2 r_{PP}$. The sum of these two connecting paths gives the phenotypic parent-offspring correlation (Wright, 1921):

$$\text{phenotypic} \quad r_{PO} = \frac{1}{2}(1 + r_{PP})\, h^2 \tag{40}$$

Again from Fig. 294, we see that the phenotypes of the two full sibs are connected by four routes: two direct and two indirect. The two direct routes are those through the genotypes of the parents without using the link $m = r_{PP}h^2$, each such route being $h(\frac{1}{2} \cdot \frac{1}{2})h = \frac{1}{4}h^2$. The two indirect routes are those through the genetic correlation between the two parents, each being $h(\frac{1}{2})(m)(\frac{1}{2})h = \frac{1}{4}(r_{PP}h^2)h^2$. The sum of these four connecting paths gives the phenotypic sib-sib correlation:

INBREEDING AND ASSORTATIVE MATING   295

$$\text{phenotypic} \quad r_{OO} = \tfrac{1}{2}(1 + r_{PP}h^2)h^2 \tag{41}$$

A term c^2 should be added to (41) if a systematic common factor in the home is effective in increasing the correlation between full sibs reared together. When the observed values of the phenotypic correlations r_{PP}, r_{PO}, and r_{OO} are available, these equations enable us to estimate the parental genetic correlation m, the environmental heritability h^2, and the influence of common home environment c^2 in an assortative-mating population. Practical applications of these results are given in the next chapter. For further development on the subject of familial resemblance under assortative mating, see Rao, Morton, and Yee (1974, 1976).

Exercises

Ex. 1. In the text we have converted the transition diagram (Fig. 284) into Table 286, upper portion, for an assortative-mating population. Now, do the reverse for a random-mating population: that is, convert Table 286, lower portion, into a transition diagram similar to Fig. 284. If you are not sure of your drawing, consult the diagram given by Li (1971).

Ex. 2. When the n pairs of genes, whether homologous or nonhomologous, are correlated with coefficient f, find the general condition under which the population variance is twice as large as that of a random-mating population with the same gene frequencies. *Hint:* Use (17') and equate $2npq = 2n(2n-1)fpq$. Solving,

$$f = \frac{1}{2n-1}, \qquad n = \frac{1+f}{2f}$$

Example, $f = 0.04$ and $n = 13$ satisfy the condition; in which case the marital correlation would be $m = 0.52$. Why? Consult (19').

Ex. 3. As stated in the text, different assumptions lead to different equilibrium compositions of assortative-mating population. Instead of using the $(1-m)$ and (m) mating model, Wright (1921) calculated the frequency of the union of the gametes that united to form the population:

Uniting gametes	(AB)	(Ab)	(aB)	(ab)		
(AB)	x	½ z	½ z	¼ w	¼$(1+f)$ =	0.30
(Ab)	½ z	y	¼ w	½ z	¼$(1-f)$ =	0.20
(aB)	½ z	¼ w	y	½ z	¼$(1-f)$ =	0.20
(ab)	¼ w	½ z	½ z	x	¼$(1+f)$ =	0.30

The totals of such uniting gametes are of course the same as those at the bottom of our Table 279. Three of the four equations of Wright are equivalent to our three equations of (29). To obtain the fourth equation, Wright assumed that the matings are such that the variance of these uniting gametes are the same (homoscedastic). The gametes *(Ab)* and *(aB)* may be combined into one group, because they have the same "measurements." From the equations (29') we already know that ¼$w = y$. Using $(+1, 0, -1)$, instead of $(2, 1, 0)$, for the gamete values, the first two rows of the table above become

(AB) +1	(Ab + aB) 0	(ab) −1	total	mean	variance
x	z	y	0.30	$\dfrac{x-y}{0.30}$	$\dfrac{x+y}{0.30} - \left(\dfrac{x-y}{0.30}\right)^2$
½ z	$2y$	½ z	0.20	0	$\dfrac{z}{0.20}$

Equating the variance of the two rows of gametes, we obtain the equation

$$\frac{x+y}{0.30} - \frac{(x-y)^2}{0.09} = \frac{z}{0.20} \qquad (42)$$

Although (42) is not the actual equation employed by Wright (1921), it is equivalent and leads to the same solution. Equation (42) and the first two equations of (30) will yield solutions for x, y, z, with $w = 4y$.

$$\begin{aligned} x + y + z &= 0.30 \\ 2y + z &= 0.20 \end{aligned} \qquad (30)$$

Subtracting, $\quad x - y = 0.10, \quad (x-y)^2 = 0.01$

Substituting in **(42)** and solving the three equations, we find the solutions:

	BB	Bb	bb	total	frequency	
AA	22	16	7	45	0.30	
Aa	16	28	16	60	0.40	(43)
aa	7	16	22	45	0.30	
				150	1.00	

The student may verify that this population also has the variance $\sigma_Y^2 = 1.60$, the same as **(35)**. Further, the correlation between single locus genotypes, (AA, Aa, aa) and (BB, Bb, bb), is the same as the correlation between uniting gametes, (AB, Ab + aB, ab) and (AB, Ab + aB, ab), both being $F = 1/3$.

Ex. 4. Correlation between genes. One of the characteristic features of assortative mating based wholly on somatic resemblance is that the genes determining the phenotype measurement are equally correlated, whether they are homologous (belonging to the same loci) or nonhomologous (belonging to different loci). Fig. 298 is intended to make this point clearer. Let P_1 and P_2 be the two parents, each possessing a number of pairs of genes (three pairs are explicitly shown in the diagram). The genetic correlation (without environmental fluctuations) between the two parents is $m = r(P_1, P_2)$. This correlation implies that the genes of the two parents are correlated. Let f' be the correlation between two homologous genes and j' be the correlation between two nonhomologous genes of the two parents, as shown in Fig. 298. But the preference for mate is not based on genotype per se but based on phenotype measurement, on which the genes A, a function exactly the same way as the genes B, b, assuming equal gene effects and equal gene frequency. Hence, $f' = j'$.

Next, let g' be the correlation between homologous genes and k' be the correlation between nonhomologous genes of the same parent. Then in the offspring individual (lower portion of Fig. 298), the correlation between two homologous genes (via uniting gametes) g is of course the same correlation between two homologous genes of the two parents. Hence, $g = f'$.

Finally, the correlation k between two nonhomologous genes in the offspring depends on whether these two genes are from the same parent or from different parents. If from the same parent, it remains k'. If from different parents, it is j' which is equal to f'. For unlinked loci, we have $k = \frac{1}{2}k' + \frac{1}{2}j'$. These are the relations among the various correlations in two successive generations. At equilibrium, we drop the primes for the parents and obtain $k = \frac{1}{2}k + \frac{1}{2}j$. Hence, $k = j = f = g$. That is, all genes are equally correlated whether homologous or nonhomologous.

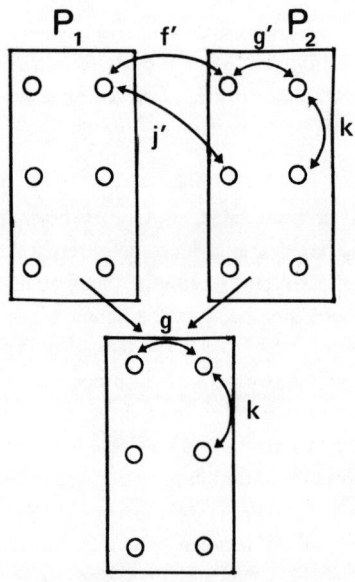

Fig. 298. Correlation between homologous and nonhomologous genes in two successive generations. Assortative mating makes $f' = j'$. In offspring, $g = f' = j'$ and $k = \frac{1}{2}k' + \frac{1}{2}j'$. At equilibrium, $k = j = f = g$. (Diagram modified from Crow and Felsenstein, 1968; notation follows Wright, 1921.)

… Chapter **10**

APPLICATIONS IN BIO-SOCIAL SCIENCES

IN THIS concluding chapter, we shall first look at a few considerations in the practical application of correlation and path coefficients, and then at a few actual examples in the field of bio-social sciences. These examples should give students an opportunity to review the methodology developed in the earlier chapters and also a "feel" for application. Needless to say, no one should take the numerical values in the examples as the established results in that field.

Standardized or Unstandardized Coefficients

The path coefficients (p_{0i}) as used in this book are standardized partial regression coefficients with respect to a causal scheme. For the same causal scheme and therefore the same regression processes, the concrete (unstandardized) partial regression coefficients are called path regression coefficients (b_{0i}). The difference between these two types of coefficients is merely a matter of units. Thus (Chapter 5):

$$p_{0i} = b_{0i}\left(\frac{\sigma_i}{\sigma_0}\right) \quad \text{or} \quad b_{0i} = p_{0i}\left(\frac{\sigma_0}{\sigma_i}\right)$$

There are no other inconsistencies between the p's and the b's. Their general properties are the same, except for the ratio of their standard deviations. Because of the absence of units, a path coefficient reduces to a correlation coefficient (a standardized coefficient) under certain simple conditions, while a concrete regression coefficient usually remains a regression coefficient with compound units, e.g., $b_{0i} = 5$ cm/gm. In certain particular cases, such as the regression of offspring on parent with respect to some quantitative trait in a random-mating population, the regression coefficient is also the correlation coefficient, $(r_{0i} = p_{0i} = b_{0i})$ as $\sigma_0 = \sigma_i$ due to equilibrium conditions.

There have been discussions about the relative merits of the standardized and concrete regression coefficients (Tukey, 1954; Turner and Stevens, 1959; Wright, 1960). We can hardly argue about tools. Is a hammer more useful than a screw driver? It depends on the nature of the task to be accomplished. In the case of standardized and unstandardized coefficients, we have even less to argue about, for it is such an easy matter to convert one type of coefficients into the other.

When the dependent variable under study is a percentage itself, there being no units involved, the path method has proved to be highly efficient. In dealing with unobservable variables (mean and variance unknown), we have to be contented with standardized coefficients, as is the case with factor analysis.

As to the nature of our objective, I find Blalock's (1967) remarks succinct as well as clear. Should there exist "causal laws," analogous to physical laws in nature (e.g., 32 feet per second per second), then these laws would be better described in terms of concrete coefficients, as they are universal and remain the same in all populations. For a hypothetical example, if our adult height-weight relationship is 5 pounds per inch, this is a physiological law on growth and is true for Italians as well as for Japanese. A standardized coefficient may vary from population to population. On the other hand, if one wishes to measure the amount of impact of one variable on another, a standardized coefficient will serve the purpose, as the method of path coefficients is essentially based on the degree of determination of the variance of the dependent variable by others. Blalock continues:

> As long as one has his purpose clearly in mind, there should be no source of confusion regarding the choice between standardized and unstandardized coefficients. If one wishes to generalize to a specific population, path coefficients are appropriate. If he wishes to compare populations to determine whether or not the underlying causal processes are basically similar, he should make use of the unstandardized coefficients.

These remarks are in perfect agreement with Wright (1931) who, in discussing the problem of borrowing coefficients, says—

> ... it is the concrete partial regression coefficient and not the path coefficient which is directly transferable (from one population to another).

Wright himself has always regarded path and regression coefficients as complementary tools rather than mutually exclusive alternatives.

Isomorphism

Certain situations in widely different fields, when expressed in an abstract form, become identical except for the terminology or names of the variables. In such a case we say these situations are "isomorphic" (Li, 1969). What is true for one situation must also be true for the other, no matter how different the subject matter or verbal descriptions may be. Population genetics, epidemiology, demography, and social sciences are admittedly different disciplines; but a number of their problems are isomorphic. Thus, an investigation in one discipline will have similar implications in the others. It is in the interest of the unity of science to mention a few examples of this nature. It happens also to be of practical value, as we may freely transfer results of one situation to another, once the isomorphism is established. The most readily understood example is one with respect to dichotomous conditions. Consider the lefthand one of the following 2 x 2 contingency tables:

		H					H		
		A	a				A	a	
W	A	0.22	0.18	0.40	W	A	0.16	0.24	0.40
	a	0.18	0.42	0.60		a	0.24	0.36	0.60
		0.40	0.60	1.00			0.40	0.60	1.00

(1)

Without revealing the meaning of the symbols H, W, A, a, we do not know what the tables are referring to. It happens that an epidemiologist is studying the possible association between husband (H) and wife (W) with respect to arthritis $(A$ = arthritis present, a = arthritis absent$)$. The lefthand table of (1) is based on a large number of husband-wife pairs within the age group 46-65. Two immediate conclusions may be drawn. The first is that the prevalence of arthritis is the same in both sexes. The second is that there is a certain amount of association between husband and wife with respect to arthritis. For, if they were not associated, the frequency of husband-wife arthritis pairs would be $(0.40)(0.40) = 0.16$, as shown in the righthand table of (1), instead of the "observed" 0.22. Assigning $A = 1$ and $a = 0$, we find the correlation between husband and wife:

$$r_{HW} = \frac{0.22 - 0.16}{(0.40)(0.60)} = 0.25$$

The 2 x 2 contingency tables above may be viewed otherwise. A population geneticist is studying the frequencies of the three genotypes *(AA, Aa, aa)* in a population and has found that the distribution (0.22, 0.36, 0.42), from which we infer that the uniting gametes must be like that shown in the lefthand table above. The marginal frequencies (0.40, 0.60) are called gene frequencies. The geneticist would say there is inbreeding in the population; that is, there is a correlation between the uniting gametes. For, if they were united at random, the distribution of the genotypes would be (0.16, 0.48, 0.36), as shown in the righthand table of (1). By the same method of calculation, the geneticist would find the inbreeding coefficient to be:

$$F = \frac{0.22 - (0.4)^2}{pq} = \frac{0.60}{0.24} = 0.25$$

Thus, we have established the correspondence for every item in epidemiology and population genetics:

Arthritis prevalence	gene frequency
Husband-wife pair	a genotype
No association	random mating
Association	inbreeding
Correlation coefficient	inbreeding coefficient

Without committing the specific meaning of the symbols of the contingency tables, they may be interpreted either way, because they are identical in form; that is, they are isomorphic. A conclusion in one situation may be transferred to the other with equal validity. Li (1969) has extended such considerations to variables with more than two conditions (multiple alleles).

The isomorphism of two situations need not be complete from beginning to end. A problem may usually be studied in stages or in various aspects. As long as two situations are isomorphic in one stage or in one aspect, the conclusions would be the same in both situations as far as that stage or that

APPLICATIONS IN BIO-SOCIAL SCIENCES

aspect is concerned. No perfect isomorphism exists in the real world. It exists only in abstract mathematics. The practical advantage of establishing a close correspondence between the situations in two different fields is to help us to see the generality of the problem and enable us to transfer methods as well as results from one field to another. There are many similar, if not strictly isomorphic, problems and methods in population genetics, epidemiology, and the social sciences. In this chapter we shall examine only the simplest examples to illustrate the principle and importance of isomorphism.

Individuals and Groups

As an application of the principle of isomorphism, let us consider the following three tables (Li, 1961, p. 84):

	I				II				III		
	(1)	(0)			(1)	(0)			(1)	(0)	
(1)	0.01	0.09	0.10	(1)	0.16	0.24	0.40	(1)	0.49	0.21	0.70
(0)	0.09	0.81	0.90	(0)	0.24	0.36	0.60	(0)	0.21	0.09	0.30
	0.10	0.90	1.00		0.40	0.60	1.00		0.70	0.30	1.00

(2)

Without identifying the specific meaning of I, II, III, and (1), (0), the tables are in an abstract form. Now, as before, let (1) = presence of arthritis, (0) = absence of arthritis, rows be wives, and columns be husbands. Further, let I = ages 36-50, II = ages 51-65, and III = ages 66-80. Again, we assume that the epidemiologist is studying the possible association between husband and wife with respect to arthritis. The reason for his study is to see if some factor(s) in the home environment and diet is contributing to the presence of arthritis. If so, husbands and wives should show some association with respect to the anomaly. From the three separate age groups of (2), the epidemiologist finds no association in any age group:

$$r_I = 0, \qquad r_{II} = 0, \qquad r_{III} = 0$$

where r_I = correlation in age group I, etc. For arithmetic convenience, let us assume that the three age groups are of equal size (say, each group is based on 1,000 husband-wife pairs). His conclusion is: there is no association between husband and wife, implying that the common home environment is not a contributing factor to arthritis. However, when he adds the three tables together to obtain an "overall" picture or a "summary" for the entire adult population, he finds something different. The following table for all age groups is obtained by $(I + II + III)/3 = T/3$.

	(1)	(0)	
(1)	0.22	0.18	0.40
(0)	0.18	0.42	0.60
	0.40	0.60	1.00

$$r_T = 0.25 \qquad (3)$$

The "summary" or "total" (T) table above shows fairly strong correlation between husbands and wives. What would the epidemiologist say now?

The correlation based on the total population, $r_T = 0.25$, is sometimes called spurious. The whole situation is sometimes described as a fallacy. A layman would think it is a paradox. But r_T has its reasons and meaning, though it does not answer our original question: is home environment a contributing factor to arthritis? We note from the tables (2) that the prevalence of arthritis increases with age, being 10% in the young group, 40% in the middle group, and 70% in the old group for both males and females.

In other words, it is age that is correlated with arthritis. As both husband and wife get older at the same time, both of them will show higher prevalence of arthritis, not because they are husband and wife but because they are both old people. Hence, r_T arises from the difference in prevalence of the three age groups, and does not answer our original question about common home environment.

Now, let us consider the three tables of (2) once more, this time as data from genetic studies of three localities or tribes. Let (1) and (0) be alleles A and a of an autosomal locus. We see that in each of the tribes or local populations (or "isolates" as geneticists call them), the mating is random $(F_i = 0, \ i = I, II, III)$. They are three separate random-mating populations. When we pool the three tribes into one large population, the resultant genotype frequencies (0.22, 0.36, 0.42) show a fairly strong inbreeding effect.

Indeed, if we disregard the tribe boundaries and treat the three random-mating local populations as a single large population, then the matings are no longer at random. Let p_i be the gene frequency of local tribe i. Then $p_1 = 0.10$, $p_2 = 0.40$, $p_3 = 0.70$ with mean $\bar{p} = 0.40$, assuming equal size of the groups. The variance of the gene frequencies is

$$\sigma_p^2 = \tfrac{1}{3}\Sigma p_i^2 - \bar{p}^2 = \tfrac{1}{3}(0.66) - (0.4)^2 = 0.22 - 0.16 = 0.06 \quad (4)$$

and the inbreeding coefficient for the total population is

$$F_T = \frac{\sigma_p^2}{\bar{p}\,\bar{q}} = \frac{0.06}{0.24} = 0.25 \quad (5)$$

which is exactly equal to $r_T = 0.25$ from table (3). The expression (5) is known as Wahlund's formula or principle in population genetics (for details see Li, 1976, Chapter 25). Expression (5) makes it clear that F_T arises from the variation of the difference in arthritis prevalence of the age groups. Again, the situation in epidemiological study of arthritis is isomorphic with that in population genetics. Any property discovered or conclusion reached from one study may be transferred to the other. For instance, geneticists use F_T as a measure of the differentiation of the local populations (i.e., how different the local populations are from each other). Similarly, we may use r_T to indicate how different the various age groups are with respect to presence of arthritis. Thus, we see that there is no paradox between the facts $r_i = F_i = 0$ and $r_T = F_T = 0.25$. The only possible source of confusion is our own loose language; we call all of them correlations.

Wright (1965 and earlier) has extended Wahlund's principle to cover hierarchical classifications and different systems of mating. Actually he has developed a system known as the F-statistics in population genetics. A review of the F-statistics would carry us too far away from our purpose here. It is mentioned here because, in view of the isomorphic nature of the problems in different fields, some of its methods and results might be useful to social research. A number of articles in the volume edited by Crawford and Workman (1973) deal with the genetic structure of anthropological populations, both empirically and analytically. They may also prove useful to other disciplines.

Diagrams for Groups and Population

The case $r_i = 0$ and $r_T > 0$ exemplified above is but one of many situations that may arise from pooling groups into one population. To avoid lengthy verbal description, a few diagrams are presented in Fig. 307 as a visual aid to understanding the various possibilities. In the upper two diagrams, a square symbolizes a group in which there is no correlation between X (abscissa) and Y (ordinate). For simplicity, only two groups (squares) are shown in each diagram. When the means $\overline{X}$ and $\overline{Y}$ of one group are both greater than (or both smaller than) those of the other, then a pooling of these two groups will create a positive correlation for the total population (upper left diagram, $r_i = 0$, r_T positive). This is the case with the example of arthritis association between husband and wife, where the prevalence (equivalent to mean) of arthritis increases with age (groups) in both husband and wife. On the other hand, if the mean of one variable in one group is larger, but that of the other variable is smaller, than those of the other group, as shown in the upper right diagram, then the pooling of these two groups will create a negative correlation in the total population ($r_i = 0$, r_T negative).

The middle two diagrams of Fig. 307 illustrate the situation in which the within-group correlations and the total-population correlation could have opposite signs. Each ellipse symbolizes a group in which there is a positive or negative correlation. The middle left diagram shows that the pooling of three groups (each r_i positive) will create a negative correlation in the total population. On the other hand, the middle right diagram shows the opposite situation: the pooling of three groups (each r_i negative) creates a positive correlation in the total population. Lest the reader think these situations are the fabrications of statisticians and would not happen in nature, the writer hastens to say that, on the contrary, these situations happen frequently; indeed, they occur whenever the mean values of the two variables vary in opposite directions.

Let us consider an extreme example for which no special knowledge of epidemiology is required. For families of any fixed size, the number of boys and girls in a family are always negatively correlated ($r_i = -1$ in this case). For instance, a study of 240 families with four children each ($s = 4$) shows the following distribution:

No. boys, B:	4	3	2	1	0	total number	
No. girls, C:	0	1	2	3	4	of families	(6)
No. families:	12	60	94	61	13	240	

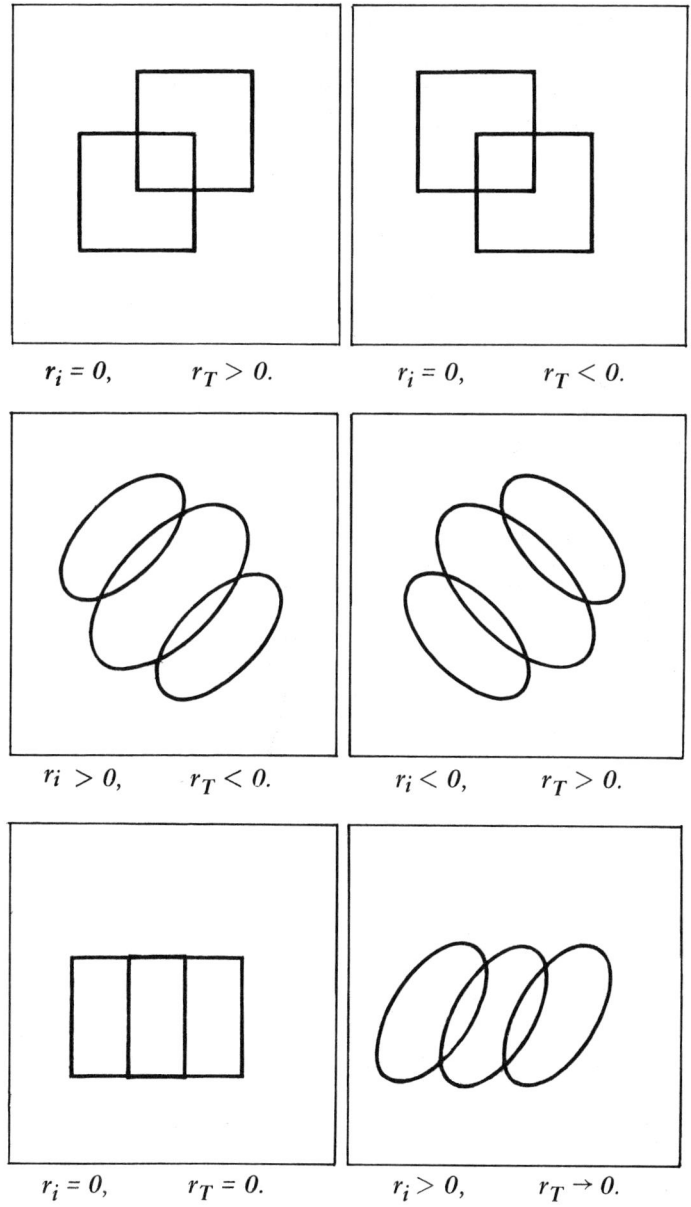

Fig. 307. A population consists of several groups. r_i = correlation within group i; r_T = correlation for the total population. Square is a group within which $r_i = 0$. Ellipse is a group within which r_i is either positive or negative, depending on the direction of its major axis.

which is nearly a symmetrical binomial distribution as expected, and $r(B, C) = -1$. For each family size (s fixed) there is the perfect *negative* correlation between B and C. However, if we look at the population as a whole, consisting of families of all sizes ($s = 0, 1, 2, 3, \ldots, 11, 12$), we will find *positive* correlation between the number of boys (B) and the number of girls (C) in a family, as shown in Table 308. The families of size $s = 4$ are marked out in Table 308, which are the numbers given in (6). Let us call the group of families of any fixed size a "slice" of the total population. In each slice, B and C are negatively correlated. When we put all the slices together, however, we find the positive correlation $r_{BC} = 0.3436$ from Table 308. This is an example of the situation depicted in the middle right diagram of Fig. 307. The marginal distributions of B and C, as well as the distribution of family size $s = B + C$ in Table 308, all have nearly negative binomial distributions (Rao, Mazumdar, Waller, and Li, 1973).

Table 308. Observed numbers of boys *(B)* and of girls *(C)* in 2133 families. (Rao, Mazumdar, Waller, and Li, 1973.)

B = number of boys in a family	\multicolumn{11}{c}{C = number of girls in a family}											Total number of families
	0	1	2	3	4	5	6	7	8	9	10	
12		1										1
10		1	1									2
9	2	1	1									4
8		1	1		2	2						6
7	1	4	2	4	3	1	3	1				19
6	3	8	4	9	4	3	4					35
5	6	19	18	10	12	6	4		1			76
4	12	37	30	30	22	9	8	3		1		152
3	43	60	67	37	18	11	9	5	1			251
2	96	105	94	46	23	13	2	3				382
1	211	206	101	61	17	6	6	5		1	1	615
0	254	183	90	44	13	3	1	2				590
Total number of families	628	626	409	241	114	54	37	19	2	2	1	2133

The lower two diagrams of Fig. 307 show the situation wherein only one variable (say, X) varies from group to group but the other variable (Y) remains the same over the groups. In such a case, if $r_i = 0$, then $r_T = 0$ also, as shown in the lower left diagram. Even if r_i is positive, the total population will show only a weak correlation, if any, as shown in the lower right diagram of Fig. 307.

Correlation between Groups

In the foregoing paragraphs some diagrams were given to show the various types of mixture that can occur in a population. The discussion was essentially centered about the within-group and the total correlations. It dealt with the situation that would be encountered if we were to study a population and find a correlation between two variables and then wonder if it were due to cause-and-effect relationship, due to some common causes, or simply due to internal heterogeneity of the population. However, there is a third kind of correlation that is frequently used by epidemiologists and sociologists; and that is the correlation obtained by regarding a group of individuals as a *unit of observation*. To illustrate, let us consider the two variables X = education and Y = income (or any other two variables). For each individual there is a pair of observations (X, Y). Suppose that we have studied a total of N individuals in Pennsylvania, of which n_1 individuals are in county 1, n_2 in county 2, etc., where $n_1 + n_2 + \ldots + n_k = N$. A correlation between X and Y may be calculated from the N individuals, ignoring their county residence; we shall call it the total individual correlation, r_T. For each county a separate correlation (r_i) between X and Y may be calculated from the n_i individuals of the ith county. The weighted average of such within-county correlations (to be denoted by r_W subsequently) will be called the within-group individual correlation. Finally, each county has its own mean values of the two variables; thus, for the ith county, the means are $(\overline{X}_i, \overline{Y}_i)$, associated with weight $n_{i.}$. There are k such pairs of means. The correlation between $\overline{X}_i$ and $\overline{Y}_i$ for the k counties will be called the between-group correlation, r_B, which is also known as the "ecological correlation."

In order to develop an algebraic relationship for r_W, r_B, and r_T, we shall first subdivide the total sum of products of deviations from the general mean into two components, one within-group and one between-group (see e.g., Li, 1964b, Chapter 14). Let $X_{i\alpha}$ and $Y_{i\alpha}$ be the measurements of the two variables for the individual α of the ith county, $\bar{X}_i$ and $\bar{Y}_i$ be the county means based on n_i individuals, and $\bar{X}$ and $\bar{Y}$ be the general means based on N individuals. Then the total sum of products (of deviations from mean) is

$$\Sigma_i \Sigma_\alpha (X_{i\alpha} - \bar{X})(Y_{i\alpha} - \bar{Y}) = \Sigma_i \Sigma_\alpha (X_{i\alpha} - \bar{X}_i)(Y_{i\alpha} - \bar{Y}_i) + \Sigma_i n_i (\bar{X}_i - \bar{X})(\bar{Y}_i - \bar{Y})$$

or $\quad T(xy) \quad = \quad W(xy) \quad + \quad B(xy)$

where, for brevity, we write $T(xy)$ for the total sum of products, $W(xy)$ for the within-group sum of products, and $B(xy)$ for the between-group sum of products. A similar subdivision of the sum of squares (of deviations from the mean) is obtained from the identity above by replacing the X's by the Y's, or vice versa. Thus,

$$\Sigma_i \Sigma_\alpha (X_{i\alpha} - \bar{X})^2 = \Sigma_i \Sigma_\alpha (X_{i\alpha} - \bar{X}_i)^2 + \Sigma_i n_i (\bar{X}_i - \bar{X})^2$$

$$\Sigma_i \Sigma_\alpha (Y_{i\alpha} - \bar{Y})^2 = \Sigma_i \Sigma_\alpha (Y_{i\alpha} - \bar{Y}_i)^2 + \Sigma_i n_i (\bar{Y}_i - \bar{Y})^2$$

which, again, may be simply written, respectively,

$$T(x^2) \quad = \quad W(x^2) \quad + \quad B(x^2)$$

$$T(y^2) \quad = \quad W(y^2) \quad + \quad B(y^2)$$

In terms of these quantities, the total, within-group, and between-group correlations mentioned above are, respectively,

$$r_T = \frac{T(xy)}{\sqrt{T(x^2)}\sqrt{T(y^2)}}, \quad r_W = \frac{W(xy)}{\sqrt{W(x^2)}\sqrt{W(y^2)}}, \quad r_B = \frac{B(xy)}{\sqrt{B(x^2)}\sqrt{B(y^2)}}$$

The total individual correlation may then be expressed in terms of the within-group and the between-group correlations. Splitting the numerator,

$$r_T = \frac{W(xy)}{\sqrt{T(x^2)}\sqrt{T(y^2)}} + \frac{B(xy)}{\sqrt{T(x^2)}\sqrt{T(y^2)}}$$

APPLICATIONS IN BIO-SOCIAL SCIENCES

Writing $W(xy) = r_W \sqrt{W(x^2)}\sqrt{W(y^2)}$ and $B(xy) = r_B \sqrt{B(x^2)}\sqrt{B(y^2)}$, we obtain

$$r_T = r_W \sqrt{\frac{W(x^2)}{T(x^2)}} \sqrt{\frac{W(y^2)}{T(y^2)}} + r_B \sqrt{\frac{B(x^2)}{T(x^2)}} \sqrt{\frac{B(y^2)}{T(y^2)}}$$

If we introduce a factor G by which the groups are formed, where G may be a quantitative or a qualitative trait, or a geographic region, then $B(x^2)/T(x^2) = \eta_{XG}^2$ is the ratio of the sum of squares of the means of groups to the total sum of squares, where η_{XG} is known as the *correlation ratio* of X on G. Similarly, $B(y^2)/T(y^2) = \eta_{YG}^2$. Then

$$r_T = r_W \sqrt{1 - \eta_{XG}^2} \sqrt{1 - \eta_{YG}^2} + r_B \, \eta_{XG} \eta_{YG}$$

which is the result of Robinson (1950).

A correlation coefficient always has the same sign as the corresponding sum of products of deviations which constitutes its numerator. A sum of products of deviations may be positive, zero, or negative. Discounting the case of zero sum of products, there are still six possibilities for the combinations of signs of the three types of correlations:

r_T	r_W	r_B	r_T	r_W	r_B
+	+	+	−	−	−
+	+	−	−	−	+
+	−	+	−	+	−

Robinson (1950) gives examples in which r_T and r_B are opposite in sign. Even when they do agree in sign, their numerical values may differ widely. To see this, let us consider an extreme case in which there are only two groups (see top row of Fig. 307). The between-group correlation will be either $r_B = +1$ or $r_B = -1$, while r_W may assume almost any value at all.

In most practical situations, the between-group correlation r_B is numerically larger than r_T and r_W. Users of r_B have long noticed that its value depends on the number of groups as well as the manner by which the groups are formed. In other words, it depends on the arbitrary subdivision of the data. Usually, the larger the groups (and therefore the fewer the groups), the larger will be the value of the between-group correlation. When it reduces to two groups, $r_B = \pm 1$, as noted previously. It becomes clear that r_B does not have the same meaning as r_W. The former describes the properties of the groups with respect to variables $\overline{X}_i$ and $\overline{Y}_i$. The latter gives the inherent association between X and Y in a homogeneous population. The example in which $r_W = 0$ and $r_B = \pm 1$, as shown in the top row of Fig. 307, should make this point obvious. The most important single conclusion from these observations is that on the basis of the between-group correlation we can make practically no inference on the nature or magnitude of the within-group correlation. The total individual correlation r_T is of course a mixture of r_W and r_B.

Segments of a Dependent Variable

A quantitative entity is usually the sum of many separate parts. The simplest example is the human stature (standing height) which is the sum of three major segments: head length, trunk length, and leg length. Each of these major segments may have its own biological properties, and its effect on the standing height may be different. Hence we write $Y = X_1 + X_2 + X_3$, where Y is the standing height and the X's are the measurements of the segments. But the standing height is influenced by heredity, nutrition, exercise, and other unknown incidental factors. A distinction must be made between these factors and the segments. The segments are parts of the standing height, while heredity, nutrition, exercise, etc., are not; they are factors influencing the standing height *via* one or more of the segments. For instance, exercises may influence the leg length but not the head length. Therefore, in evaluating the relative importance of the various factors on stature, we may either make comparisons among the segments (e.g., leg length is more important than head length), or among the exogenous factors (e.g., exercise is more important than nutrition), but not between the segments and the exogenous factors. Thus, a statement like "nutrition is more important than leg length" has no meaning. The correct path diagrams for such a situation are given in Fig. 313 in which A, B, C, are the exogenous factors, X_i are the segments and U_i are the residuals.

APPLICATIONS IN BIO-SOCIAL SCIENCES 313

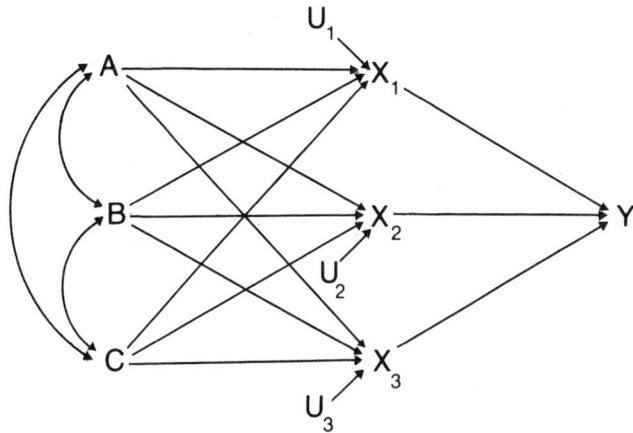

Fig. 313. Segments of a composite variable: $Y = X_1 + X_2 + X_3$ by definition, where $A, B, C,$ are exogenous factors. The diagram shows A, B, C influencing Y via the segments. The diagram involving each segment X_i is a self-contained system by itself.

The diagram shows that each segment may be influenced by the exogenous factors. In such a situation, the study of any one of the segments is a self-contained study by itself. The composite variable Y is merely the sum of the separate studies. Although the discussion above concerns the standing height for ease of understanding, the principle involved here is a general one. In the following pages, "segments" will be adopted as a general term for all similar situations with respect to a composite variable.

The factors of a composite variable may be multiplicative rather than additive. Again, we shall consider an obvious example. Agronomists and plant breeders, in comparing the yields of several strains of barley, wish to know in what respects the yields differ; they may differ in weight per grain, number of grains per head, number of heads per plant, or number of plants per plot. A plot is a small piece of experimental land being used as the unit-area. Thus the yield from a plot may be written as a product of several factors (n. = number of):

$$\frac{\text{yield}}{\text{plot}} = \frac{\text{yield}}{\text{n. grains}} \times \frac{\text{n. grains}}{\text{n. heads}} \times \frac{\text{n. heads}}{\text{n. plants}} \times \frac{\text{n. plants}}{\text{plot}}$$

Taking the logarithms (natural or common) of both sides renders the relationship additive. Let $Y = \log(\text{yield/plot})$, $X_1 = \log(\text{yield/grains})$, etc. Then the relationship above becomes

$$Y = X_1 + X_2 + X_3 + X_4$$

so that our discussions of the additive "segments" in the previous paragraphs would apply. The yield is influenced by exogenous factors such as amount of precipitation, fertilizers, cultivation, plant diseases, and unknown incidental factors. Such factors should be treated as A, B, C, in Fig. 313 as these factors influence the composite yield via one or more of the segments.

The number of segments of a composite variable is usually arbitrary. In the example of standing height, we may study only two segments: the sitting height and the leg length, instead of three. In the example of barley yield, we may increase or decrease the number of segments, depending on our interest and the nature of the segments. For instance, the segments X_2 derived from (n. grains per head) may be split into two segments, as (n. grains per head) = (n. grains per row) (n. rows per head).

Duncan (1966) cited a study of population density in Chicago by Winsborough, using the same decomposition method as in barley yield. The population density may be written:

$$\frac{\text{population}}{\text{area}} = \frac{\text{population}}{\text{dwelling units}} \times \frac{\text{dwelling units}}{\text{structures}} \times \frac{\text{structures}}{\text{area}}$$

A logarithmic transformation makes the segments additive: $Y = X_1 + X_2 + X_3$. But the population density is influenced by exogenous factors such as distance from the center of downtown, recency of growth, etc. These factors should be treated as A, B, C, in Fig. 313, influencing the various segments of the composite density. Duncan rightly pointed out that in studying and interpreting population growth, the exogenous factors and the segments cannot be compared on equal footing. In social research of this nature, a path analysis corresponding to Fig. 313 will go a long way toward clarification and interpretation of the results.

Environment and Heredity

At the outset the reader should note that the topic under discussion is environment *and* heredity, not environment *versus* heredity. The latter phrase has no meaning to me, as every organism must have received the hereditary material (chromosomes and possibly other particles in the cytoplasm) from its parents and must have developed under certain environmental conditions. When environment is broadly defined as all factors other than the hereditary material derived from parents, then it may be formally stated that the characteristics of an individual are entirely (by definition) determined by heredity *and* environment. The complication lies in the way heredity and environment interact with each other—they act both separately and jointly, depending on the particular trait under consideration. The phrase "the relative importance of environment and heredity" is also meaningless, unless a particular trait is specified. Thus, with respect to color blindness, heredity is 100% responsible and environment plays no role. With respect to a quantitative trait, such as height, both heredity and environment are involved. The problem is to evaluate their relative importance on the variation of height.

The piebald pattern of guinea-pigs. The first example chosen to illustrate the method of evaluating the relative importance of heredity and environment is the piebald pattern in guinea-pigs (Wright, 1920). The reader may think this is too remote from human traits of interest. Yes, it is; it is also the reason that it is chosen, so that we can concentrate on methodology rather than being diverted by the subject matter. This example is also of historical importance, as it is one of the earliest attempts to evaluate the relative importance of heredity and environment in the development of a quantitative trait—one of the oldest problems of animal and plant breeders. Wright, in 1920, was working at the Bureau of Animal Industry, United States Department of Agriculture.

A description of the strains of guinea-pigs and their crosses would be out of place here and would confuse those not in the field of animal breeding. Suffice it to say that the piebald trait is a type of coat pattern, consisting of colored spots on a white ground. The trait under consideration is the percentage of white in the dorsal areas of the animal. This trait has never been fixed in many generations of close inbreeding—a fact indicating that the colored spotting is influenced by developmental or environmental factors. The observed correlations for two stocks are as follows:

relationship	Random bred		Inbred	
	number	correlation	number	correlation
sire-dam	105	0.019 ± 0.066	73	0.029 ± 0.079
parent-offspring	1962	0.211 ± 0.015	942	0.014 ± 0.022
litter mates	1233	0.214 ± 0.018	579	0.069 ± 0.028

The negligible correlations between sire and dam show that the matings are virtually at random with respect to the piebald pattern. In the inbred stock, there is practically no correlation between parent and offspring; the correlation between litter mates (0.069) is of doubtful significance. But the parent-offspring and litter mates correlations in the random-bred stock are significant. Adopting the arguments and path diagrams in Chapter 8 for random-mating populations, the following equations may be set up:

			Random	Inbred
r_{PO}	$= \frac{1}{2}b^2$	$=$	0.211,	0.014
r_{OO}	$= \frac{1}{2}b^2 + c^2$	$=$	0.214,	0.069

where b^2 is the environmental heritability, and c^2 is the contribution from factors common to the litter (analogous to the common home environment for full sibs reared together). The contribution from random environmental factors is e^2, where $b^2 + c^2 + e^2 = 1$. Here it is assumed there is no dominance, the dominance heritability being $g^2 = 1$. Solving the equations above, we obtain

			Random	Inbred
environmental heritability,	b^2	=	0.422	0.028
common litter factors,	c^2	=	0.003	0.055
random factors	e^2	=	0.575	0.917
total determination			1.000	1.000

In the inbred stock, where the genetic background is comparatively uniform, the coat pattern is essentially determined by random environmental factors ($e^2 = 0.917$). In the random stock about 42% of the spotting pattern is determined by heredity and 58% by random environmental factors. There are practically no common litter factors involved that would influence the development of the piebald pattern.

The analysis may be carried one step further by using the actual variances of the two stocks. The variance in the random-bred stock is found to be $\sigma_O^2 = 0.643$ and that of the inbred stock is $\sigma_I^2 = 0.364$. The former is larger than the latter. The random-bred stock contains both genetic and environmental variations, while the inbred stock (being genetically uniform) contains only environmental variations. Hence, $e^2 \sigma_O^2$ should be approximately equal to σ_I^2, if these two types of variations are additive. Thus, $(0.58)(0.643) = 0.373$ is in good agreement with the observed 0.364, considering the sampling variations. This shows that the genetic and environmental variabilities are essentially additive in determining the piebald pattern of the guinea-pig coat.

Multiple Regression, Partial Correlation, and Path Coefficient

While still on the general subject of environment and heredity, we now switch to an example about a human quantitative trait—the intelligence quotient or IQ scores. The current explosion of publications on the heritability of human IQ were initiated by Jensen (1969, 1972, 1973). For our present purpose we shall again choose a historical example. Wright (1931), using the data of Miss D. S. Burks, carried out a path analysis of the relative importance of environment and heredity in determining human IQ. The choice of this particular example is not because Burks' data are better than current ones, nor because it is free of contemporary personalities, but because it illustrates the importance of constructing a meaningful causal scheme *before* the analysis. The example to be discussed will show the difference between the results obtained by multiple regression, partial correlation, and path analysis.

The data of Burks consist of IQ test scores of some 100 families (parents and children) in California and a scale (a grading system) for measuring their home environment. The data also consist of test scores of foster parents and some 200 children they adopted at an average age of three months. The two groups of parents were closely similar with respect to IQ and home environment. For simplicity of analysis, the midparent IQ scores are used. A midparent score is the mean of the two parents' scores. In actual calculation we may use the parents' total score, saving the labor of dividing it by 2). The basic data yield the following correlations:

	natural parents	foster parents
midparent, home environment	$r_{PE} = 0.86$	($r_{FE} = 0.86$)
midparent, child	$r_{PO} = 0.61$	$r_{FC} = 0.23$
child, home environment,	$r_{OE} = 0.49$	$r_{CE} = 0.29$

In the above, P and O denote natural parent and offspring, while F and C denote foster parent and adopted child. The correlation between parents and home environment is observed to be 0.86 for the group of natural parents, but unobserved for the group of foster parents. It is reasonable to assume that the correlation is the same as that of the natural parent group, as it is essentially the correlation between a home and its occupants. The data presented above will be analyzed in three different ways.

The first model is of the type of the ordinary multiple regression (Chapter 5); viz., expressing the child's IQ as a linear function of parents' IQ and home environment. The causal diagrams corresponding to this model are shown in Fig. 319 for both groups. The equations for the path coefficients for the two groups are:

natural parents

$r_{OP} = p_{OP} + 0.86\, p_{OE} = 0.61$
$r_{OE} = 0.86\, p_{OP} + p_{OE} = 0.49$

foster parents

$r_{CF} = p_{CF} + 0.86\, p_{CE} = 0.23$
$r_{CE} = 0.86\, p_{CF} + p_{CE} = 0.29$

Solving each set of equations separately, we obtain the path coefficients:

natural group

$p_{OP} = 0.724$
$p_{OE} = -0.133$

foster group

$p_{CF} = -0.074$
$p_{CE} = 0.354$

which are indicated in their respective path diagrams (Fig. 319). In the natural parent group, the path coefficient from home environment to child's IQ

APPLICATIONS IN BIO-SOCIAL SCIENCES   319

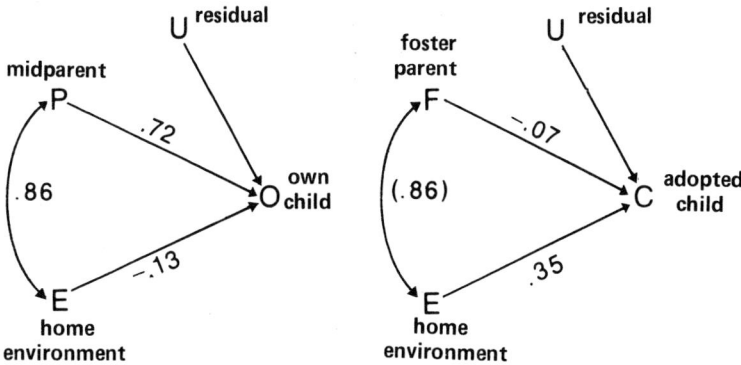

Fig. 319. Path diagrams equivalent to multiple regression with residual *U*. *Left:* natural parent (midparent, *P*) and offspring *(O)*. *Right:* foster parent (midparent, *F*) and adopted child *(C)*. *E* = home environment, as measured by a certain grading system. Redrawn from Wright, 1931. These diagrams are to be compared with those of Fig. 321.

is $p_{OE} = -0.135$. This would mean that a better home lowers the child's IQ! The corresponding concrete partial regression coefficient is of course also negative. It apparently does not permit any interpretation, as it is contrary to our general observation. In the foster-parent group, the path coefficient from parent to adopted children is $p_{CF} = -0.074$, which may be dismissed as nonsignificant; but at its face value, it is equally difficult to understand. In fact, there is little meaning for the direct path from a foster parent to an adopted child, as they have no biological relationship. Although there is no interpretation to these results, they may be used for the purpose of prediction, as the multiple regression is intended to do.

The reader must not think that this is an isolated example, chosen for the meaningless (or even misleading) negative path coefficient from home environment to child's IQ. Similar data, when treated the same way, will give similar results. For instance, Jencks, et al (1972, p. 340) use the multiple regression model for child's IQ on father's education, occupation, and IQ. Sure enough, the path coefficient from father's education to child's IQ was found to be -0.114, and the authors concluded: "Unfortunately, the results are not at all plausible." Then they speculated on the possible reasons for such a result. The fact is that no interpretation is possible or necessary. The multiple regression was obtained by minimizing the discrepancy between the observed and the calculated values of the dependent variable and thus yield the closest prediction of the dependent variable. It was not intended to mean that the influence of the father's education is negative on the child's IQ.

The second approach is to use the partial correlation. What would be the correlation between home environment and child's IQ, eliminating the effect of his natural parents? By the usual formula for partial correlation, we have

$$r_{OE.P} = \frac{r_{OE} - r_{OP} r_{EP}}{\sqrt{1 - r_{OP}^2}\sqrt{1 - r_{EP}^2}} = \frac{0.49 - (0.61)(0.86)}{\sqrt{1 - 0.61^2}\sqrt{1 - 0.86^2}} = -0.086$$

Again, it is negative and apparently admits no interpretation. There is no reason why a good home environment should produce children with lower IQ, assuming the parents are the same. This shows that neither the multiple regression model nor the partial correlation method yield meaningful results.

The third approach is to reconstruct the causal schemes so that they conform more closely with biological facts. The reconstructed causal schemes are shown in Fig. 321. Ignoring the numerical values associated with the paths, for the time being, and concentrating on the structure of the diagrams, we see several important differences between Fig. 319 and Fig. 321. As mentioned previously, when we define environment broadly to include all factors other than heredity, then an individual's IQ score (or any other trait) is completely determined by heredity and environment. Consequently, there is no residual factor to the child's IQ in Fig. 321, as there was in Fig. 319.

Another important feature of Fig. 321 is that there is no direct path from midparent's IQ to child's IQ, not even for natural parent and child. In a purely regression problem, we can always find the regression of child on parent for purposes of description or prediction. But, for biological interpretation, there is no direct connection between a parent's phenotype and his child's phenotype. The correlation between the two phenotypes has to go through the hereditary mechanism which is the underlying cause

APPLICATIONS IN BIO-SOCIAL SCIENCES 321

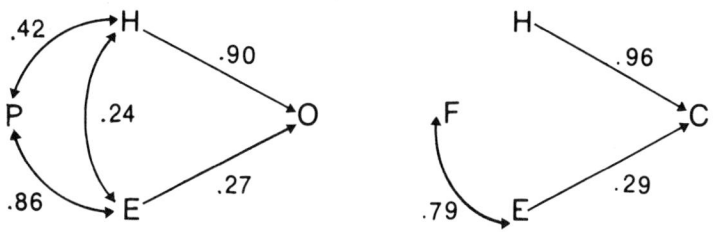

Fig. 321. New path diagrams based on the same data as for Fig. 319. The main feature is the introduction of child's heredity (H). Left: natural parent and offspring (P and O). Right: foster parent and adopted child (F and C). Complete determination of child's phenotype by heredity and environment is assumed; there is no residual variable. Redrawn from Wright, 1931.

for their phenotypic correlation. Hence a new variable, the child's heredity (H), absent in Fig. 319, has been introduced in Fig. 321. Incidentally, this provides another example of using an unobserved variable in a causal scheme.

We shall analyze the adopted children group (right-hand diagram of Fig. 321) first, as it is so simple. It will be recalled that for this group, the correlation between midparent IQ and home environment has not been observed. The two actually observed correlations and the situation of complete determination give us three equations:

environment-child, $r_{CE} = p_{CE} = 0.29$, $p_{CE} = 0.29$

midparent-child, $r_{FC} = r_{FE} p_{CE} = 0.23$, $r_{FE} = 0.23/0.29 = 0.793$

complete determination, $p_{CE}^2 + p_{CH}^2 = 1.00$, $p_{CH} = \sqrt{1 - 0.29^2} = 0.957$

These values are then entered into the right-hand diagram of Fig. 321. It can be seen that the results of this analysis are not only consistent with observed data, but admit interpretation. There is no negative effect from the midparent to the adopted child. The effect of the midparent on the adopted child is through the home environment, being $(0.793)(0.29) = 0.23$.

For the natural parents and children group (left diagram of Fig. 321), four equations may be written down immediately: three from the observed correlations and one from the complete determination of the dependent variable (child's IQ).

$$r_{PE} = 0.86$$
$$r_{OP} = p_{OH} r_{HP} + p_{OE} r_{EP} = 0.61$$
$$r_{OE} = p_{OH} r_{HE} + p_{OE} = 0.49$$
$$p_{OE}^2 + p_{OH}^2 + 2 p_{OE} (r_{HE}) p_{OH} = 1.00$$

Since there are five unknowns to be determined in the diagram, the path method brings out the fact that the observations are inadequate to yield solutions to the problem as formulated by the causal diagram (Fig. 321). This is one of the merits of the path method.

In the present case, a fifth equation may be obtained by making an extra assumption. Comparing the two diagrams of Fig. 321, we see that in each diagram there are two paths leading to child's IQ: one from the child's heredity (H) and one from home environment (E). In the adopted children group we have already found $p_{CH} = 0.957$ and $p_{CE} = 0.290$. Since these two groups of parents are very similar with respect to their home environment and other characteristics, we assume that in the natural children group the corresponding two path coefficients maintain the same proportion (see pp. 133-134). This will provide the needed fifth equation:

$$\frac{h}{e} = \frac{p_{OH}}{p_{OE}} = \frac{p_{CH}}{p_{CE}} = \frac{0.957}{0.290} = 3.30, \text{ or, } h = 3.30e$$

where $h = p_{OH}$ and $e = p_{OE}$ for brevity. Note that we merely assumed the same proportion for the two path coefficients, because path coefficients are not transferable from one set of data to another with a different causal scheme.

The five equations are easily solved. Since $r_{PE} = 0.86$, there are actually four nonlinear equations with four unknowns. By the last equation, we may eliminate $h = p_{OH}$ so that there are only three equations to be solved.

APPLICATIONS IN BIO-SOCIAL SCIENCES

For brevity, writing $r = r_{HP}$ and $s = r_{HE}$, we obtain the solutions (consult Chapter 2, Ex. 5, if necessary):

$h = p_{OH} = 0.90,$ $\qquad e = p_{OE} = 0.27$

$r = r_{HP} = 0.42,$ $\qquad s = r_{HE} = 0.24$

These values are then entered into the left diagram of Fig. 321, completing the analysis. It is seen now that home environment does not have a negative effect on child's IQ. Its direct effect is 0.27 and indirect effect is $0.90(0.24) = 0.22$, the total correlation being $r_{OE} = 0.27 + 0.22 = 0.49$. Wright (1931) carried the analysis further, but the example above should suffice for readers not in genetics. The main purpose of citing this example in some detail is to illustrate the point that we must have a meaningful causal scheme in order to obtain meaningful solutions. Knowledge of the subject matter is indispensable in the formulation of a system of relationships.

Assortative Mating in Human Populations

Human matings are essentially at random with respect to most genetic traits that have no visible symptoms, such as blood groups, enzyme activities, etc. However, with respect to conspicuous physical traits, whether hereditary or not, the matings are selective, such as albinos with albinos, dwarfs with dwarfs, etc. It has long been known that there is assortative mating with respect to human stature (standing height), the correlation between husband and wife being approximately + 0.30. Sociologists now are studying assortative matings with respect to a number of social variables such as level of education, economic brackets, etc. The degree of selection or assortativeness of mating with respect to IQ scores seems high. Jencks, et al (1972) reported that the husband-wife correlation with respect to IQ is 0.57 (their p. 273), the parent-offspring correlation is 0.48 (their p. 274), and that between full siblings reared together is 0.52 (their p. 289). In the last section of Chapter 9, it was shown that $m = r_{PP} h^2$, where m is the (unobserved) genetic correlation between husband and wife, r_{PP} is the observed phenotypic (IQ) correlation between them, and h^2 is the environmental heritability, assuming additive gene effects. The three observed correlations cited above yield the following three equations:

husband-wife,	r_{PP}	$= m/h^2$		$= 0.57$
parent-offspring,	r_{PO}	$= \frac{1}{2}(1 + r_{PP})\,h^2$		$= 0.48$
sib-sib,	r_{OO}	$= \frac{1}{2}(1 + r_{PP}h^2)\,h^2 + c^2$		$= 0.52$

where c^2 is the contribution to correlation from the common home environment that may affect the children reared together. Solving the equations, we obtain

genetic correlation between mates,	m	$= 0.35$
common home environment,	c^2	$= 0.11$
environmental heritability,	h^2	$= 0.61$

and $e^2 = 1 - h^2 - c^2 = 0.28$ is due to random- or nonsystematic-environmental effects. The analysis of Jencks, et al (1972) is far more complicated than indicated above; in fact, they have not obtained any reasonable estimates of these parameters. Considering the parsimony of our model, estimates obtained above may serve as starting points for further refinement.

Education, Occupation, and Child Intelligence

Duncan's (1968 and later) studies on ability and achievement (essentially income) are too comprehensive to be cited as an illustrative example for methodology here. However, we shall consider a subsystem of his that involves only five variables. The following observed correlations and the causal diagram (Fig. 325) are based on Jencks, et al (1972, Appendix B).

	father education X_1	father occupation X_2	child early IQ X_0	child education D	child adult IQ Q
X_1	(1)	0.509	0.300	0.382	0.305
X_2		(1)	0.300	0.420	0.314
X_0			(1)	0.550	0.830
D				(1)	0.630

APPLICATIONS IN BIO-SOCIAL SCIENCES 325

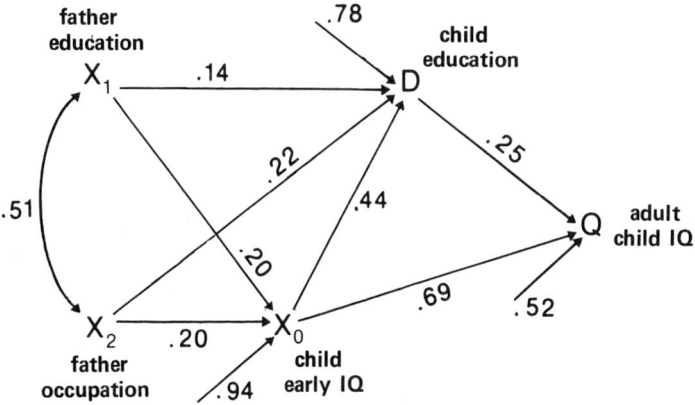

Fig. 325. Path relationships between father's education and occupation and child's early IQ, education, and adult IQ of native white nonfarm males aged 25-64 in 1962. (Excerpted from Jencks, et al, 1972, p. 339.)

The path coefficients shown in Fig. 325 are found in successive stages by identifying self-contained subsystems. This example is suitable to illustrate the procedure. An inspection of Fig. 325 shows that the three variables X_1, X_2, and X_0 constitute a self-contained system by themselves—a multiple regression of X_0 on X_1 and X_2 plus a residual. Hence, we find the path coefficients p_{01} and p_{02} first, using the correlation r_{12} = 0.51.

equations   solutions

$$\begin{cases} r_{01} = p_{01} + 0.51\, p_{02} = 0.30 \\ r_{02} = 0.51\, p_{01} + p_{02} = 0.30 \end{cases} \quad \begin{cases} p_{01} = 0.20 \\ p_{02} = 0.20 \end{cases}$$

$$R^2_{0(12)} = p_{01} r_{10} + p_{02} r_{20} = 0.12; \quad u = \sqrt{1 - 0.12} = 0.938$$

where u is the path coefficient from the residual to X_0, the child's early IQ. These values are then entered into the left part of the diagram (Fig. 325).

Next, we see that the child's education (D) is determined by X_0, X_1, X_2, and a residual, forming another complete system by itself. The equations for path coefficients and their solutions are as follows:

equations

$$\begin{cases} r_{D0} = p_{D0} + p_{D1}r_{10} + p_{D2}r_{20} = 0.550, \\ r_{D1} = p_{D0}r_{01} + p_{D1} + p_{D2}r_{21} = 0.382, \\ r_{D2} = p_{D0}r_{02} + p_{D1}r_{12} + p_{D2} = 0.420, \end{cases}$$

solutions

$$\begin{cases} p_{D0} = 0.44 \\ p_{D1} = 0.14 \\ p_{D2} = 0.22 \end{cases}$$

where all the r_{ij}'s have been observed. In the process of solving these equations, it is advisable to retain one or two more digits and then round them off to two significant figures in the final solutions. The three path coefficients above are then entered into the diagram (Fig. 325). The residual path is calculated in the usual way:

$$R^2_{D(012)} = p_{D0}r_{0D} + p_{D1}r_{1D} + p_{D2}r_{2D} = 0.39$$

$$u = \sqrt{1 - 0.39} = 0.78$$

as indicated in the diagram. Finally, the child's adult IQ (Q) is determined by his education (D) and early IQ (X_0) plus a residual. There are only two paths and two equations:

$$\begin{cases} r_{QD} = p_{QD} + p_{Q0}r_{0D} = 0.630, \\ r_{Q0} = p_{QD}r_{D0} + p_{Q0} = 0.830, \end{cases} \quad \begin{cases} p_{QD} = 0.25 \\ p_{Q0} = 0.69 \end{cases}$$

$$R^2_{Q(D0)} = p_{QD}r_{DQ} + p_{Q0}r_{0Q} = 0.733, \quad u = \sqrt{0.267} = 0.52$$

These complete the entries in Fig. 325. We have found all the relevant path coefficients in the diagram. Several features of the analysis are worth noting. One is the large residual path to X_0, the child's early IQ scores. This means that neither the father's education (X_1) nor the father's occupation (X_2) play any important role in determining the child's early IQ. Education and occupation are probably two of the most important factors in sociology, and yet they have so little direct influence on child's IQ score. In Fig. 325, no genetic factors leading to X_0 have been included, and thus the large residual path. If the IQ scores of the parents were included in the study, the residual coefficient to child's early IQ would be much diminished.

Another feature of Fig. 325 is the effect of child's education on his adult IQ. The direct path from education *(D)* to adult IQ is 0.25, much lower than 0.69, the direct effect of early IQ (X_0) on adult IQ.

Finally, we note that the path coefficient from early IQ to education is 0.44, which is higher than that from education to adult IQ. The implication seems to be that it is the children with higher IQ who go to school rather than that schooling improves children's IQ. The indirect effect from early IQ to adult IQ via education is $(0.44)(0.25) = 0.11$.

In the analysis above, we have used 8 of the 10 observed correlations. The two correlations not used are $r_{Q1} = 0.305$ and $r_{Q2} = 0.314$. On the basis of Fig. 325, two additional equations may be written down for these two correlations. These are called *excess* equations and the system is said to be over-determined or over-identified. General comments on this problem will be found in a later paragraph. For the time being, we take the path values indicated in Fig. 325 as valid estimates of the respective influences and ask, if so, what should be the correlation between X_1 and Q? The connecting chains between these two variables and their values are:

connecting paths	value	
$Q - D - X_1$	$(0.25)(0.14)$	$= 0.035$
$Q - D - X_0 - X_1$	$(0.25)(0.44)(0.20)$	$= 0.022$
$Q - D - X_2 - X_1$	$(0.25)(0.22)(0.51)$	$= 0.028$
$Q - D - X_0 - X_2 - X_1$	$(0.25)(0.44)(0.20)(0.51)$	$= 0.011$
$Q - X_0 - X_1$	$(0.69)(0.20)$	$= 0.138$
$Q - X_0 - X_2 - X_1$	$(0.69)(0.20)(0.51)$	$= 0.070$
	Total	$= 0.304$

According to the diagram, the expected correlation between X_1 and Q is 0.304, while the observed correlation is $r_{Q1} = 0.305$, in spite of the sampling and rounding off of errors. A similar situation exists for r_{Q2}. We conclude that the path coefficients indicated in Fig. 325 are consistent with the data.

Urbanization and Income

Blalock (1961) considered various possible relationships among five variables in 150 randomly selected southern counties taken from the 1950 census. This is a typical example of studying the between-group correlations (r_B) rather than the within-group correlations (r_W), as discussed earlier in the chapter. However, we shall use this example solely for the purpose of illustrating the method of path coefficients and will not argue about the validity of the numerical conclusions. In this particular case, there is no way to study the within-group correlation based on census tabulations. It can be studied only when the individual data are available.

The five variables chosen for study are: X_1 = an index of urbanization, X_2 = the percentage of nonwhites in the county, X_3 = income of white people, X_4 = an index of nonwhite educational level, and X_5 = income of nonwhite people. All indices are crude. The nonwhite educational level of a county is the percentage of males age 25+ with more than six years of schooling. The index for income is the percentage of families with annual income of $1,500 or more. The ten observed correlations among the five variables are as follows:

		% nonwhite X_2	white income X_3	nonwhite education X_4	nonwhite income X_5
urbanization	X_1	−0.389	0.670	0.264	0.736
% nonwhite	X_2		0.067	−0.531	−0.440
white income	X_3			0.042	0.599
nonwhite education	X_4				0.386

Blalock's (1961) third model may be expressed in the form of a path diagram (Fig. 329) as was done by Boudon (1965) omitting the residuals for simplicity. A glance at Fig. 329 shows that it is essentially a study about

APPLICATIONS IN BIO-SOCIAL SCIENCES 329

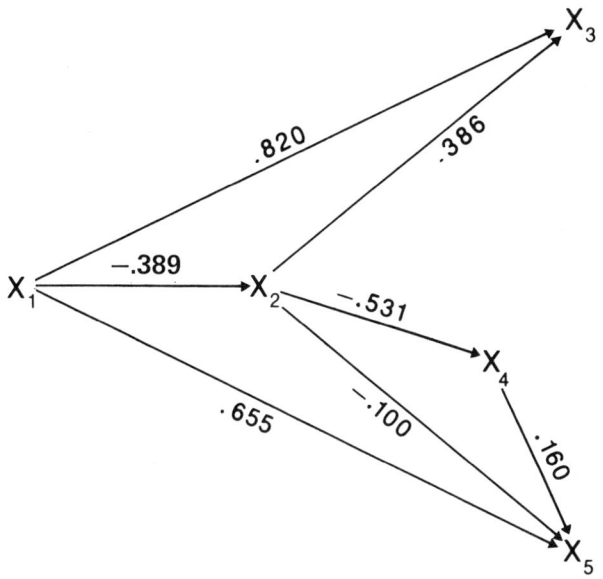

Fig. 329. A model of causal relations between urbanization (X_1) and white income (X_3) and nonwhite income (X_5). The intermediate variables are: X_2 = percentage of nonwhite population in a county; X_4 = educational level of nonwhites. (Redrawn from Blalock, 1961, and Boudon, 1965.)

urbanization and income. The diagram consists of two subsystems; the variables X_1, X_2, X_3 form a self-contained system concerning the white income. The variables X_1, X_2, X_4, X_5 form another system concerning the nonwhite income. The existence of the subsystems implies that the equations involving path coefficients fall into groups and can be solved separately for each group. First, we note that there is only one connection between X_1 and X_2, and between X_2 and X_4. Hence,

$$r_{21} = p_{21} = -0.389$$

$$r_{42} = p_{42} = -0.531$$

Next, the following two equations form a group by themselves:

$$r_{31} = p_{31} + p_{32}r_{21} = 0.670 \qquad p_{31} = 0.820$$

$$r_{32} = p_{31}r_{12} + p_{32} = 0.067 \qquad p_{32} = 0.386$$

Similarly, the following three equations form a group:

$$r_{51} = p_{51} + p_{52}r_{21} + p_{54}r_{41} = 0.736$$

$$r_{52} = p_{51}r_{12} + p_{52} + p_{54}r_{42} = -0.440$$

$$r_{54} = p_{51}r_{14} + p_{52}r_{24} + p_{54} = 0.386$$

Substituting the observed values of the r's and solving, we obtain

$$p_{51} = 0.655, \qquad p_{52} = -0.100, \qquad p_{54} = 0.160$$

Thus we have obtained the values of the seven path coefficients. These numerical values are indicated in Fig. 329. Since all ten correlations among the five variables have been observed and there are only seven path coefficients in the diagram, the system is over-determined. The extra three correlations may be used to see if they are consistent with the path coefficients indicated in the diagram. A prime is used to indicate the calculated values of the correlations:

	calculated		observed
$r'_{41} = p_{42}p_{21} =$	0.207,	$r_{41} =$	0.264
$r'_{34} = p_{42}r_{23} =$	-0.036,	$r_{34} =$	0.042
$r'_{35} = p_{31}r_{15} + p_{32}r_{25} =$	0.434,	$r_{35} =$	0.599

Although the results are in general agreement, we note that the calculated correlations are all lower than the observed. The addition of a direct path from urbanization (X_1) to nonwhite educational level (X_4) in Fig. 329 would give a new equation:

APPLICATIONS IN BIO-SOCIAL SCIENCES

$$r_{41} = p_{41} + p_{42}p_{21} = p_{41} + (-0.531)(-0.389) = 0.264$$

yielding

$$p_{41} = 0.264 - 0.207 = 0.057$$

But this does not account for the comparatively high correlation of the white and nonwhite income, $r_{35} = 0.599$. The causal scheme of Fig. 329 has only one basic common factor (X_1) for X_3 and X_5. This seems to suggest that some other relevant common factors for the white and nonwhite incomes are not included in Fig. 329.

Excess Equations and Estimation

The discussions above concentrate on the consistency of the causal scheme. Obviously, the "agreement" or "disagreement" between an observed correlation and a calculated one from an excess equation is admittedly subjective and arbitrary. When there are excess equations to a system (over-determination or over-identification), Hauser and Goldberger (1971) have applied the method of maximum likelihood to obtain estimates of path coefficients from all relevant equations. This is, of course, statistically sound when the causal scheme has been accepted and it becomes merely a matter of estimation. However, if the disagreement between an observed and a calculated (from an excess equation) correlation suggests that the causal scheme itself should be revised, then the improved estimates (a compromise among the relevant equations) would only serve to minimize the discrepancy and possibly lead to delaying the revision of the causal scheme itself. Thus, we see that in the matter of searching for causal relationships, the element of subjectivity is (unfortunately) more difficult to overcome than in a purely statistical problem.

Utilization of Physician Service

There is a paucity of examples of the application of path coefficients in epidemiology and medical care problems. Lilienfeld (1966), an epidemiologist, apparently had a clear understanding of causal diagrams but did not elaborate. The inclusion of one example in this field may stimulate other workers to pursue the subject further. A study of medical care and utilization of physician service, by Kalimo and Bice (1973), will serve as our example. The four variables of this study were: X_1 = age, classified into three categories (15–44, 45–64, and 65 + years); X_2 = presence or absence of chronic illness or physical impairment; X_3 = number of sick days, in bed or not working, in the last two weeks (0 = none, 1 = one or two days, 2 = three or more days); X_4 = visit to physician (0 or 1). The causal scheme proposed is shown in Fig. 333. The following six observed correlations among the four variables are based on nearly 35,000 individual respondents.

		chronic problem X_2	sick days X_3	physician use X_4
age,	X_1	0.25	0.11	0.04
chronic problem	X_2		0.21	0.08
sick days,	X_3			0.30

The five equations for the five path coefficients in Fig. 333 fall into three groups. There is only one connection between X_1 and X_2, and there are two connecting paths between other pairs of variables. The equations and solutions are as follows:

equations

$r_{21} = p_{21} = 0.25$

$r_{31} = p_{31} + p_{32}r_{21} = 0.11$
$r_{32} = p_{31}r_{12} + p_{32} = 0.21$

$r_{42} = p_{42} + p_{43}r_{32} = 0.08$
$r_{43} = p_{42}r_{23} + p_{43} = 0.30$

solutions

$p_{21} = 0.25$

$p_{31} = 0.061$
$p_{32} = 0.195$

$p_{42} = 0.018$
$p_{43} = 0.296$

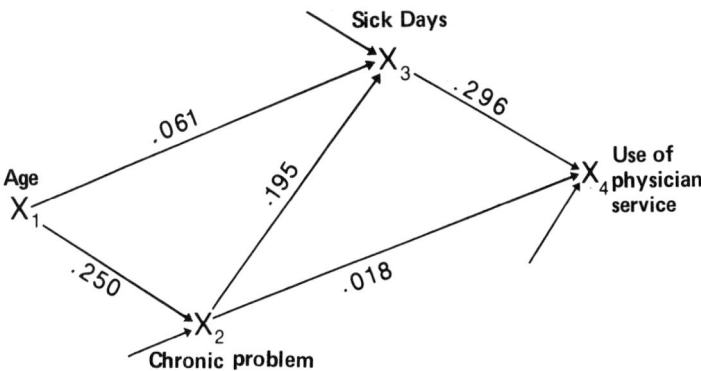

Fig. 333. A causal model for physician use. X_1 = age, X_2 = chronic condition, X_3 = sick days, X_4 = visit to physician. The small unlabeled arrows are residual factors. (Redrawn from Kalimo and Bice, 1973.)

These p values are entered into the corresponding paths of Fig. 333. Since there are five path coefficients and six observed correlations, the system is over-identified by one equation. The problem and method of adjusting these estimates have just been mentioned. Suppose that we wish to know if these p values are consistent with the unused equation concerning r_{41}. The calculated value of correlation between X_1 and X_4 is indicated by a prime:

$$r'_{41} = p_{42}r_{21} + p_{43}r_{31} = 0.037$$

which agrees with the observed $r_{41} = 0.04$. From Fig. 333 two conclusions are obvious. The first is that sick-days is a more important cause for physician use than chronic problems. The second is that physician use is largely due to factors not indicated in the diagram. The degree of determination of X_4 (physician use) is

$$p_{43}^2 + p_{42}^2 + 2p_{43}p_{42}r_{23} = (0.296)^2 + (0.018)^2 + 2(0.296)(0.018)(0.21) = 0.09$$

leaving 91% of the variance of physician use unaccounted for (i.e., due to residual factors). The characteristic feature of the causal scheme (Fig. 333) is that there is no direct connection between age (X_1) and physician use (X_4). One does not visit a physician simply because he has had another birthday. He sees his physician because of his illness and therefore the effect of age is an indirect one.

Epidemiology

Like the social sciences, epidemiology usually involves many factors whose causal relationships are not clear and may vary from one geographical region to another, making the construction of path diagrams more difficult than in biology. Nevertheless, epidemiologists now would like to explore the possibilities of using path analysis in their studies. Goldsmith and Berglund (1974) presented tentative path diagrams for (i) childhood asthma and bronchitis, (ii) adult bronchitis and emphysema, (iii) respiratory functional impairment in children and adolescents, and (iv) cancer of the lung (bronchogenic carcinoma), as proposals for further research. Their goal was "a more comprehensive use of numerical analysis to describe the relationship and pathways by which environmental exposures affect health." In other words, one of the aims of epidemiological studies is to identify the etiologic factors of human diseases. In applying any statistical method in epidemiology, Lilienfeld's (1965) pragmatic concept of causality—"a factor may be defined as a cause of a disease if the incidence is diminished when exposure to this factor is likewise diminished"—gives us a general guide to the construction of path diagrams relating factors and diseases.

Other examples and conclusion

There are many other examples of the application of path analysis in the social sciences, of which only a few may be mentioned, not necessarily in any systematic order. Following Duncan's (1966) first exposition, Land's (1969) treatment concentrated on the mathematical and statistical aspects of path coefficients. Heise (1969) reiterated the basic assumptions or requirements for path analysis and warned against overly-simple applications of the path technique. As to research on social variables, there is an interesting early study by Sewell and Shah (1967) on socioeconomic status, intelligence, and the attainment of higher education. Quite comprehensive are Featherman's (1972) studies on achievement orientations and socioeconomic career attainments and Hauser's (1973) monograph on socioeconomic background and educational performance.

The applications of path analysis have also begun to crop up in other sciences. Analogous to Duncan's "sociological examples" (1966), Wert and Linn (1970) provided some "psychological examples." They applied path analysis to a set of kindred problems, including the correlation of true scores over time, allowing specific measurement errors to be correlated, the "bouncing beta" in stepwise regression analysis, the specification of the number and interrelationships of latent variables in a set of data, adjustment for covariate unreliability in the analysis of covariance, and the analysis of multitrait-multimethod matrices. These topics are obviously too technical for a general reader but they should be helpful to educational psychologists. The economist Gintis (1971) reported on a path analysis of education and worker productivity. Griliches and Mason (1972) employed some path analysis in a study of education and income, treating aptitude tests as indicators of an unobservable variable representing ability.

Path analysis may be regarded as a special form of "structural equations" employed by other authors in other fields. Two pioneer examples are Wold (1954) in economics and Simon's "models of man" (1957). The volume edited by Blalock (1971) is a collection of articles dealing with causal models and is a good source of references to related methods.

Of special interest is Goldberger's (1972) review of structural equation methods in the social sciences. Students of genetics and agriculture have long known the early work of Wright (1925) on corn and hog correlations. Goldberger, an economist, a social scientist, and a statistician, regards Wright's 1925 analysis of the relations between corn price, hog price, and other variables as a pioneer example of using structural equations in the field of economics (at least in agricultural economics). He then relates a fascinating story about the personal and professional friendship between the pioneer economist Henry Schultz and the geneticist Sewall Wright. The latter used the path method to estimate a supply-demand model for hogs. He also fitted a supply-demand model for potatoes with lagged price as the instrument, and developed a dynamic version of the model. Wright acknowledged: "I am indebted to my colleague, Professor Henry Schultz, for data on the quantity and price of potatoes marketed annually from 1896-1914 and the suggestion that it would be interesting material for analysis by this method." In Schultz's own work, however, Wright's method was barely mentioned but not used. Hence the name Wright and the path method remained unknown to econometricians for another 45 years. Goldberger (1974) remarked:

> It is reasonable to expect that, writing in 1938, Schultz would draw on Wright's work.... It is difficult to find an explanation for Schultz' failure to recognize, let alone utilize, Wright's powerful approach to the formulation and estimation of structural equation models. What makes the matter particularly puzzling is that Schultz and Wright were more than casually acquainted.

The explanations given by personalities directly involved in an event are often different from those given by detached persons (e.g., historians with hindsight). Wright's own explanation (personal communication to Goldberger, 1971) is: "Henry was so committed to a completely formal objective approach in multivariate analysis that I could never get him to see the possible usefulness of my somewhat informal subjective mode of approach." While this is undoubtedly true, one wonders why other economists also failed to pick up the trail when it was still visible?

To a geneticist, the story of the belated recognition of the path method in economic analysis is reminiscent of the belated recognition of Mendel's law of heredity (1866), until the scientific atmosphere was ready for it in 1900. Apparently, the 1930's are the period of growth of formal multivariate analysis; the subjective structural method was born too early. Now, there seems to be a stampede for path analysis by the proselytes.

It may be prudent to end this book with a warning. One serious misconception about the path method is that it tells us the cause and effect relationship among the variables. That this is not the case has been repeatedly stated in the previous chapters on path analysis. It has been emphasized that the path diagram represents a particular (subjective) point of view concerning the relationships of the variables. Whether it makes sense or not depends much upon our subject matter knowledge. Once a particular viewpoint has been taken, the analysis must be performed accordingly, so that we may ascertain the logical consequences of such a structure. In Wright's own words (1934, 1960, 1968, respectively):

> The method of path coefficients is not intended to accomplish the impossible task of deducing causal relations from the values of the correlation coefficients.
>
> The purpose of path analysis is to determine whether a proposed set of interpretations is consistent throughout.
>
> The (path) method is not a mill from which interpretations can be arrived at automatically.

This book attempts to introduce the method of path coefficients but does not advocate its use for all occasions to the exclusion of other methods.

* * *

REFERENCES

Blalock, H. M. 1961. *Causal inferences in nonexperimental research.* Chapel Hill, N.C.: Univ. of North Carolina Press.

———. 1961. Correlation and causality: the multivariate case. *Social Forces* 39: 246-251.

———. 1962. Four-variable causal models and partial correlations. *Am. J. Sociol.* 68: 182-194.

———. 1963. Making causal inferences for unmeasured variables from correlations among indicators. *Am. J. Sociol.* 69: 53-62.

———. 1967. Path coefficients versus regression coefficients. *Am. J. Sociol.* 72: 675-676.

———., ed. 1971. *Causal models in the social sciences.* Chicago: Aldine-Atherton.

———, and Blalock, A. B., eds. 1968. *Methodology in social research.* New York: McGraw-Hill.

Borgatta, E. F., ed. 1969. *Sociological methodology.* San Francisco: Jossey-Bass.

Boudon, R. 1965. A method of linear causal analysis: dependence analysis. *Am. Social Rev.* 30: 365-374.

Brownlee, K. A. 1965. *Statistical theory and methodology in science and engineering.* New York: John Wiley.

Burt, C. 1971. Quantitative genetics in psychology. *British J. Math. & Stat. Psychol.* 24: 1-21.

———, and Howard, M. 1956. The multifactorial theory of inheritance and its application to intelligence. *British J. Stat. Psychol.* 9: 95-131.

Cancro, R., ed. 1971. *Intelligence: genetic and environmental influences.* New York: Grune & Stratton.

Cockerham, C. C. 1954. An extension of the concept of partitioning hereditary variance for analysis of covariance among relatives when epistasis is present. *Genetics* 39: 859-882.

Crawford, M. H., and Workman, P. L., eds. 1973. *Methods and theories of anthropological genetics.* Albuquerque, N.M.: Univ. of New Mexico Press.

Crow, J. F., and Felsenstein, J. 1968. The effect of assortative mating on the genetic composition of a population. *Eugenics Quar.* 15: 85-97.

Cruz-Coke, R.; Nagel, R.; and Etcheverry, R. 1964. Effects of locus MN on diastolic blood pressure in a human population. *Annals of Human Genetics* 28: 39-48.

REFERENCES

Dobzhansky, Th. 1937. *Genetics and the origin of species.* New York: Columbia Univ. Press.

Draper, N. R., and Smith, H. 1966. *Applied regression analysis.* New York: John Wiley.

Duncan, O. D. 1966. Path analysis—sociological examples. *Am. J. Sociol.* 72: 1-16.

———. 1968. Ability and achievement. *Eugenics Quarterly* 15: 1-11.

———. 1969. Contingencies in constructing causal models. In *Sociological methodology,* E. F. Borgatta, ed., pp. 74-112. San Francisco: Jossey-Bass.

———. 1969. Inheritance of poverty or inheritance of race? In *On understanding poverty,* D. P. Moynihan, ed. New York: Basic Books.

———. 1975. *Introduction to structural equation models.* New York: Academic Press.

Emigh, T. H. 1974. "Statistical methodology of the nature-nurture controversy in human intelligence." Master's thesis, Iowa State Univ.

Ezekiel, M., and Fox, K. A. 1959. *Methods of correlation and regression analysis.* 3rd. ed. New York: John Wiley.

Falconer, D. S. 1963. Quantitative inheritance. In *Methodology in mammalian genetics,* ed. W. J. Burdette, pp. 193-216. San Francisco: Holden-Day.

Featherman, D. L. 1972. Achievement orientations and socioeconomic career attainments. *Amer. Sociological Rev.* 37: 131-143.

Fisher, R. A. 1918. The correlation between relatives on the supposition of Mendelian inheritance. *Trans. Roy. Soc. Edinb.* 52(2): 399-433.

Gintis, H. 1971. Education, technology, and the characteristics of worker productivity. *Amer. Economic Rev.* 61: 266-279.

Goldberger, A. S. 1972. Structural equation methods in the social sciences. *Econometrica* 40: 979-1001.

———, and Duncan, eds. 1973. *Structural equation models in the social sciences.* New York and London: Seminar Press.

Goldsmith, J. R., and Berglund, K. 1974. Epidemiological approach to multiple factor interactions in pulmonary disease: the potential usefulness of path analysis. *Annals N.Y. Acad. Sci.* 221: 361-375.

Griliches, Z., and Mason, W. 1972. Education, income, and ability. *J. Political Economy* 80: S74-103.

Hauser, R. M. 1973. *Socioeconomic background and educational performance.* Amer. Sociological Assoc., Washington, D.C.

———, and Goldberger, A. S. 1971. The treatment of unobservable variables in path analysis. In *Sociological methodology,* ed. H. L. Costner, pp. 81-117. San Francisco: Jossey-Bass.

Heise, D. R. 1969. Problems in path analysis and causal inference. In *Sociological methodology*, ed. E. F. Borgatta, pp. 38-73. San Francisco: Jossey-Bass.

Holt, S. B. 1968. *The genetics of dermal ridges.* Springfield, Illinois: Charles C. Thomas.

Jencks, C. et al. 1972. *Inequality, a reassessment of the effect of family and schooling in America.* New York: Basic Books.

Jensen, A. R. 1969. How much can we boost IQ and scholastic achievement? *Harvard Educational Rev.* 39: 1-123.

———. 1972. *Genetics and education.* New York: Harper & Row.

———. 1973a. *Educability and group differences.* New York: Harper & Row.

———. 1973b. *Educational differences.* London: Methuen & Co.

Kalimo, E., and Bice, T. W. 1973. Causal analysis and ecological fallacy in cross-national epidemiological research. *Scand. J. Soc. Med.* 1: 17-24.

Keeler, C. 1968. Some oddities in the delayed appreciation of "Castle's law." *J. Heredity* 59: 110-112.

Kempthorne, O. 1954. The correlations between relatives in a random mating population. *Proc. Royal Soc. London B*, 143: 103-113.

———. 1955a. The theoretical values of correlations between relatives in random mating populations. *Genetics* 40: 153-167.

———. 1955b. The correlations between relatives in random mating populations. *Cold Spring Harbor Symposium* 20: 60-78.

———. 1957. *An introduction to genetic statistics.* New York: John Wiley.

———. 1971. The statistical treatment of data with genetics structure. In *Craniofacial growth in man,* ed. R. E. Moyers and W. M. Krogman, pp. 163-182. Oxford: Pergamon Press.

Land, K. C. 1969. Principles of path analysis. In *Sociological methodology,* ed. E. F. Borgatta, pp. 3-37. San Francisco: Jossey-Bass.

Li, C. C. 1948. *An introduction to population genetics.* National Peking Univ. Press.

———. 1953. A direct proof of the relation between genotypic mating correlation and gametic uniting correlation in equilibrium populations. *J. Heredity* 44: 39-40.

———. 1954. The correlation between parents and offspring in a random mating population. *Am. J. Human Genetics* 6: 383-386.

———. 1955a. A diagrammatic representation of the sum of squares and products. *J. Am. Stat. Assoc.* 50: 1056-1063.

———. 1955b. *Population genetics.* Chicago: Univ. of Chicago Press.

REFERENCES

———. 1956. The concept of path coefficient and its impact on population genetics. *Biometrics* 12: 190-210.

———. 1961. *Human genetics: principles and methods.* New York: McGraw-Hill. (Paperbacks available from author.)

———. 1964a. Two additional views of linear regression coefficients. *The Amer. Statistician* 18: 27-28.

———. 1964b. *Introduction to experimental statistics.* New York: McGraw-Hill.

———. 1967. Genetic equilibrium under selection. *Biometrics* 23: 397-484.

———. 1968. Fisher, Wright, and path coefficients. *Biometrics* 24: 471-483.

———. 1969. Population subdivision with respect to multiple alleles. *Annals of Human Genetics* 33: 23-29.

———. 1970. Human genetic adaptation. In *Essays in evolution and genetics in honor of Theodosius Dobzhansky*, ed. M. K. Hecht and W. C. Steere, pp. 545-577. New York: Appleton-Century-Crofts.

———. 1971. A tale of two thermos bottles: properties of a genetic model for human intelligence. In *Intelligence: genetic and environmental influences*, ed. R. Cancro, pp. 162-181. New York: Grune & Stratton.

———. 1975. Assortative mating in man. *Proc. 2nd meeting of Mexico Society of Genetics.*

———. 1976. *First course in population genetics.* Pacific Grove, Calif.: Boxwood Press.

———; Mazumdar, S.; and Rao, B. R. 1975. Partial correlation in terms of path coefficients. *The Amer. Statistician* 29: 89-90.

———, and Sacks, L. 1954. The derivation of joint distribution and correlation between relatives by the use of stochastic matrices. *Biometrics* 10: 347-360.

Lilienfeld, A. M. 1966. Epidemiologic methods and inferences. In *Chronic diseases and public health*, ed. A. M. Lilienfeld and A. J. Gifford, pp. 99-112. Baltimore: Johns Hopkins Univ. Press.

Mather, K., and Jinks, J. L. 1971. *Biometrical genetics.* Ithaca, N.Y.: Cornell Univ. Press.

Moran, P. A. P. 1961. Path coefficients reconsidered. *Australian J. Stat.* 3: 87-93.

———, and Smith, C. A. B. 1966. Commentary on R. A. Fisher's paper on the correlation between relatives on the supposition of Mendelian inheritance. *Eugenics Laboratory Memoirs s1.* London: Cambridge Univ. Press.

Penrose, L. S. 1971. Notes on the interpretation of intrafamilial correlation coefficients. *Annals of Human Genetics* 34: 291-293.

Rao, B. R.; Mazumdar, S.; Waller, J. H.; and Li, C. C. 1973. Correlation between the numbers of two types of children in a family. *Biometrics* 29: 271-279.

Rao, D. C.; Morton, N. E.; and Yee, S. 1974. Analysis of family resemblance. II. A linear model for familial correlation. *Am. J. Human Genetics* 26: 331-359.

———. 1976. Resolution of cultural and biological inheritance by path analysis. *Am. J. Human Genetics* 28: 228-242.

Roberts, R. C. 1967. Some concepts and methods in quantitative genetics. In *Behavior-genetic analysis*, ed. J. Hirsch, pp. 214-257. New York: McGraw-Hill.

Robinson, W. S. 1950. Ecological correlations and the behavior of individuals. *Amer. Sociological Rev.* 15: 351-357.

Schultz, J. 1973. Human values and human genetics. *Amer. Naturalist* 107: 585-597.

Sewell, W. H., and Shah, V. P. 1967. Socioeconomic status, intelligence, and the attainment of higher education. *Sociology of Education* 40: 1-23.

———; Haller, A. O.; and Ohlendorf, G. W. 1970. The educational and early occupational status attainment process: replication and revision. *Amer. Sociological Rev.* 35: 1014-1027.

Simon, H. A. 1957. *Models of man.* New York: John Wiley.

Trustrum, G. B. 1961. The correlations between relatives in a random mating diploid population. *Proc. Camb. Phil. Soc.* 57: 315-320.

Tukey, J. W. 1954. Causation, regression, and path analysis. In *Statistics and mathematics in biology*, ed. Kempthorne, et al., pp. 35-66. Ames, Iowa: Iowa State Univ. Press.

Turner, M. E., and Stevens, C. E. 1959. The regression analysis of causal paths. *Biometrics* 15: 236-258.

Werts, C. E., and Linn, R. L. 1970. Path analysis: psychological examples. *Psychological Bulletin* 74: 193-212.

Wold, H. 1954. Causality and econometrics. *Econometrics* 22: 162-177.

Wright, S. 1917. The average correlation within subgroups of a population. *J. Wash. Acad. Sci.* 7: 532-535.

———. 1918. On the nature of size factors. *Genetics* 3: 367-374.

———. 1920. The relative importance of heredity and environment in determining the piebald pattern of guinea pigs. *Proc. Nat. Acad. Sci.* 6: 320-332.

———. 1921a. Correlation and causation. *J. Agric. Res.* 20: 557-585.

———. 1921b. Systems of matings, I, II, III, IV, V. *Genetics* 6: 111-178.

———. 1923. The theory of path coefficients—a reply to Niles' criticism. *Genetics* 8: 239-255.

———. 1931a. Evolution in Mendelian populations. *Genetics* 16: 97-159.

REFERENCES

———. 1931b. Statistical methods in biology. *Proc. Amer. Stat. Assoc.* 26: 155-163.

———. 1932. General, group, and special size factors. *Genetics* 17: 603-619.

———. 1934. The method of path coefficients. *Ann. Math. Stat.* 5: 161-215.

———. 1952. The genetics of variability. In *Quantitative inheritance,* ed. Agric. Res. Council, pp. 5-41. London: Her Majesty's Stationery Office.

———. 1954. The interpretation of multivariate systems. In *Statistics and mathematics in biology,* ed. Kempthorne, et al., pp. 11-33. Ames, Iowa: Iowa State Univ. Press.

———. 1960a. Path coefficients and path regression: alternative or complementary concepts? *Biometrics* 16: 189-202.

———. 1960b. The treatment of reciprocal interaction, with or without lag, by path analysis. *Biometrics* 16: 423-445.

———. 1963. Genic interaction. In *Methodology in mammalian genetics,* ed. W. J. Burdette, pp. 159-192. San Francisco: Holden-Day.

———. 1965. The interpretation of population structure by F-statistics with special regard to systems of mating. *Evolution* 19: 395-420.

———. 1968-1969. *Genetic and biometric foundations.* Evolution and the genetics of populations, vol. I. Chicago: Univ. of Chicago Press.

———. 1968-1969. *The theory of gene frequencies.* Evolution and the genetics of populations, vol. II. Chicago: Univ. of Chicago Press.

Yule, G. U., and Kendall, M. G. 1968. *An introduction to the theory of statistics.* 14th ed. New York: Hafner.

INDEX

A

Acquired characters, 9
Allele, 190-191
Analysis of variance, 222
Analytical geometry, 106
Annual rings, 2
Arthritis prevalence, 302
Assortative mating, 270
    IQ, with respect to, 323
    mating system, 279
    parent-offspring diagram, 284, 289
    phenotypic correlation, 293
Autosomes, 188

B

Behavior traits, 8
Bilinear form, 157
Blending inheritance, 187
Breed true, 192

C

Castle's law, 198
Cause and effect, 1-2
Causes, correlated, 145
Chain of causes, 137
Chromosomes, 188
Closed system, 1
Coefficient of determination, 106
Common causes, 141
Common factors, proportion, 160
Connection rule, 107-108
Contingency tables, pooled, 303
Contraceptive devices, 3
Correlated causes, 145
Correlation coefficient, 50
    multiple, 83
    partial, 94
Correlation between
    boys and girls, 308
    groups, 309
    linear functions, 153, 159
    mates, 266
    nonhomologous genes, 273, 297
    uniting gametes, 267

Correlation, ecological, 309
Correlation, relatives
    additive scale, 217, 219, 227
    ancestors and descendants, 288
    assortative mating, with, 294, 295
    dominance, with, 238, 240, 242
    environmental effects, with, 251, 254, 260
    inbreeding, 267, 269
    midparent and children, 228
    one parent and children, 260
    path diagrams, 256, 259
Correlation ratio, 311
Correlation, sum of connecting paths, 110, 149
Correlation, unanalyzed, 147
Covariance, 59

D

Dependence analysis, 16
Determinant, 24, 85
Determination coefficient, 106
Determination, joint, 116
    partial, 114
Diagrams, equivalent, 185
Diploids, 188
Dominance, 192, 229
    degree of, 233, 236
    overdominance, 194
Dominance heritability, g^2, 235
Dominance ratio, 235

E

Ecological correlation, 309
Economics, 335
Education and intelligence, 324
Environmental effects, 245, 291
    common home, 260, 323
    interaction with heredity, 246
    random and systematic, 247
Environmental heritability, 252, 254
Environmentalist, 11
Epidemiology, 334
Epistasis, 244
Equations, linear, 21

INDEX

Equations, nonlinear, 28
Equilibrium state, 198
Equivalent diagrams, 185
Estimates of heritability, 255
Excess equations, 331

F

Fertilization, 190
Fertilizer, 11
Finger ridge count, 221
Flower color, 192

G

g^2, dominance heritability, 235
    table of, 236
Gadgets, electric, 3
Gametes, 189
Gene, 187
Gene effects, additive, 214
    with dominance, 229
Gene frequency, 196
Genetic correlations
    dominance, with, 237
    graphs of, 242
    midparent-child, 228
    parent-child, 218
    sib-sib, 219
Genetic variance, 230
    dominance component, 232
    families, between and within, 223
    linear component, 231
Genetics
    Mendelian, 187
    population, 195
Genotype, 189, 192
Geology, experimental, 5
Groups and population, 306
Guinea pigs, 12, 315

H

Half-sibs, 209
Hardy-Weinberg law, 198
Hereditarian, 10-11
Heritability, 8-9, 245
    dominance, 235
    environmental, 252, 254

Heterosis, 194
Heterozygote, 191
Homology, 188
Homozygote, 191
Hypothesis, disproof of, 7

I

Identity, algebraic, 17
Identity of genes, 203
Inbreeding, 266
    genotype distribution, 270
Income and urbanization, 328
Infant learning program, 7
Inheritance, Mendelian, 190
Intelligence,
    assortative mating, 323
    occupation and education, 324
Isomorphism, 301
ITO matrices, 206
    T decomposition, 217

J

Johannsen, 10
Joint determination, 116

L

Law of inheritance, 191
Least squares, 39, 77
Linear equations, 21
Linear regression, 37
    list of relationships, 60
Locus (of genes), 188
Lysenko, 9

M

Mating system, 196
Mating type, 191
Mendelian law, 191
MN blood type, 211
Multiple correlation, 83
    as a path, 150-151
Multiple loci, 243
Multiple regression, 317
    on two variables, 75

on three variables, 90
on four variables, 124
standardized, 108, 110

N

Nonlinear equations, 28
Normal equations, 40, 77, 91
   reduced form, 81
   standardized, 110, 161

O

Observation unit, 309
Occam's razor, 6
Occupation and intelligence, 324
Overdetermined system, 178
Overdominance (heterosis), 194

P

Parent-child combinations, 201, 218
   with dominance, 239
Parent-child correlation, 218
   environmental effects, 251
   with inbreeding, 269
   (*also see* genetic correlation)
Partial correlation, 94, 317
   path diagram of, 99, 127, 173
   relation with path, 129, 170
Partial determination, 114
Path analysis, genetic, 224
   half-sibs, 227
   parent-offspring, 226
   sib-sib, 227
Path analysis, summary, 178
Path coefficient, 103, 317
   history, 12
   ratio of, 133-134
   relation with partial correlation, 99, 129, 170
   relation with partial correlation, 129, 170
Path diagrams, 106
   multiple regression, 113, 119, 124
   rules of reading, 161
Parsimony, principle of, 6
Partition of sum of squares, 42
   geometrical representation, 44

Pea plant, 194
Phenotype, 192
   measurement of, 248
Physician utilization, 331
Piebald pattern, 12, 315
Plausible interpretation, 3-4
Pleiotropism, 194
Pooled contingency tables, 303
Population genetics, 195
Population and groups, 306
Psychology, 5
Pure line, 10

Q

Quadratic form, 157
Quantitative trait, 193, 214

R

R^2 components, 85
Rainfall, 2
Random deviation and covariance, 250
Random mating, 197
Random union of gametes, 198
Ratio of path coefficients, 133-134
Regression coefficient, 41
   regression ssq, 45
   residual ssq, 45
Regression equation, 39, 75, 90, 124
   choosing of, 121
Relatives, 206
   (*also see* correlation between relatives)
Remote factors, 146
Reproductive cells, 189
Round trip, 115

S

Scatter diagram, 38
Scattering of points, 66, 249
Segments of a variable, 312
Segregating families, 192
Sex chromosomes, 188
Sib-pair distribution, 202, 203
   components of, 213, 219
   conditional probabilities, 208
Sib-sib correlation, 219

with dominance, 240
　　　environmental effects, 254
Slope, weight of, 72
ssq = sum of squares, 45
ssq due to regression, 43-45, 80, 82, 91
　　　relation with *r*, 54
　　　relation with *R*, 84
Standard deviation, 59
Standardized variables, 100
　　　regression of, 102
Stationary state, 198
Sum of products, 19, 21
Sum of squares, 20, 21
　　　partition by regression, 42
　　　(also see ssq)

T

Transfer of path coefficients, 300

U

Unanalyzed correlation, 147
Underdetermined system, 178
Unit of observation 309
Unobserved variable, 175
　　　diagram of, 177
Urbanization and income, 328

V

Variable segments 312
Variance, 58
Viewpoint and interpretation, 165

W

Weight of slope, 72
Weinberg law, 198
Wheat yield, 2

XYZ

Zygote, 190